Other Titles in This Series

573 **Wayne Aitken,** An arithmetic Riemann-Roch theorem for singular arithmetic surfaces, 1996

572 **Ole H. Hald and Joyce R. McLaughlin,** Inverse nodal problems: Finding the potential from nodal lines, 1996

571 **Henry L. Kurland,** Intersection pairings on Conley indices, 1996

570 **Bernold Fiedler and Jürgen Scheurle,** Discretization of homoclinic orbits, rapid forcing and "invisible" chaos, 1996

569 **Eldar Straume,** Compact connected Lie transformation groups on spheres with low cohomogeneity, I, 1996

568 **Raúl E. Curto and Lawrence A. Fialkow,** Solution of the truncated complex moment problem for flat data, 1996

567 **Ran Levi,** On finite groups and homotopy theory, 1995

566 **Neil Robertson, Paul Seymour, and Robin Thomas,** Excluding infinite clique minors, 1995

565 **Huaxin Lin and N. Christopher Phillips,** Classification of direct limits of even Cuntz-circle algebras, 1995

564 **Wensheng Liu and Héctor J. Sussmann,** Shortest paths for sub-Riemannian metrics on rank-two distributions, 1995

563 **Fritz Gesztesy and Roman Svirsky,** (m)KdV solitons on the background of quasi-periodic finite-gap solutions, 1995

562 **John Lindsay Orr,** Triangular algebras and ideals of nest algebras, 1995

561 **Jane Gilman,** Two-generator discrete subgroups of $PSL(2, R)$, 1995

560 **F. Tomi and A. J. Tromba,** The index theorem for minimal surfaces of higher genus, 1995

559 **Paul S. Muhly and Baruch Solel,** Hilbert modules over operator algebras, 1995

558 **R. Gordon, A. J. Power, and Ross Street,** Coherence for tricategories, 1995

557 **Kenji Matsuki,** Weyl groups and birational transformations among minimal models, 1995

556 **G. Nebe and W. Plesken,** Finite rational matrix groups, 1995

555 **Tomás Feder,** Stable networks and product graphs, 1995

554 **Mauro C. Beltrametti, Michael Schneider, and Andrew J. Sommese,** Some special properties of the adjunction theory for 3-folds in $\mathbb{P}^5$, 1995

553 **Carlos Andradas and Jesús M. Ruiz,** Algebraic and analytic geometry of fans, 1995

552 **C. Krattenthaler,** The major counting of nonintersecting lattice paths and generating functions for tableaux, 1995

551 **Christian Ballot,** Density of prime divisors of linear recurrences, 1995

550 **Huaxin Lin,** C^*-algebra extensions of $C(X)$, 1995

549 **Edwin Perkins,** On the martingale problem for interactive measure-valued branching diffusions, 1995

548 **I-Chiau Huang,** Pseudofunctors on modules with zero dimensional support, 1995

547 **Hongbing Su,** On the classification of C*-algebras of real rank zero: Inductive limits of matrix algebras over non-Hausdorff graphs, 1995

546 **Masakazu Nasu,** Textile systems for endomorphisms and automorphisms of the shift, 1995

545 **John L. Lewis and Margaret A. M. Murray,** The method of layer potentials for the heat equation on time-varying domains, 1995

544 **Hans-Otto Walther,** The 2-dimensional attractor of $x'(t) = -\mu x(t) + f(x(t-1))$, 1995

(Continued in the back of this publication)

MEMOIRS
of the
American Mathematical Society

Number 573

An Arithmetic Riemann-Roch Theorem for Singular Arithmetic Surfaces

Wayne Aitken

March 1996 • Volume 120 • Number 573 (first of 4 numbers) • ISSN 0065-9266

American Mathematical Society
Providence, Rhode Island

1991 *Mathematics Subject Classification.*
Primary 14G40, 11G30.

Library of Congress Cataloging-in-Publication Data

Aitken, Wayne, 1963–
An arithmetic Riemann-Roch theorem for singular arithmetic surfaces / Wayne Aitken.
p. cm. – (Memoirs of the American Mathematical Society, ISSN 0065-9266; no. 573)
"March 1996, volume 120, number 573 (first of 4 numbers)."
Includes bibliographical references.
ISBN 0-8218-0407-3 (alk. paper)
1. Arithmetical algebraic geometry. 2. Riemann-Roch theorems. I. Title. II. Series.
QA3.A57 no. 573
[QA242.5]
510 s–dc20
[512′.74]

95-52304
CIP

Memoirs of the American Mathematical Society

This journal is devoted entirely to research in pure and applied mathematics.

Subscription information. The 1996 subscription begins with Number 568 and consists of six mailings, each containing one or more numbers. Subscription prices for 1996 are $391 list, $313 institutional member. A late charge of 10% of the subscription price will be imposed on orders received from nonmembers after January 1 of the subscription year. Subscribers outside the United States and India must pay a postage surcharge of $25; subscribers in India must pay a postage surcharge of $43. Expedited delivery to destinations in North America $30; elsewhere $92. Each number may be ordered separately; *please specify number* when ordering an individual number. For prices and titles of recently released numbers, see the New Publications sections of the *Notices of the American Mathematical Society*.

Back number information. For back issues see the *AMS Catalog of Publications*.

Subscriptions and orders should be addressed to the American Mathematical Society, P. O. Box 5904, Boston, MA 02206-5904. *All orders must be accompanied by payment.* Other correspondence should be addressed to Box 6248, Providence, RI 02940-6248.

Memoirs of the American Mathematical Society is published bimonthly (each volume consisting usually of more than one number) by the American Mathematical Society at 201 Charles Street, Providence, RI 02904-2213. Second-class postage paid at Providence, Rhode Island. Postmaster: Send address changes to Memoirs, American Mathematical Society, P. O. Box 6248, Providence, RI 02940-6248.

This publication is indexed in *Science Citation Index®, SciSearch®, Research Alert®, CompuMath Citation Index®, Current Contents®/Physical, Chemical & Earth Sciences.*

Printed in the United States of America.

∞ The paper used in this book is acid-free and falls within the guidelines established to ensure permanence and durability.

♻ Printed on recycled paper.

10 9 8 7 6 5 4 3 2 1 00 99 98 97 96

Contents

Abstract

The following gives a development of Arakelov theory general enough to handle not only regular arithmetic surfaces but also a large class of arithmetic surfaces whose generic fiber has singularities. This development culminates in an arithmetic Riemann-Roch theorem for such arithmetic surfaces.

The first half of this work gives a treatment of Deligne's functorial intersection theory tailored to the needs of this paper. This treatment is intended to satisfy three requirements. The first is that it be general enough to handle families of singular curves. The second is that it be reasonably self-contained. For example, this treatment bypasses Knudson and Mumford's development of the determinant of cohomology and instead gives a direct and concrete approach; all arising sign problems are handled directly. This treatment also develops much of the needed duality theory. The third requirement is that the constructions given be readily adaptable to the process of adding norms and metrics such as is done in the second half of this paper.

The second half of this paper is devoted to developing a class of intersection functions for singular curves which behaves analogously to the canonical Green's functions introduced by Arakelov for smooth curves. We call these functions *intersection functions* since they give a measure of intersection over the infinite places of a number field; the intersection over finite places can be defined in terms of the standard apparatus of algebraic geometry. There are major differences between my intersection functions and Arakelov's canonical Greens functions. For example, for a given non-singular curve, Arakelov's canonical Green's function is unique; however, for a given singular curve, the set of intersection functions is in general parameterized by a finite dimensional real vector space. We give the dimension of this space. Another difference is that for any given non-singular curve its canonical Green's function is bounded from below (or above, depending on the convention used). For any given singular curve, an associated intersection function has no such bounds, but in fact exhibits certain asymptotics as the points approach the singularities of the curve.

Finally, using the above mentioned intersection functions together with our treatment of Deligne's functorial intersection theory, we define an intersection theory for arithmetic surfaces which includes a large class of singular arithmetic surfaces. This culminates in a proof of the arithmetic Riemann-Roch theorem.

Introduction

The present work is divided into two parts. The first part, Chapters 1 through 3, is a self-contained treatment of Deligne's functorial intersection theory which defines an intersection pairing between invertible sheaves on a family of curves. The family is not assumed to be smooth over the base nor is the family itself assumed to be regular. We do assume an integral base scheme. Most of the results here are due to Deligne, or are slight generalizations of Deligne's results (see SGA4 XVIII and [D]). However, the method I use to derive the results is different in several important ways. My development follows the philosophy of L. Moret-Bailly in [MB] which is to define the intersection pairing in terms of the determinant of cohomology, and then to prove that this pairing has all the desired properties such as bilinearity. The present treatment is intended to be concrete and relatively self-contained. For example, we do not use Knudson and Mumford's development of the determinant of cohomology, but instead give a direct and concrete approach; all arising sign problems are handled directly. In addition, much of the needed duality theory is developed (in Chapter 3). One main requirement of this treatment is that the constructions given be readily adaptable to the process of adding norms and metrics as is done in Chapter 5.

The second part, Chapters 4 and 5, develops an arithmetic Riemann-Roch theorem for an important class of non-regular arithmetic surfaces. An arithmetic surface can be regarded as a family of curves over a one-dimensional base scheme, where this base scheme is the affine scheme associated with the ring of integers of some number field. I consider the case where the generic fiber of this surface is allowed to be singular, and is not required to be irreducible.

The general idea is as follows. Since the base scheme of an arithmetic surface is affine, in order to develop an intersection theory it is necessary to "complete" the base scheme and then to "complete" the arithmetic surface over the completed base scheme. When we do this, we study not invertible sheaves on the base scheme, but "metrized" invertible sheaves which are invertible sheaves with some additional structure. I extend all the constructions given in the first part of this work to yield not just invertible sheaves on the base, but actually metrized invertible sheaves. I show that the Riemann-Roch isomorphism is not only an isomorphism of invertible sheaves on the base scheme, but an isomorphism of metrized invertible sheaves. One property of these metrized invertible sheaves is that they have a well-defined degree. The Riemann-Roch theorem is obtained by taking degrees of both sides of the Riemann-Roch isomorphism.

The intended audience for this paper includes not only those who are interested in generalizations of Faltings' arithmetic Riemann-Roch theorem [F1], but also those who are looking for an introduction to Arakelov theory from the "functorial" point of view. With this in mind, I have tried to make this account as accessible and self-contained as possible. This is why, for instance, I give a brief account of the theory of the canonical Green's function for the non-singular case before developing the generalization to the singular case. Those who are interested in seeing an exposition of Arakelov theory and are mainly interested in the case where the arithmetic surfaces are regular are advised to skip the last six sections of Chapter 4; those who are interested in my generalization of the arithmetic Riemann-Roch theorem, but are already familiar with the use of the "functorial" point of view in Arakelov Theory, are advised to concentrate on the last two chapters.

This work grew out of my 1991 Ph.D. thesis. I first learned Arakelov theory from Barry Mazur, who then became my thesis advisor. I would like to thank him for his encouragement and support. I would also like to express my thanks to John Tate and Serge Lang for their help and encouragement.

Wayne Aitken
October, 1993

Received by the editor November 10, 1993, and in revised form November 2, 1994.

Chapter 1

The Intersection Pairing for One-Dimensional Schemes

In SGA4 XVIII and [D], Deligne defines an intersection pairing which assigns to any pair of line bundles on a smooth projective family of curves a line bundle on the base scheme. In this context, he interpreted the Grothendieck-Riemann-Roch theorem not as an equality between certain elements of the Picard group of the base, but as a canonical isomorphism between the corresponding line bundles on the base.

In Part I of this work I give a development of Deligne's intersection pairing and the associated Riemann-Roch isomorphism. I do so not only for smooth families of curves, but for a larger class of families of curves. I do not assume that the generic fiber is non-singular or even irreducible. My development of the intersection pairing was inspired by the approach of L. Moret-Bailly in [MB] where he developed the pairing for certain types of generically smooth families (of total dimension 2). His method was to *define* the intersection pairing in terms of the determinant of cohomology, and then to prove that this pairing has all the desired properties. The reader should also be aware of generalizations of Deligne's work by R. Elkik [E1] and J. Frank [Fr].

I have divided the development of the intersection pairing into two stages: this chapter is concerned with the "absolute" case, i.e., a projective one-dimensional scheme over a field; the next chapter is concerned with the "relative" case, i.e., a flat, projective family of one-dimensional schemes over an arbitrary integral base scheme. This seems to be a good division of labor: in the first chapter we don't have to worry about the more technical issues such as flatness criteria or relative cohomology and their determinants. The basic ideas can be presented more clearly. Then in the second chapter we extend the constructions to the case when the base scheme is any integral scheme.

1. Preliminaries

(1.1) Fractional Notation for Line Bundles. Throughout this work the terms *line bundle* and *invertible sheaf* will be used synonymously.

- For L and M line bundles on a scheme X we will use the following notation:

$$\frac{L}{M} \stackrel{\text{def}}{=} L \otimes M^\vee \qquad \text{where } M^\vee \text{ is the dual line bundle to } M.$$

- Given a pair of isomorphisms of line bundles, $\lambda\colon L \xrightarrow{\sim} L'$ and $\mu\colon M \xrightarrow{\sim} M'$, define

$$\frac{\lambda}{\mu} : \frac{L}{M} \xrightarrow{\sim} \frac{L'}{M'}$$

to be the isomorphism $\lambda \otimes (\mu^\vee)^{-1}$ where $\mu^\vee$ is the dual isomorphism from $(M')^\vee$ to $M^\vee$ induced by μ.

- Let L and M be line bundles on a scheme X with sections, l of L and m of M, on an open subscheme U. Also assume that m is non-zero at every point of U. We use the notation

$$\frac{l}{m} \stackrel{\text{def}}{=} l \otimes m^\vee$$

where $m^\vee$ is the homomorphism from $M|_U$ to $\mathcal{O}_X|_U$ sending m to 1. So $\dfrac{l}{m}$ is a section of $\dfrac{L}{M}$ on U.

(1.2) Meromorphic Sections. We will now establish the terminology and basic properties concerning meromorphic sections of line bundles and other related matters. This terminology will be used throughout this work.

We will always assume that the schemes we deal with are Noetherian and separated. Such a scheme X has a finite number of *associated points*, where, by definition, $x \in X$ is an associated point if and only if the maximal ideal of the local ring $\mathcal{O}_{X,x}$ is an associated prime ideal of $\mathcal{O}_{X,x}$. In other words, $x \in X$ is an associated point if and only if every element of the maximal ideal of the local ring $\mathcal{O}_{X,x}$ is a zero divisor in this ring. On an affine scheme, $X = \operatorname{Spec} A$, a function $f \in A$ is a zero divisor of A if and only if the zero set of f contains at least one associated point of X. A flat morphism between two schemes sends associated points to associated points.

The closure of any associated point is called an *associated component*. In particular, each component of X is an associated component. Those associated components which are not components of X are called *embedded components* and their generic points are called *embedded points*. In many common situations there are no embedded points. For example, if a scheme is reduced it has no embedded points. Likewise, if a scheme is Cohen-Macaulay it has no embedded points. Conversely, if a scheme is one-dimensional and contains no embedded points, then it is Cohen-Macaulay. Zero-dimensional schemes are, of course, always Cohen-Macaulay.

Let L be a line bundle on a scheme X. Intuitively, a *meromorphic section* l of L is a section of L which is defined and nowhere vanishing on an open set U which contains every associated point of X; in other words, l is defined on a dense open subscheme U which intersects every embedded component. In particular, if X is reduced or Cohen-Macaulay, we only need to require that U be dense. To formulate the definition more precisely, consider the set of pairs (U, l) where U is an open subscheme of X containing all the associated points of X and where l is a section of L on U which is not zero anywhere on U. We define an equivalence relation on such pairs by saying (U, l) and (U', l') are equivalent if and only if $l|_{U''} = l'|_{U''}$ for some open subscheme U'' of $U \cap U'$ also containing all the associated points of X. A meromorphic section of L is defined to be an equivalence class of such pairs. The need to work with equivalence classes is eliminated by the fact that any meromorphic section of L uniquely determines a "maximal" pair (U, l) representing it. This pair is maximal in the sense that, for any other pair (U', l') representing the same meromorphic section, U' is contained in U and $l|_{U'} = l'$. The proof of this fact is based on the principle that if l and l' are two global sections (not necessarily meromorphic) of a line bundle L, and if $l|_W = l'|_W$ for some open subscheme W containing all the associated points of X, then $l = l'$.

Note that the definition of meromorphic given here is stronger than the other common definition: we require not only that the section be defined but also that it be nowhere zero on some open set containing all the associated points of the scheme.

An *invertible meromorphic function* of X is defined to be a meromorphic section of $\mathcal{O}_X$.

Any isomorphism between line bundles takes meromorphic sections to meromorphic sections in a natural way; moreover, such an isomorphism is uniquely determined by a single meromorphic section of the first line bundle and its image under the isomorphism. Any meromorphic section l of a line bundle L determines a Cartier divisor $D = (l)$, and there is an isomorphism between L and $\mathcal{O}_X(D)$ characterized by the rule $l \mapsto \mathbf{1}_D$. Here $\mathbf{1}_D$ is the canonical meromorphic section of $\mathcal{O}_X(D)$. Thus, given two line bundles with meromorphic sections such that the Cartier divisors associated to each of the meromorphic sections are equal, then there is a unique isomorphism between the line bundles which takes the one meromorphic section to the other.

Note that if X is projective over an affine scheme, then, given a line bundle L and a finite set $x_1, \ldots, x_m$ of closed points on X, there is a meromorphic section of L defined and non-zero at each of the points $x_1, \ldots, x_m$. A corollary to this is that every line bundle on such a scheme is isomorphic to $\mathcal{O}_X(D)$ for some Cartier divisor D.

2. The Determinant of Cohomology Functor

(2.1) Determinants. Let V be a vector space of dimension n over a field K. We define the determinant of V, written $\operatorname{Det} V$, to be $\bigwedge^n V$, the nth order exterior product of V. The determinant of V is a one-dimensional K-vector space. When V is zero-dimensional, $\operatorname{Det} V$ is canonically isomorphic to the field K.

For V and W two K-vector spaces and α an isomorphism between them, we define $\operatorname{Det}\alpha$, the associated determinant map, to be the isomorphism from $\operatorname{Det} V$ and $\operatorname{Det} W$ induced by α.

(2.2) The Determinant Map of a Sequence. Given an exact sequence

$$(\xi) \qquad 0 \longrightarrow V_1 \xrightarrow{\beta_1} V_2 \xrightarrow{\beta_2} \cdots \xrightarrow{\beta_{n-1}} V_n \xrightarrow{\beta_n} 0,$$

we define the determinant map of this sequence $\operatorname{Det}\xi$ to be the isomorphism

$$\frac{\bigotimes_{i \text{ even}} \operatorname{Det} V^i}{\bigotimes_{i \text{ odd}} \operatorname{Det} V^i} \longrightarrow K$$

defined as follows:

For $i = 2, \ldots, n$, let $b_{i,1}, \ldots, b_{i,m_i}$ be a basis of the kernel of β_i. For each $b_{i,j}$, choose an element $c_{i-1,j}$ of V_{i-1} which maps to $b_{i,j}$. Clearly $b_{i,1}, \ldots, b_{i,m_i}, c_{i,1}, \ldots, c_{i,m_{i+1}}$ is a basis of V_i, and so

$$d_i \overset{\text{def}}{=} b_{i,1} \wedge \ldots \wedge b_{i,m_i} \wedge c_{i,1} \wedge \ldots \wedge c_{i,m_{i+1}}$$

is a non-zero element of $\operatorname{Det} V_i$. We define $\operatorname{Det}\xi$ by the rule

$$d = \frac{\bigotimes_{i \text{ even}} d_i}{\bigotimes_{i \text{ odd}} d_i} \mapsto 1.$$

By the basic properties of exterior products, d_i is independent of the choice of elements $c_{i,1}, \ldots, c_{i,m_{i+1}}$ mapping to $b_{i+1,1}, \ldots, b_{i+1,m_{i+1}}$. In other words, d_i depends only on the choice of basis $(b_{i+1,j})$, if $1 \le i < n$, and the basis $(b_{i,j})$, if $1 < i \le n$. Although both d_{i-1} and d_i (for $1 < i \le n$) depend on the choice of $(b_{i,j})$, it follows, from basic properties of exterior products, that the "ratio"

$$\frac{d_{i-1}}{d_i} \in \frac{\operatorname{Det} V^{i-1}}{\operatorname{Det} V^i},$$

and hence d itself, is independent of this choice. Thus $\operatorname{Det}\xi$ is well-defined.

(2.3) The Determinant of Cohomology. Let Y be a scheme proper over a field K, and let $\mathcal{F}$ be a coherent sheaf of Y. Define the *determinant of cohomology* of $\mathcal{F}$, denoted $\mathbf{D}(\mathcal{F})$, to be the one-dimensional K-vector space

$$\mathbf{D}(\mathcal{F}) \overset{\text{def}}{=} \frac{\bigotimes_{i \text{ even}} \operatorname{Det} H^i(\mathcal{F})}{\bigotimes_{i \text{ odd}} \operatorname{Det} H^i(\mathcal{F})}.$$

Given an isomorphism α between coherent sheaves $\mathcal{F}$ and $\mathcal{G}$, define $\mathbf{D}(\alpha)$ to be the isomorphism between $\mathbf{D}(\mathcal{F})$ and $\mathbf{D}(\mathcal{G})$ given by the formula

$$\mathbf{D}(\alpha) \overset{\text{def}}{=} \frac{\bigotimes_{i \text{ even}} \operatorname{Det} H^i(\alpha)}{\bigotimes_{i \text{ odd}} \operatorname{Det} H^i(\alpha)},$$

where $H^i(\alpha)$ is the isomorphism between $H^i(\mathcal{F})$ and $H^i(\mathcal{G})$ induced by α, and $\operatorname{Det} H^i(\alpha)$ is the determinant of this isomorphism.

(2.4) The Isomorphism Associated to a Long Exact Sequence. Let Y be a scheme proper over a field K. Suppose we have a short exact sequence of coherent sheaves on Y:

$$(\eta) \qquad 0 \longrightarrow \mathcal{F}_1 \longrightarrow \mathcal{F}_2 \longrightarrow \mathcal{F}_3 \longrightarrow 0\,.$$

Consider the associated long exact sequence

$$(\eta') \quad 0 \longrightarrow H^0(\mathcal{F}_1) \longrightarrow H^0(\mathcal{F}_2) \longrightarrow H^0(\mathcal{F}_3) \longrightarrow \ldots \longrightarrow H^d(\mathcal{F}_3) \longrightarrow 0,$$

where d is the dimension of Y. We have the determinant isomorphism, Det (η'), as defined in (2.2) above which maps a certain tensor product of one dimensional vector spaces to the field K. By bringing some of these spaces to the right hand side of the isomorphism, and then rearranging terms, so to speak, we obtain naturally the isomorphism

$$\mathbf{D}(\mathcal{F}_2) \xrightarrow{\sim} \mathbf{D}(\mathcal{F}_1) \otimes \mathbf{D}(\mathcal{F}_3).$$

We call this isomorphism $\mathbf{D}(\eta)$.

(2.5) Compatibility for Three-by-Three Diagrams. As before, let Y be a scheme proper over a field K. Suppose we have a commutative diagram of coherent sheaves with exact rows and columns:

$$\begin{array}{ccccccccc}
 & & (\zeta_1) & & (\zeta_2) & & (\zeta_3) & & \\
 & & 0 & & 0 & & 0 & & \\
 & & \downarrow & & \downarrow & & \downarrow & & \\
(\eta_1) & 0 \longrightarrow & \mathcal{F}_{1,1} & \longrightarrow & \mathcal{F}_{1,2} & \longrightarrow & \mathcal{F}_{1,3} & \longrightarrow & 0 \\
 & & \downarrow & & \downarrow & & \downarrow & & \\
(\eta_2) & 0 \longrightarrow & \mathcal{F}_{2,1} & \longrightarrow & \mathcal{F}_{2,2} & \longrightarrow & \mathcal{F}_{2,3} & \longrightarrow & 0 \\
 & & \downarrow & & \downarrow & & \downarrow & & \\
(\eta_3) & 0 \longrightarrow & \mathcal{F}_{3,1} & \longrightarrow & \mathcal{F}_{3,2} & \longrightarrow & \mathcal{F}_{3,3} & \longrightarrow & 0 \\
 & & \downarrow & & \downarrow & & \downarrow & & \\
 & & 0 & & 0 & & 0 & &
\end{array}$$

(2.5.1) Theorem. *In the above situation, the following diagram commutes*

$$\begin{array}{ccccc}
\mathbf{D}(\mathcal{F}_{2,2}) & \xrightarrow{\mathbf{D}(\eta_2)} & \mathbf{D}(\mathcal{F}_{2,1})\otimes\mathbf{D}(\mathcal{F}_{2,3}) & \xrightarrow{\mathbf{D}(\zeta_1)\otimes\mathbf{D}(\zeta_3)} & \mathbf{D}(\mathcal{F}_{1,1})\otimes\mathbf{D}(\mathcal{F}_{3,1})\otimes\mathbf{D}(\mathcal{F}_{1,3})\otimes\mathbf{D}(\mathcal{F}_{3,3}) \\
\| & & & & \downarrow i \\
\mathbf{D}(\mathcal{F}_{2,2}) & \xrightarrow{\mathbf{D}(\zeta_2)} & \mathbf{D}(\mathcal{F}_{1,2})\otimes\mathbf{D}(\mathcal{F}_{3,2}) & \xrightarrow{\mathbf{D}(\eta_1)\otimes\mathbf{D}(\eta_3)} & \mathbf{D}(\mathcal{F}_{1,1})\otimes\mathbf{D}(\mathcal{F}_{1,3})\otimes\mathbf{D}(\mathcal{F}_{3,1})\otimes\mathbf{D}(\mathcal{F}_{3,3})
\end{array}$$

where i is $(-1)^\pi$ times the canonical isomorphism which switches the two innermost terms, and where π is an integer which will be defined in the course of the proof.

Note It is possible to adjust, by a sign, the definition given in (2.4) above in order to make the above diagram actually commute (see [A] for more details). To do this we also need to adjust, by a sign, the natural isomorphism which switches the order of the tensor product of two determinants of cohomology.

Proof. For $1 \le i, j \le 3$, let $H^k_{i,j}$ denote the cohomology space $H^k(\mathcal{F}_{i,j})$, which is, by convention, the zero space when $k < 0$. For general integral values of i and j, we define $H^k_{i,j}$ by the following rule: we require that $H^k_{i+3,j} = H^k_{i,j+3} = H^{k+1}_{i,j}$.

The above 3×3 diagram of exact sequences gives linear maps $H^k_{i,j} \to H^k_{i+1,j}$ and $H^k_{i,j} \to H^k_{i,j+1}$ which are, depending on i and j, either maps induced by the cohomology functors, or are boundary maps coming from the exact sequences of the above diagram. In particular for all i,j and k the sequences

$$H^k_{i-1,j} \to H^k_{i,j} \to H^k_{i+1,j} \qquad \text{and} \qquad H^k_{i,j-1} \to H^k_{i,j} \to H^k_{i,j+1}$$

are exact.

(2.5.2) Lemma. *In the above situation, the following diagram commutes:*

$$\begin{array}{ccc} H^k_{i,j} & \longrightarrow & H^k_{i,j+1} \\ \downarrow & & \downarrow \\ H^k_{i+1,j} & \longrightarrow & H^k_{i+1,j+1} \end{array}$$

unless $i = j = 3$ (mod 3), in which case the above diagram is "anti-commutative", i.e., for any vector v of $H^k_{i,j}$, its images under the compositions

$$H^k_{i,j} \to H^k_{i,j+1} \to H^k_{i+1,j+1} \qquad \text{and} \qquad H^k_{i,j} \to H^k_{i+1,j} \to H^k_{i+1,j+1}$$

differ by a factor of -1.

Proof. The first statement is a basic fact of cohomology theory (in fact, it is, by definition, true of any δ-functor). So we can assume $i = j = 3$, and, of course, that $k \ge 0$. In other words, we must show that the following diagram of boundary maps is anti-commutative:

$$\begin{array}{ccc} H^k(\mathcal{F}_{3,3}) & \longrightarrow & H^{k+1}(\mathcal{F}_{3,1}) \\ \downarrow & & \downarrow \\ H^{k+1}(\mathcal{F}_{1,3}) & \longrightarrow & H^{k+2}(\mathcal{F}_{1,1}) \end{array}$$

Let $\mathcal{U}$ be a finite cover of Y by affine open sets. For any coherent sheaf $\mathcal{F}$, let $\mathcal{C}^\bullet(\mathcal{F})$ denote the Čech complex of $\mathcal{F}$ with respect to this cover.

Let ϕ be an element of $H^k(\mathcal{F}_{3,3})$, and let $f^k_{3,3}$ be an element of $\mathcal{C}^k(\mathcal{F}_{3,3})$ representing ϕ. We have the following commutative diagram of complexes with exact rows and columns:

$$\begin{array}{ccccccccc}
 & & 0 & & 0 & & 0 & & \\
 & & \downarrow & & \downarrow & & \downarrow & & \\
0 & \longrightarrow & \mathcal{C}^\bullet(\mathcal{F}_{1,1}) & \longrightarrow & \mathcal{C}^\bullet(\mathcal{F}_{1,2}) & \longrightarrow & \mathcal{C}^\bullet(\mathcal{F}_{1,3}) & \longrightarrow & 0 \\
 & & \downarrow & & \downarrow & & \downarrow & & \\
0 & \longrightarrow & \mathcal{C}^\bullet(\mathcal{F}_{2,1}) & \longrightarrow & \mathcal{C}^\bullet(\mathcal{F}_{2,2}) & \longrightarrow & \mathcal{C}^\bullet(\mathcal{F}_{2,3}) & \longrightarrow & 0 \\
 & & \downarrow & & \downarrow & & \downarrow & & \\
0 & \longrightarrow & \mathcal{C}^\bullet(\mathcal{F}_{3,1}) & \longrightarrow & \mathcal{C}^\bullet(\mathcal{F}_{3,2}) & \longrightarrow & \mathcal{C}^\bullet(\mathcal{F}_{3,3}) & \longrightarrow & 0 \\
 & & \downarrow & & \downarrow & & \downarrow & & \\
 & & 0 & & 0 & & 0 & &
\end{array}$$

Let $f^k_{2,2}$ be an element of $\mathcal{C}^k(\mathcal{F}_{2,2})$ mapping to $f^k_{3,3}$. Under the boundary map d of the Čech complex, $f^k_{3,3}$ is sent to 0 and $f^k_{2,2}$ is sent to an element $df^k_{2,2}$ of $\mathcal{C}^{k+1}(\mathcal{F}_{2,2})$.

Let $f^{k+1}_{3,2}$ be the image of $df^k_{2,2}$ in $\mathcal{C}^{k+1}(\mathcal{F}_{3,2})$; likewise, let $f^{k+1}_{2,3}$ be the image of $df^k_{2,2}$ in $\mathcal{C}^{k+1}(\mathcal{F}_{2,3})$. By exactness, $f^{k+1}_{3,2}$ comes from an element $f^{k+1}_{3,1}$ of $\mathcal{C}^{k+1}(\mathcal{F}_{3,1})$; likewise, $f^{k+1}_{2,3}$ comes from an element $f^{k+1}_{1,3}$ of $\mathcal{C}^{k+1}(\mathcal{F}_{1,3})$. We point out that the cohomology class associated to $f^{k+1}_{3,1}$ is the image of ϕ under the boundary map $\mathrm{H}^k(\mathcal{F}_{3,3}) \to \mathrm{H}^{k+1}(\mathcal{F}_{3,1})$, and that the class of $f^{k+1}_{1,3}$ is the image of ϕ under the boundary map $\mathrm{H}^k(\mathcal{F}_{3,3}) \to \mathrm{H}^{k+1}(\mathcal{F}_{1,3})$.

Let $f^{k+1}_{2,1}$ be an element of $\mathcal{C}^{k+1}(\mathcal{F}_{2,1})$ mapping to $f^{k+1}_{3,1}$, and let α be its image in $\mathcal{C}^{k+1}(\mathcal{F}_{2,2})$. Note that the image of $df^k_{2,2} - \alpha$ in $\mathcal{C}^{k+1}(\mathcal{F}_{3,2})$ is 0; therefore, there is an element $f^{k+1}_{1,2}$ of $\mathcal{C}^{k+1}(\mathcal{F}_{1,2})$ whose image β in $\mathcal{C}^{k+1}(\mathcal{F}_{2,2})$ is $df^k_{2,2} - \alpha$, i.e.,

$$df^k_{2,2} = \alpha + \beta.$$

Note that the image of $f^{k+1}_{1,2}$ in $\mathcal{C}^{k+1}(\mathcal{F}_{1,3})$ is $f^{k+1}_{1,3}$.

Under the boundary maps d of the Čech complexes, the elements $f^{k+1}_{1,3}$, $f^{k+1}_{3,1}$, and $df^k_{2,2}$ are sent to zero. Hence we have the equation

$$d\alpha + d\beta = 0,$$

and $df^{k+1}_{2,1}$ and $df^{k+1}_{1,2}$ come from elements, which we will call α' and β' respectively, of $\mathcal{C}^{k+2}(\mathcal{F}_{1,1})$. The image of $\alpha' + \beta'$ in $\mathcal{C}^{k+2}(\mathcal{F}_{2,2})$ is $d\alpha + d\beta = 0$, so, by injectivity,

$$\alpha' + \beta' = 0.$$

But α' represents the image of ϕ under the composition of boundary maps

$$H^k(\mathcal{F}_{3,3}) \to H^{k+1}(\mathcal{F}_{3,1}) \to H^{k+2}(\mathcal{F}_{1,1}),$$

and β' represents the image of ϕ under the composition

$$H^k(\mathcal{F}_{3,3}) \to H^{k+1}(\mathcal{F}_{1,3}) \to H^{k+2}(\mathcal{F}_{1,1}).$$

The lemma follows.

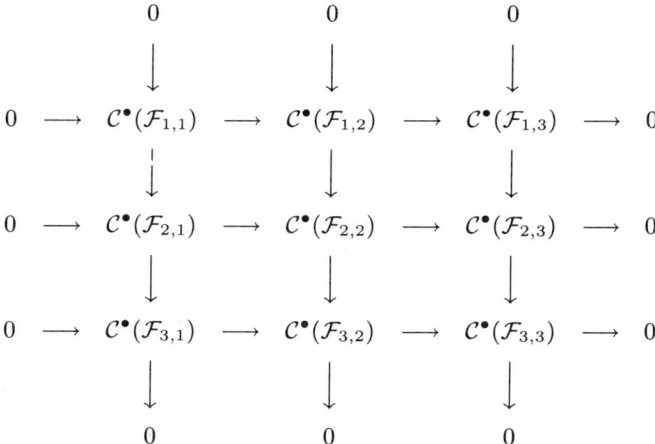

(2.5.3) We now continue the proof of Theorem 2.5.1. Let $H^k_{i,j}$ be as above. The next step will be to define a sequence of subspaces for each $H^k_{i,j}$:

$$A^k_{i,j} \subseteq B^k_{i,j} \subseteq C^k_{i,j} \subseteq D^k_{i,j} \subseteq H^k_{i,j}.$$

Define $A^k_{i,j}$ to be the image of $H^k_{i-1,j-1}$ under the composition map

$$\begin{array}{ccc} H^k_{i-1,j-1} & \longrightarrow & H^k_{i-1,j} \\ \downarrow & & \downarrow \\ H^k_{i,j-1} & \longrightarrow & H^k_{i,j} \end{array}$$

Define $B^k_{i,j}$ to be the intersection of the kernels of the two maps

$$H^k_{i,j} \to H^k_{i+1,j} \qquad \text{and} \qquad H^k_{i,j} \to H^k_{i,j+1}.$$

Define $C^k_{i,j}$ to be the subspace of $H^k_{i,j}$ generated by the images of the two maps

$$H^k_{i-1,j} \to H^k_{i,j} \qquad \text{and} \qquad H^k_{i,j-1} \to H^k_{i,j}.$$

Finally, define $D^k_{i,j}$ to be the kernel of the composition map

$$\begin{array}{ccc} H^k_{i,j} & \longrightarrow & H^k_{i,j+1} \\ \downarrow & & \downarrow \\ H^k_{i+1,j} & \longrightarrow & H^k_{i+1,j+1} \end{array}$$

(2.5.4) Lemma. *For all i, j, and k, the map $H^k_{i,j} \to H^k_{i+1,j}$ induces an isomorphism*

$$D^k_{i,j}/C^k_{i,j} \xrightarrow{\sim} B^k_{i+1,j}/A^k_{i+1,j}.$$

Likewise, the map $H^k_{i,j} \to H^k_{i,j+1}$ induces an isomorphism

$$D^k_{i,j}/C^k_{i,j} \xrightarrow{\sim} B^k_{i,j+1}/A^k_{i,j+1}.$$

Proof. Easy.

(2.5.5) The above lemma results in a cycle of six isomorphic spaces:

$$\begin{aligned} D^k_{i,j}/C^k_{i,j} &\xrightarrow{\sim} B^k_{i,j+1}/A^k_{i,j+1} \xleftarrow{\sim} D^k_{i-1,j+1}/C^k_{i-1,j+1} \xrightarrow{\sim} B^k_{i-1,j+2}/A^k_{i-1,j+2} \\ &\xleftarrow{\sim} D^k_{i-2,j+2}/C^k_{i-2,j+2} \xrightarrow{\sim} B^k_{i-2,j+3}/A^k_{i-2,j+3} \\ &\xleftarrow{\sim} D^k_{i-3,j+3}/C^k_{i-3,j+3} = D^k_{i,j}/C^k_{i,j}. \end{aligned}$$

The following three lemmas show that these isomorphisms are well behaved. In particular, the composition (using the inverses of the second, fourth, and sixth isomorphisms) of the above six isomorphisms is, up to sign, the identity map. Each lemma treats one of the three possible cases.

(2.5.5.1) Lemma. *Consider the cycle of spaces* $D^{k-1}_{3,3}$, $B^k_{3,1}$, $D^k_{2,1}$, $B^k_{2,2}$, $D^k_{1,2}$, *and* $B^k_{1,3}$. *Let* $x_{3,3}$ *be an element of* $D^{k-1}_{3,3}$. *It is possible to find elements* $x_{3,1}$, $x_{2,1}$, $x_{2,2}$, $x_{1,2}$, *and* $x_{1,3}$ *of* $B^k_{3,1}$, $D^k_{2,1}$, $B^k_{2,2}$, $D^k_{1,2}$, *and* $B^k_{1,3}$ *respectively, such that, under the maps*

$$D^{k-1}_{3,3} \to B^k_{3,1} \qquad D^k_{2,1} \to B^k_{2,2} \qquad \text{and} \qquad D^k_{1,2} \to B^k_{1,3},$$

we have

$$x_{3,3} \mapsto x_{3,1} \qquad x_{2,1} \mapsto x_{2,2} \qquad \text{and} \qquad x_{1,2} \mapsto x_{1,3},$$

and, under the maps

$$D^{k-1}_{3,3} \to B^k_{1,3} \qquad D^k_{1,2} \to B^k_{2,2} \qquad \text{and} \qquad D^k_{2,1} \to B^k_{3,1},$$

we have

$$x_{3,3} \mapsto -x_{1,3} \qquad x_{1,2} \mapsto -x_{2,2} \qquad \text{and} \qquad x_{2,1} \mapsto -x_{3,1}.$$

Proof. Fix, once and for all, an affine open cover of Y. Let $\mathcal{C}^\bullet_{i,j}$ be the Čech complex associated with $\mathcal{F}_{i,j}$ and the above open cover of Y.

Let $a_{3,3}$ be an element of $\mathcal{C}^{k-1}_{3,3}$ which represents $x_{3,3}$. Choose an element $a_{2,2}$ of $\mathcal{C}^{k-1}_{2,2}$ which is mapped to $a_{3,3}$. Let $a_{3,2}$ and $a_{2,3}$ be the images of $a_{2,2}$ in $\mathcal{C}^{k-1}_{3,2}$ and $\mathcal{C}^{k-1}_{2,3}$ respectively.

The boundary map, $\mathcal{C}^{k-1}_{3,3} \to \mathcal{C}^k_{3,3}$, sends $a_{3,3}$ to 0, so the boundary element $\mathrm{d}(a_{3,2})$ in $\mathcal{C}^k_{3,2}$ is the image of an element $b_{3,1}$ of $\mathcal{C}^k_{3,1}$. Note that the cohomology class of $b_{3,1}$ is the image of $x_{3,3}$ under the boundary map

$$\mathrm{H}^{k-1}(\mathcal{F}_{3,3}) \to \mathrm{H}^k(\mathcal{F}_{3,1}).$$

Likewise, there is an element $c_{1,3}$ of $\mathcal{C}^k_{1,3}$ which maps to $\mathrm{d}(a_{2,3})$, and so represents the image of $x_{3,3}$ under the boundary map

$$\mathrm{H}^{k-1}(\mathcal{F}_{3,3}) \to \mathrm{H}^k(\mathcal{F}_{1,3}).$$

Let $b_{2,1}$ be an element of $\mathcal{C}^k_{2,1}$ which maps to $b_{3,1}$. The fact that $\mathrm{dd}(a_{3,2}) = 0$ implies that $\mathrm{d}(b_{3,1}) = 0$, which in turn implies that $\mathrm{d}(b_{2,1})$ is the image of an element ϵ of $\mathcal{C}^{k+1}_{1,1}$. The class represented by ϵ is the image of $x_{3,3}$ in $\mathrm{H}^{k+1}(\mathcal{F}_{1,1})$ under the composition of the two boundary maps. So the assumption that $x_{3,3}$ is in $D^{k-1}_{3,3}$ implies that ϵ is $\mathrm{d}(\delta_{1,1})$ for some element $\delta_{1,1}$ of $\mathcal{C}^k_{1,1}$. Let $\delta_{2,1}$ be the image of $\delta_{1,1}$ in $\mathcal{C}^k_{2,1}$. Let $b'_{2,1} = b_{2,1} - \delta_{2,1}$. Hence $b'_{2,1}$ is an an element of $\mathcal{C}^k_{2,1}$ which maps to $b_{3,1}$ and such that $\mathrm{d}b'_{2,1} = 0$. Let $b'_{2,2}$ be the image of $b'_{2,1}$ in $\mathcal{C}^k_{2,2}$.

Likewise, construct an element $c_{1,2}$ of $\mathcal{C}^k_{1,2}$ which maps to $c_{1,3}$ and such that $\mathrm{d}(c_{1,2}) = 0$. Let $c_{2,2}$ be the image of $c_{1,2}$ in $\mathcal{C}^k_{2,2}$. Observe that the element

$$e_{2,2} \stackrel{\text{def}}{=} c_{2,2} + b'_{2,2} - \mathrm{d}(a_{2,2})$$

maps to 0 under the maps

$$\mathcal{C}^k_{2,2} \to \mathcal{C}^k_{2,3} \qquad \text{and} \qquad \mathcal{C}^k_{2,2} \to \mathcal{C}^k_{3,2},$$

and so there is an element $e_{1,1}$ of $\mathcal{C}^k_{1,1}$ which maps to $e_{2,2}$ under the composition

$$\begin{array}{ccc} \mathcal{C}^k_{1,1} & \longrightarrow & \mathcal{C}^k_{1,2} \\ \downarrow & & \downarrow \\ \mathcal{C}^k_{2,1} & \longrightarrow & \mathcal{C}^k_{2,2} \end{array}$$

Let $e_{1,2}$ be the image of $e_{1,1}$ in $\mathcal{C}^k_{1,2}$. Let $c'_{1,2} = c_{1,2} - e_{1,2}$. Note that $d(e_{2,2}) = 0$ which implies that $d(e_{1,2}) = 0$ and hence $d(c'_{1,2}) = 0$. Note also that the image of $c'_{1,2}$ in $\mathcal{C}^k_{1,3}$ is $c_{1,3}$. Let $c'_{2,2}$ be the image of $c'_{1,2}$ in $\mathcal{C}^k_{2,2}$, so $c'_{2,2} = da_{2,2} - b'_{2,2}$. Thus the cohomology classes of $c'_{2,2}$ and $b'_{2,2}$ in $H^k(\mathcal{F}_{2,2})$ satisfy the formula $[c'_{2,2}] = -[b'_{2,2}]$.

Let $x_{3,1}$ in $B^k_{3,1} \subseteq H^k(\mathcal{F}_{3,1})$ be the cohomology class represented by $b_{3,1}$.

Let $x_{2,1}$ in $D^k_{2,1} \subseteq H^k(\mathcal{F}_{2,1})$ be the cohomology class represented by $-b'_{2,1}$.

Let $x_{2,2}$ in $B^k_{2,2} \subseteq H^k(\mathcal{F}_{2,2})$ be the cohomology class represented by $-b'_{2,2}$ (or $c'_{2,2}$).

Let $x_{1,2}$ in $D^k_{1,2} \subseteq H^k(\mathcal{F}_{1,2})$ be the cohomology class represented by $-c'_{1,2}$.

Let $x_{1,3}$ in $B^k_{1,3} \subseteq H^k(\mathcal{F}_{1,3})$ be the cohomology class represented by $-c_{1,3}$.

(2.5.5.2) Lemma. *Consider the cycle of spaces $D^k_{2,2}$, $B^k_{2,3}$, $D^k_{1,3}$, $B^{k+1}_{1,1}$, $D^k_{3,1}$, and $B^k_{3,2}$. Let $x_{2,2}$ be an element of $D^k_{2,2}$. It is possible to find elements $x_{2,3}$, $x_{1,3}$, $x_{1,1}$, $x_{3,1}$, and $x_{3,2}$ of $B^k_{2,3}$, $D^k_{1,3}$, $B^{k+1}_{1,1}$, $D^k_{3,1}$, and $B^k_{3,2}$ respectively, such that, under the maps*

$$D^k_{1,3} \to B^{k+1}_{1,1} \qquad D^k_{3,1} \to B^k_{3,2} \qquad \text{and} \qquad D^k_{2,2} \to B^k_{2,3},$$

we have

$$x_{1,3} \mapsto x_{1,1} \qquad x_{3,1} \mapsto x_{3,2} \qquad \text{and} \qquad x_{2,2} \mapsto x_{2,3},$$

and, under the maps

$$D^k_{1,3} \to B^k_{2,3} \qquad D^k_{3,1} \to B^{k+1}_{1,1} \qquad \text{and} \qquad D^k_{2,2} \to B^k_{3,2},$$

we have

$$x_{1,3} \mapsto -x_{2,3} \qquad x_{3,1} \mapsto -x_{1,1} \qquad \text{and} \qquad x_{2,2} \mapsto -x_{3,2}.$$

Proof. Fix, once and for all, an affine open cover of Y. Let $\mathcal{C}^\bullet_{i,j}$ be the Čech complex associated with $\mathcal{F}_{i,j}$ and this open cover of Y.

Let $a_{2,2}$ be an element of $\mathcal{C}^k_{2,2}$ which represents $x_{2,2}$. Let $a_{2,3}$ and $a_{3,2}$ be the images of $a_{2,2}$ in $\mathcal{C}^k_{2,3}$ and $\mathcal{C}^k_{3,2}$ respectively. Since $x_{2,2}$ is in $D^k_{2,2}$, the image of $a_{2,2}$ in $\mathcal{C}^k_{3,3}$ is $\mathrm{d}(b_{3,3})$ for some element $b_{3,3}$ in $\mathcal{C}^{k-1}_{3,3}$.

Let $b_{2,2}$ be an element of $\mathcal{C}^{k-1}_{2,2}$ whose image in $\mathcal{C}^{k-1}_{3,3}$ is $b_{3,3}$. Let $b_{2,3}$ and $b_{3,2}$ be the images of $b_{2,2}$ in $\mathcal{C}^{k-1}_{2,3}$ and $\mathcal{C}^{k-1}_{3,2}$ respectively.

Define $c_{2,3}$ to be $a_{2,3} - \mathrm{d}(b_{2,3})$ in $\mathcal{C}^k_{2,3}$. The image of $c_{2,3}$ in $\mathcal{C}^k_{3,3}$ is 0, hence there is an element $c_{1,2}$ of $\mathcal{C}^k_{1,2}$ whose image in $\mathcal{C}^k_{2,3}$ is $c_{2,3}$. Let $c_{2,2}$ and $c_{1,3}$ be the images of $c_{1,2}$ in $\mathcal{C}^k_{2,2}$ and $\mathcal{C}^k_{1,3}$ respectively.

Likewise, define $e_{3,2}$ to be $a_{3,2} - \mathrm{d}(b_{3,2})$ in $\mathcal{C}^k_{3,2}$. The image of $e_{3,2}$ in $\mathcal{C}^k_{3,3}$ is 0, hence there is an element $e_{2,1}$ of $\mathcal{C}^k_{2,1}$ whose image in $\mathcal{C}^k_{3,2}$ is $e_{3,2}$. Let $e_{2,2}$ and $e_{3,1}$ be the images of $e_{2,1}$ in $\mathcal{C}^k_{2,2}$ and $\mathcal{C}^k_{3,1}$ respectively.

Let $\epsilon_{2,2} = a_{2,2} - e_{2,2} - c_{2,2} - \mathrm{d}(b_{2,2})$. The image of $\epsilon_{2,2}$ in $\mathcal{C}^k_{2,3}$ is 0; likewise, its image in $\mathcal{C}^k_{3,2}$ is 0. Therefore, there is an element, $\epsilon_{1,1}$ of $\mathcal{C}^k_{1,1}$ whose image in $\mathcal{C}^k_{2,2}$ is $\epsilon_{2,2}$. Let $\epsilon_{1,2}$ and $\epsilon_{2,1}$ be the images of $\epsilon_{1,1}$ in $\mathcal{C}^k_{1,2}$ and $\mathcal{C}^k_{2,1}$ respectively.

Since $\mathrm{d}(e_{3,2}) = \mathrm{d}(a_{3,2}) = 0$, it follows that $\mathrm{d}(e_{3,1}) = 0$. Therefore, $\mathrm{d}(e_{2,1})$ is the image of an element $f_{1,1}$ of $\mathcal{C}^{k+1}_{1,1}$. Likewise, $\mathrm{d}(c_{1,2})$ is the image of an element $g_{1,1}$ of $\mathcal{C}^{k+1}_{1,1}$. Let $f_{2,2}$ and $g_{2,2}$ be the images in $\mathcal{C}^{k+1}_{2,2}$ of $f_{1,1}$ and $g_{1,1}$ respectively. Note that the cohomology class of $f_{1,1}$ is the image under the boundary map,

$$H^k(\mathcal{F}_{3,1}) \to H^{k+1}(\mathcal{F}_{1,1})$$

of the class represented by $e_{3,1}$. Likewise, the cohomology class of $g_{1,1}$ is the image under the boundary map,

$$H^k(\mathcal{F}_{1,3}) \to H^{k+1}(\mathcal{F}_{1,1})$$

of the class represented by $c_{1,3}$. Observe that

$$\begin{aligned} \mathrm{d}(\epsilon_{2,2}) &= \mathrm{d}(a_{2,2}) - \mathrm{d}(e_{2,2}) - \mathrm{d}(c_{2,2}) - \mathrm{dd}(b_{2,2}) \\ &= 0 - f_{2,2} - g_{2,2} - 0. \end{aligned}$$

It follows that $\mathrm{d}(\epsilon_{1,1}) = -f_{1,1} - g_{1,1}$, i.e., the cohomology class of $f_{1,1}$ in $H^{k+1}(\mathcal{F}_{1,1})$ is equal to the class of $-g_{1,1}$.

Let $x_{2,3}$ in $B^k_{2,3} \subseteq H^k(\mathcal{F}_{2,3})$ be the cohomology class represented by $a_{2,3}$ (or $c_{2,3}$).

Let $x_{1,3}$ in $D^k_{1,3} \subseteq H^k(\mathcal{F}_{1,3})$ be the cohomology class represented by $-c_{1,3}$.

Let $x_{1,1}$ in $B^{k+1}_{1,1} \subseteq H^{k+1}(\mathcal{F}_{1,1})$ be the cohomology class represented by $-g_{1,1}$ (or $f_{1,1}$).

Let $x_{3,1}$ in $D^k_{3,1} \subseteq H^k(\mathcal{F}_{3,1})$ be the cohomology class represented by $-e_{3,1}$.

Let $x_{3,2}$ in $B^k_{3,2} \subseteq H^k(\mathcal{F}_{3,2})$ be the cohomology class represented by $-a_{3,2}$ (or $-e_{3,2}$).

(2.5.5.3) Lemma. *Consider the cycle of spaces* $D^k_{1,1}$, $B^k_{1,2}$, $D^{k-1}_{3,2}$, $B^{k-1}_{3,3}$, $D^{k-1}_{2,3}$, *and* $B^k_{2,1}$. *Let* $x_{1,1}$ *be an element of* $D^k_{1,1}$. *It is possible to find elements* $x_{1,2}$, $x_{3,2}$, $x_{3,3}$, $x_{2,3}$, *and* $x_{2,1}$ *of* $B^k_{1,2}$, $D^{k-1}_{3,2}$, $B^{k-1}_{3,3}$, $D^{k-1}_{2,3}$, *and* $B^k_{2,1}$ *respectively, such that, under the maps*

$$D^k_{1,1} \to B^k_{1,2} \qquad D^{k-1}_{3,2} \to B^{k-1}_{3,3} \qquad \text{and} \qquad D^{k-1}_{2,3} \to B^k_{2,1},$$

we have

$$x_{1,1} \mapsto x_{1,2} \qquad x_{3,2} \mapsto x_{3,3} \qquad \text{and} \qquad x_{2,3} \mapsto x_{2,1},$$

and, under the maps

$$D^k_{1,1} \to B^k_{2,1} \qquad D^{k-1}_{3,2} \to B^k_{1,2} \qquad \text{and} \qquad D^{k-1}_{2,3} \to B^{k-1}_{3,3},$$

we have

$$x_{1,1} \mapsto x_{2,1} \qquad x_{3,2} \mapsto x_{1,2} \qquad \text{and} \qquad x_{2,3} \mapsto x_{3,3}.$$

Proof. Fix, once and for all, an affine open cover of Y. Let $\mathcal{C}^\bullet_{i,j}$ be the Čech complex associated with $\mathcal{F}_{i,j}$ and this open cover of Y.

Let $a_{1,1}$ be an element of $\mathcal{C}^k_{1,1}$ which represents $x_{1,1}$. Let $a_{1,2}$ and $a_{2,1}$ be the images of $a_{1,1}$ in $\mathcal{C}^k_{1,2}$ and $\mathcal{C}^k_{2,1}$ respectively. Since $x_{1,1}$ is in $D^k_{1,1}$, the image of $a_{1,1}$

in $\mathcal{C}^k_{2,2}$ is $\mathrm{d}(b_{2,2})$ for some element $b_{2,2}$ in $\mathcal{C}^{k-1}_{2,2}$. Let $b_{2,3}$, $b_{3,2}$, and $b_{3,3}$ be the images of $b_{2,2}$ in $\mathcal{C}^{k-1}_{2,3}$, $\mathcal{C}^{k-1}_{3,2}$ and $\mathcal{C}^{k-1}_{3,3}$ respectively.

Note that the cohomology class of $a_{2,1}$ is the image under the boundary map,

$$H^{k-1}(\mathcal{F}_{2,3}) \to H^k(\mathcal{F}_{2,1})$$

of the class represented by $b_{2,3}$. Likewise, the cohomology class of $a_{1,2}$ is the image under the boundary map,

$$H^{k-1}(\mathcal{F}_{3,2}) \to H^k(\mathcal{F}_{1,2})$$

of the class represented by $b_{3,2}$.

Let $x_{1,2}$ in $B^k_{1,2} \subseteq H^k(\mathcal{F}_{1,2})$ be the cohomology class represented by $a_{1,2}$.

Let $x_{3,2}$ in $D^{k-1}_{3,2} \subseteq H^{k-1}(\mathcal{F}_{3,2})$ be the cohomology class represented by $b_{3,2}$.

Let $x_{3,3}$ in $B^{k-1}_{3,3} \subseteq H^{k-1}(\mathcal{F}_{3,3})$ be the cohomology class represented by $b_{3,3}$.

Let $x_{2,3}$ in $D^{k-1}_{2,3} \subseteq H^{k-1}(\mathcal{F}_{2,3})$ be the cohomology class represented by $b_{2,3}$.

Let $x_{2,1}$ in $B^k_{2,1} \subseteq H^k(\mathcal{F}_{2,1})$ be the cohomology class represented by $a_{2,1}$.

(2.5.6) We now continue the proof of Theorem 2.5.1. Recall that in (2.5.3), we defined, for each i, j, and k, a sequence of subspaces of $H^k_{i,j}$:

$$A^k_{i,j} \subseteq B^k_{i,j} \subseteq C^k_{i,j} \subseteq D^k_{i,j} \subseteq H^k_{i,j}.$$

Let $a^k_{i,j}$, $b^k_{i,j}$, and $d^k_{i,j}$ be the dimensions of $A^k_{i,j}$, $B^k_{i,j}/A^k_{i,j}$, and $D^k_{i,j}/C^k_{i,j}$ respectively.

For each $1 \le i, j \le 3$ and each k, choose a basis of $A^k_{i,j}$:

$$\alpha^k_{i,j}(1),\ \alpha^k_{i,j}(2),\ \ldots,\ \alpha^k_{i,j}(a^k_{i,j}).$$

For each such $\alpha^k_{i,j}(l)$, choose an element $\epsilon^k_{i-1,j-1}(l)$ of $H^k_{i-1,j-1}$ which maps to $\alpha^k_{i,j}(l)$ via the composition

$$H^k_{i-1,j-1} \to H^k_{i-1,j} \to H^k_{i,j}.$$

For each such $\epsilon^k_{i-1,j-1}(l)$, let $\gamma^k_{i-1,j}(l)$ be its image under $H^k_{i-1,j-1} \to H^k_{i-1,j}$. Similarly, let $\sigma^k_{i,j-1}(l)$ be the image of $\epsilon^k_{i-1,j-1}(l)$ under $H^k_{i-1,j-1} \to H^k_{i,j-1}$. Using the equalities $H^k_{i+3,j} = H^{k+1}_{i,j}$ and $H^k_{i,j+3} = H^{k+1}_{i,j}$, we get a basis $\left\{\alpha^k_{i,j}(l)\right\}$ of $A^k_{i,j}$ and sets $\left\{\epsilon^k_{i,j}(l)\right\}$, $\left\{\gamma^k_{i,j}(l)\right\}$, and $\left\{\sigma^k_{i,j}(l)\right\}$ for all integers i,j, and k.

Note that the image of

$$\epsilon^k_{i,j}(1),\ \epsilon^k_{i,j}(2),\ \ldots,\ \epsilon^k_{i,j}\left(a^k_{i+1,j+1}\right)$$

in $H^k_{i,j}/D^k_{i,j}$ forms a basis. Note also that

$$\left\{\gamma^k_{i,j}(1),\ \gamma^k_{i,j}(2),\ \ldots,\ \gamma^k_{i,j}(a^k_{i+1,j})\right\} \bigcup \left\{\sigma^k_{i,j}(1),\ \sigma^k_{i,j}(2),\ \ldots,\ \sigma^k_{i,j}(a^k_{i,j+1})\right\}$$

is contained in $C^k_{i,j}$. In fact, it is not to hard to see that the image of this set in $C^k_{i,j}/B^k_{i,j}$ forms a basis.

For all $i = j = 1, \ldots, 3$ and all k, choose a set of elements

$$\delta^k_{i,i}(1),\ \delta^k_{i,i}(2),\ \ldots,\ \delta^k_{i,i}(d^k_{i,i})$$

of $D^k_{i,i}$ whose image in $D^k_{i,i}/C^k_{i,i}$ forms a basis. By Lemmas 2.5.5.1 – 2.5.5.3, we obtain elements

$$\delta^k_{i,j}(1),\ \delta^k_{i,j}(2),\ \ldots,\ \delta^k_{i,j}(d^k_{i,j})$$

in $D^k_{i,j}$ and elements

$$\beta^k_{i,j}(1),\ \beta^k_{i,j}(2),\ \ldots,\ \beta^k_{i,j}(b^k_{i,j})$$

in $B^k_{i,j}$ for all $1 \le i,j \le 3$, and k (and then for all integers i,j, and k using the convention $H^k_{i+3,j} = H^{k+1}_{i,j}$ and $H^k_{i,j+3} = H^{k+1}_{i,j}$), chosen to satisfy the mapping properties set forth in these lemmas. In particular, this implies that, for every i, j, and k, the image of

$$\delta^k_{i,j}(1),\ \delta^k_{i,j}(2),\ \ldots,\ \delta^k_{i,j}(d^k_{i,j})$$

in $D^k_{i,j}/C^k_{i,j}$ forms a basis, and the image of

$$\beta^k_{i,j}(1),\ \beta^k_{i,j}(2),\ \ldots,\ \beta^k_{i,j}(b^k_{i,j})$$

in $B^k_{i,j}/A^k_{i,j}$ likewise forms a basis.

To summarize, for all integers i, j and k we have the following:

- We have a series of subspaces $A^k_{i,j} \subseteq B^k_{i,j} \subseteq C^k_{i,j} \subseteq D^k_{i,j} \subseteq H^k_{i,j}$.
- We have a basis $\{\alpha^k_{i,j}(l)\}$ of $A^k_{i,j}$.
- We have a subset $\{\beta^k_{i,j}(l)\}$ of $B^k_{i,j}$ whose image in $B^k_{i,j}/A^k_{i,j}$ forms a basis.
- We have a subset $\{\gamma^k_{i,j}(l)\} \bigcup \{\sigma^k_{i,j}(l)\}$ of $C^k_{i,j}$ whose image in $C^k_{i,j}/B^k_{i,j}$ forms a basis.
- We have a subset $\{\delta^k_{i,j}(l)\}$ of $D^k_{i,j}$ whose image in $D^k_{i,j}/C^k_{i,j}$ forms a basis.
- We have a subset $\{\epsilon^k_{i,j}(l)\}$ of $H^k_{i,j}$ whose image in $H^k_{i,j}/D^k_{i,j}$ forms a basis.

The above facts imply that, for any fixed i, j, and k, the union

$$\{\alpha^k_{i,j}(l)\} \bigcup \{\beta^k_{i,j}(l)\} \bigcup \{\gamma^k_{i,j}(l)\} \bigcup \{\sigma^k_{i,j}(l)\} \bigcup \{\delta^k_{i,j}(l)\} \bigcup \{\epsilon^k_{i,j}(l)\}$$

is a basis of $H^k_{i,j}$.

This basis is well-behaved with respect to the map $H^k_{i,j} \to H^k_{i,j+1}$. Note the following:

- The set $\{\alpha^k_{i,j}(l)\} \bigcup \{\beta^k_{i,j}(l)\} \bigcup \{\gamma^k_{i,j}(l)\}$ forms a basis of the kernel.
- If $(i,j) \ne (1,3) \pmod 3$, the (ordered) set $\{\sigma^k_{i,j}(l)\}$ is sent to $\{\alpha^k_{i,j+1}(l)\}$.
- If $i = 1, j = 3 \pmod 3$, the (ordered) set $\{\sigma^k_{i,j}(l)\}$ is sent to $\{-\alpha^k_{i,j+1}(l)\}$.
- The (ordered) set $\{\delta^k_{i,j}(l)\}$ is sent to $\{\beta^k_{i,j+1}(l)\}$.
- The (ordered) set $\{\epsilon^k_{i,j}(l)\}$ is sent to $\{\gamma^k_{i,j+1}(l)\}$.

This basis is likewise well-behaved with respect to the map $H^k_{i,j} \to H^k_{i+1,j}$:

- The set $\{\alpha^k_{i,j}(l)\} \bigcup \{\beta^k_{i,j}(l)\} \bigcup \{\sigma^k_{i,j}(l)\}$ forms a basis of the kernel.
- The (ordered) set $\{\gamma^k_{i,j}(l)\}$ is sent to $\{\alpha^k_{i+1,j}(l)\}$.
- If $i+j \neq 2 \pmod 3$, the (ordered) set $\{\delta^k_{i,j}(l)\}$ is sent to $\{-\beta^k_{i+1,j}(l)\}$.
- If $i+j = 2 \pmod 3$, the (ordered) set $\{\delta^k_{i,j}(l)\}$ is sent to $\{\beta^k_{i+1,j}(l)\}$.
- The (ordered) set $\{\epsilon^k_{i,j}(l)\}$ is sent to $\{\sigma^k_{i+1,j}(l)\}$.

Recall that, in the 3×3 commutative diagram in (2.5), we labelled the vertical exact sequences ζ_1, ζ_2, and ζ_3, and the horizontal exact sequences η_1, η_2, and η_3. First we study the maps $\mathbf{D}(\eta_i)$.

If we look at the definitions of (2.2) and (2.4), we see that the map $\mathbf{D}(\eta_i)$ is defined in terms of ordered bases of the kernels of the maps $H^k_{i,j} \to H^k_{i,j+1}$. For each $1 \leq i, j \leq 3$ and each k we choose the (ordered) basis

$$\{\alpha^k_{i,j}(l)\} \bigcup \{\beta^k_{i,j}(l)\} \bigcup \{\gamma^k_{i,j}(l)\}$$

unless $i = j = 1$, in which case we choose the (ordered) basis

$$\{-\alpha^k_{i,j}(l)\} \bigcup \{\beta^k_{i,j}(l)\} \bigcup \{\gamma^k_{i,j}(l)\}.$$

The conventions $H^k_{i,j+3} = H^{k+1}_{i,j}$ and $H^k_{i+3,j} = H^{k+1}_{i,j}$ define a basis for all i and j.

Now to use the definitions of (2.2) and (2.4) we must choose, for each $1 \leq i, j \leq 3$ and k, a sequence of elements in $H^k_{i,j}$ which maps (in sequential order) to the above chosen ordered basis of the kernel of $H^k_{i,j+1} \to H^k_{i,j+2}$. According to the facts mentioned above, the following (ordered) set of elements will work:

$$\{\sigma^k_{i,j}(l)\} \bigcup \{\delta^k_{i,j}(l)\} \bigcup \{\epsilon^k_{i,j}(l)\}.$$

When we append this list to the above chosen ordered basis of the kernel of the map $H^k_{i,j} \to H^k_{i,j+1}$, we get an ordered basis of $H^k_{i,j}$ which determines an element $x^k_{i,j}$ of $\operatorname{Det} H^k_{i,j}$.

The definitions of (2.2) and (2.4) together imply that the map

$$\mathbf{D}(\eta_i)\colon \mathbf{D}(\mathcal{F}_{i,2}) \to \mathbf{D}(\mathcal{F}_{i,1}) \otimes \mathbf{D}(\mathcal{F}_{i,3})$$

is defined by the requirement that $x_{i,2} \mapsto x_{i,1} \otimes x_{i,3}$ where

$$x_{i,j} \stackrel{\text{def}}{=} \frac{\bigotimes_{k \text{ even}} x^k_{i,j}}{\bigotimes_{k \text{ odd}} x^k_{i,j}}.$$

Now we investigate $\mathbf{D}(\zeta_j)$. We start by choosing an ordered basis for the kernel of $H^k_{i,j} \to H^k_{i+1,j}$ for each $0 \leq i, j \leq 3$ and each k: if $i + j \neq 0 \pmod 3$ we choose the (ordered) basis

$$\{\alpha^k_{i,j}(l)\} \bigcup \{-\beta^k_{i,j}(l)\} \bigcup \{\sigma^k_{i,j}(l)\},$$

but if $i + j = 0 \pmod 3$ we choose the (ordered) basis

$$\{\alpha^k_{i,j}(l)\} \bigcup \{\beta^k_{i,j}(l)\} \bigcup \{\sigma^k_{i,j}(l)\}.$$

The conventions $H^k_{i,j+3} = H^{k+1}_{i,j}$ and $H^k_{i+3,j} = H^{k+1}_{i,j}$ define a basis for all i and j.

Now we choose, for each $1 \leq i, j \leq 3$ and k, a sequence of elements in $H^k_{i,j}$ which maps (in sequential order) to the above chosen ordered basis of the kernel of $H^k_{i+1,j} \to H^k_{i+2,j}$. According to the facts mentioned above, the following (ordered) set of elements will work:

$$\{\gamma^k_{i,j}(l)\} \bigcup \{\delta^k_{i,j}(l)\} \bigcup \{\epsilon^k_{i,j}(l)\}.$$

When we append this list to the above chosen ordered basis of the kernel of the map $H^k_{i,j} \to H^k_{i+1,j}$, we get an ordered basis of $H^k_{i,j}$ which determines an element $y^k_{i,j}$ of $\mathrm{Det} H^k_{i,j}$.

The definitions of (2.2) and (2.4) together imply that the map

$$\mathbf{D}(\zeta_j)\colon \mathbf{D}(\mathcal{F}_{2,j}) \ \to \ \mathbf{D}(\mathcal{F}_{1,j}) \otimes \mathbf{D}(\mathcal{F}_{3,j})$$

is defined by the requirement that $y_{2,j} \mapsto y_{1,j} \otimes y_{3,j}$ where

$$y_{i,j} \stackrel{\text{def}}{=} \frac{\bigotimes_{k \text{ even}} y^k_{i,j}}{\bigotimes_{k \text{ odd}} y^k_{i,j}}.$$

The two elements $x_{i,j}$ and $y_{i,j}$ of $\mathbf{D}(\mathcal{F}_{i,j})$ are defined in terms of the same basis, up to sign, but ordered differently, therefore they are equal, up to sign:

$$x_{i,j} = (-1)^{\pi_{i,j}}\, y_{i,j}$$

for some integer $\pi_{i,j}$. For future reference we give a formula for $\pi_{i,j}$. For each i, j, and k, let $a^k_{i,j}$ and $b^k_{i,j}$ be as above: $a^k_{i,j}$ is the dimensions of $A^k_{i,j}$ which is the number of elements in $\{\alpha^k_{i,j}(l)\}$, and $b^k_{i,j}$ is the dimensions of $B^k_{i,j}/A^k_{i,j}$ which is the number of elements in $\{\beta^k_{i,j}(l)\}$. In addition, let $c^k_{i,j}$ be the number of elements of $\{\gamma^k_{i,j}(l)\}$, i.e., $c^k_{i,j} = a^k_{i+1,j}$; finally, let $s^k_{i,j}$ be the number of elements of $\{\sigma^k_{i,j}(l)\}$, i.e., $s^k_{i,j} = a^k_{i,j+1}$.

By looking at the differences between the basis used to define $x^k_{i,j}$ and the basis used to define $y^k_{i,j}$, we see that if $i = j = 1$, then

$$\pi_{i,j} = \sum_k \left(c^k_{i,j} s^k_{i,j} + b^k_{i,j} + a^k_{i,j}\right) \qquad \text{mod } 2;$$

if $i + j = 0 \pmod 3$, then

$$\pi_{i,j} = \sum_k c^k_{i,j} s^k_{i,j} \qquad \text{mod } 2;$$

otherwise

$$\pi_{i,j} = \sum_k \left(c^k_{i,j} s^k_{i,j} + b^k_{i,j}\right) \qquad \text{mod } 2.$$

Now we are in a position to prove the main result. We do so by keeping track of the images of $x_{2,2}$ under the two compositions. Under the first composition,

$$\begin{aligned}
x_{2,2} \ &\mapsto \ x_{2,1} \otimes x_{2,3} \\
&= \ (-1)^{\pi_{2,1}+\pi_{2,3}} \left(y_{2,1} \otimes y_{2,3}\right) \\
&\mapsto \ (-1)^{\pi_{2,1}+\pi_{2,3}} \left(y_{1,1} \otimes y_{3,1} \otimes y_{1,3} \otimes y_{3,3}\right) \\
&= \ (-1)^{\pi_{2,1}+\pi_{2,3}+\pi_{1,1}+\pi_{3,1}+\pi_{1,3}+\pi_{3,3}} \left(x_{1,1} \otimes x_{3,1} \otimes x_{1,3} \otimes x_{3,3}\right).
\end{aligned}$$

Under the second composition,

$$\begin{aligned} x_{2,2} &= (-1)^{\pi_{2,2}}(y_{2,2}) \\ &\mapsto (-1)^{\pi_{2,2}}(y_{1,2}\otimes y_{3,2}) \\ &= (-1)^{\pi_{2,2}+\pi_{1,2}+\pi_{3,2}}(x_{1,2}\otimes x_{3,2}) \\ &\mapsto (-1)^{\pi_{2,2}+\pi_{1,2}+\pi_{3,2}}(x_{1,1}\otimes x_{1,3}\otimes x_{3,1}\otimes x_{3,3}). \end{aligned}$$

So we have the desired result, where

$$\begin{aligned} \pi &= (\pi_{2,1}+\pi_{2,3}+\pi_{1,1}+\pi_{3,1}+\pi_{1,3}+\pi_{3,3})+(\pi_{2,2}+\pi_{1,2}+\pi_{3,2}) \\ &= \sum_{1\le i,j\le 3}\pi_{i,j} \\ &= \sum_{k;\,1\le i,j\le 3} c_{i,j}^k s_{i,j}^k + \sum_k a_{1,1}^k + \sum_k (b_{1,1}^k + b_{1,3}^k + b_{2,2}^k + b_{2,3}^k + b_{3,1}^k + b_{3,2}^k) \\ &= \sum_{k;\,1\le i,j\le 3} c_{i,j}^k s_{i,j}^k + \sum_k a_{1,1}^k + \sum_k (b_{1,1}^k + b_{2,2}^k) \pmod 2. \end{aligned}$$

The last step follows from the fact that $b_{2,3}^k = b_{3,2}^k$ and $b_{1,3}^k = b_{3,1}^k$; see Lemmas 2.5.5.2 and 2.5.5.1.

3. The Norm Functor for Zero-Dimensional Schemes

Throughout this section, Y will be a scheme finite over a field K. It follows that Y is affine; i.e., $Y = \mathrm{Spec}(A)$ for some ring A which is a finite dimensional K-vector space. In this section we define and study the norm functor on Y.

(3.1) The Norm of an Element. Let a be an element of A, and consider the associated K-linear transformation $T_a\colon A \to A$ sending an element x to ax. We define the *norm* of a, written $\mathrm{Norm}_Y(a)$ or $\mathrm{Norm}_A(a)$, to be the determinant of T_a.

Note that, for all a_1 and a_2 in A,

$$\mathrm{Norm}_A(a_1\cdot a_2) = \mathrm{Norm}_A(a_1)\cdot \mathrm{Norm}_A(a_2).$$

(3.2) The Norm Functor. Let L be a line bundle on Y. We define the *norm* of L, written $\mathbf{N}_Y(L)$ or simply $\mathbf{N}(L)$, to be the one-dimensional K-vector space given by the formula

$$\mathbf{N}_Y(L) \stackrel{\text{def}}{=} \frac{\mathbf{D}(L)}{\mathbf{D}(\mathcal{O}_Y)} = \mathbf{D}(L)\otimes \mathbf{D}(\mathcal{O}_Y)^{-1}$$

where $\mathcal{O}_Y$ is the structure sheaf of Y.

The line bundle L is the sheaf associated to a K-vector space V_L which is a locally free A-module of rank 1. Since Y consists of a finite number of closed points, V_L is a free A-module of rank 1. Since Y is a zero-dimensional scheme,

$$\mathbf{D}(\mathcal{O}_Y) = \mathrm{Det}(A) \qquad \text{and} \qquad \mathbf{D}(L) = \mathrm{Det}(V_L).$$

Thus

$$\mathbf{N}(L) = \frac{\mathrm{Det}(V_L)}{\mathrm{Det}(A)}.$$

If σ is an isomorphism between two line bundles L and M, then let $\mathbf{N}(\sigma)$ be the following isomorphism

$$\mathbf{N}(\sigma) \stackrel{\text{def}}{=} \frac{\mathbf{D}(\sigma)}{\mathbf{D}(I)}\colon\ \mathbf{N}(L) = \frac{\mathbf{D}(L)}{\mathbf{D}(\mathcal{O}_Y)} \longrightarrow \mathbf{N}(M) = \frac{\mathbf{D}(M)}{\mathbf{D}(\mathcal{O}_Y)}$$

where I is the identity map. Thus $\mathbf{N}(\bullet)$ gives a covariant functor from the category whose objects are line bundles on Y and whose morphisms are isomorphisms between line bundles, to the category of 1-dimensional K-vector spaces.

(3.3) Definition. Let L be a line bundle on Y, and let V_L be its associated free, rank 1 A-module. For any global section l of L, i.e., element of V_L, we define $\mathbf{N}(l) = \mathbf{N}_Y(l)$ to be the element of $\mathbf{N}(L)$ determined as follows: let $a_1, \dots, a_n$ be a K-basis of A, so $a_1 \wedge \dots \wedge a_n$ is a non-zero vector of $\mathrm{Det}(A)$ and $a_1\, l \wedge \dots \wedge a_n\, l$ is a vector of $\mathrm{Det}(V_L)$. Define

$$\mathbf{N}(l) \stackrel{\text{def}}{=} \frac{(a_1\, l \wedge \dots \wedge a_n\, l)}{(a_1 \wedge \dots \wedge a_n)}.$$

Note that, if l is a *non-vanishing* section of L, i.e., if l generates V_L as an A-module, then $\mathbf{N}(l)$ is a *non-zero* vector of $\mathbf{N}(L)$.

Now we will show that $\mathbf{N}(l)$ is well-defined, i.e., it is independent of the choice of $a_1, \dots, a_n$. Let $a'_1, \dots, a'_n$ be another K-basis of A, and let $[\alpha_{i,j}]$ be the $n \times n$ matrix of elements of K such that, for each a'_i, we have $a'_i = \sum \alpha_{i,j} a_j$. So, for each a'_i, we have $a'_i\, l = \sum \alpha_{i,j}(a_j\, l)$. Thus

$$(a'_1 \wedge \dots \wedge a'_n) = \mathrm{Det}[\alpha_{i,j}]\ (a_1 \wedge \dots \wedge a_n),$$

and

$$(a'_1\, l \wedge \dots \wedge a'_n\, l) = \mathrm{Det}[\alpha_{i,j}]\ (a_1\, l \wedge \dots \wedge a_n\, l).$$

Hence

$$\frac{(a'_1\, l \wedge \dots \wedge a'_n\, l)}{(a'_1 \wedge \dots \wedge a'_n)} = \frac{(a_1\, l \wedge \dots \wedge a_n\, l)}{(a_1 \wedge \dots \wedge a_n)}.$$

Finally, we note that, if σ is an isomorphism between two line bundles L and M of Y, and if l is a global section of L, then the map $\mathbf{N}(\sigma)$ sends $\mathbf{N}(l)$ to $\mathbf{N}(\sigma(l))$.

(3.4) Proposition. *Let L be a line bundle on Y, and let V_L be its associated free, rank 1 A-module. Let l be a global section of L, i.e., an element of V_L, and let a be an element of A. Then*

$$\mathbf{N}(a \cdot l) = \mathrm{Norm}_A(a) \cdot \mathbf{N}(l).$$

Proof. Let $a_1, \dots, a_n$ be a K-basis of A, and assume, for now, that l generates V_L as an A-module. The matrix representing the action of a on V_L with respect to the K-basis $a_1\, l, \dots, a_n\, l$ is the same as that representing the action of a on A with respect to the K-basis $a_1, \dots, a_n$. The determinant of this matrix is, by definition, $\mathrm{Norm}_A(a)$. Hence

$$(a(a_1\, l) \wedge \dots \wedge a(a_n\, l)) = \mathrm{Norm}_A(a)\ (a_1\, l \wedge \dots \wedge a_n\, l).$$

Therefore,

$$\begin{aligned}\mathbf{N}(a\,l) &\stackrel{\text{def}}{=} \frac{(a_1(a\,l)\wedge\ldots\wedge a_n(a\,l))}{(a_1\wedge\ldots\wedge a_n)}\\ &= \frac{(a(a_1\,l)\wedge\ldots\wedge a(a_n\,l))}{(a_1\wedge\ldots\wedge a_n)}\\ &= \operatorname{Norm}_A(a)\cdot\frac{(a_1\,l\wedge\ldots\wedge a_n\,l)}{(a_1\wedge\ldots\wedge a_n)}\\ &\stackrel{\text{def}}{=} \operatorname{Norm}_A(a)\cdot\mathbf{N}(l)\end{aligned}$$

Now for an arbitrary global section l, we can write $l = a'l'$ for some $a' \in A$ and some l' generating V_L as an A-module. So

$$\begin{aligned}\mathbf{N}(a\,l) &= \mathbf{N}(a\,a'\,l')\\ &= \operatorname{Norm}_A(a\,a')\cdot\mathbf{N}(l')\\ &= \operatorname{Norm}_A(a)\cdot\operatorname{Norm}_A(a')\cdot\mathbf{N}(l')\\ &= \operatorname{Norm}_A(a)\cdot\mathbf{N}(a'\,l')\\ &= \operatorname{Norm}_A(a)\cdot\mathbf{N}(l).\end{aligned}$$

(3.5) Proposition. *Let L_1 and L_2 be line bundles on Y. There is a canonical isomorphism*

$$\mathbf{N}(L_1\otimes L_2)\xrightarrow{\sim}\mathbf{N}(L_1)\otimes\mathbf{N}(L_2)$$

characterized by the requirement that for any choice of global sections, l_1 of L_1 and l_2 of L_2, we have

$$\mathbf{N}(l_1\otimes l_2)\mapsto\mathbf{N}(l_1)\otimes\mathbf{N}(l_2).$$

Proof. Let l_1' be a fixed non-vanishing global section of L_1, and let l_2' be a fixed non-vanishing global section of L_2. Then $\mathbf{N}(l_1')$, $\mathbf{N}(l_2')$, and $\mathbf{N}(l_1'\otimes l_2')$ are non-zero elements of $\mathbf{N}(L_1')$, $\mathbf{N}(L_2')$, and $\mathbf{N}(L_1'\otimes L_2')$ respectively. Therefore, the rule

$$\mathbf{N}(l_1'\otimes l_2')\mapsto\mathbf{N}(l_1')\otimes\mathbf{N}(l_2')$$

defines uniquely an isomorphism

$$\phi\colon \mathbf{N}(L_1\otimes L_2)\xrightarrow{\sim}\mathbf{N}(L_1)\otimes\mathbf{N}(L_2).$$

This isomorphism has the desired property for at least one choice of global sections, l_1' and l_2'. Now for arbitrary global sections, l_1 of L_1 and l_2 of L_2, there are elements a_1 and a_2 of A such that $l_1 = a_1\cdot l_1'$ and $l_2 = a_2\cdot l_2'$, and consequently

$$l_1\otimes l_2 = a_1a_2\cdot l_1'\otimes l_2'.$$

So, under the map ϕ,

$$\begin{aligned}\mathbf{N}(l_1\otimes l_2) &= \mathbf{N}(a_1a_2\cdot l_1'\otimes l_2')\\ &= \operatorname{Norm}_A(a_1\,a_2)\cdot\mathbf{N}(l_1'\otimes l_2')\\ &\mapsto \operatorname{Norm}_A(a_1\,a_2)\cdot\mathbf{N}(l_1')\otimes\mathbf{N}(l_2')\\ &= \operatorname{Norm}_A(a_1)\cdot\operatorname{Norm}_A(a_2)\cdot\mathbf{N}(l_1')\otimes\mathbf{N}(l_2')\\ &= \mathbf{N}(a_1\cdot l_1')\otimes\mathbf{N}(a_2\cdot l_2')\\ &= \mathbf{N}(l_1)\otimes\mathbf{N}(l_2).\end{aligned}$$

(3.6) Proposition. *Suppose Y is the disjoint union of Y_1 and Y_2, where Y_1 and Y_2 are finite over K. For every line bundle L on Y there is a canonical isomorphism*

$$\mathbf{N}_Y(L) \xrightarrow{\sim} \mathbf{N}_{Y_1}(L|_{Y_1}) \otimes \mathbf{N}_{Y_2}(L|_{Y_2})$$

characterized by

$$\mathbf{N}_Y(l) \mapsto \mathbf{N}_{Y_1}(l|_{Y_1}) \otimes \mathbf{N}_{Y_2}(l|_{Y_2})$$

for any global section l of L.

Proof. Let A_1 and A_2 be the associated rings, of finite dimension over K, such that $Y_1 = \mathrm{Spec}(A_1)$, $Y_2 = \mathrm{Spec}(A_2)$, and $Y = \mathrm{Spec}(A_1 \times A_2)$. Let l_0 be a fixed nowhere-vanishing global section of L; hence, $l_0|_{Y_1}$ is a nowhere-vanishing global section of $L|_{Y_1}$ and $l_0|_{Y_2}$ is a nowhere-vanishing global section of $L|_{Y_2}$. It follows that $\mathbf{N}_Y(l_0)$, $\mathbf{N}_{Y_1}(l_0|_{Y_1})$, and $\mathbf{N}_{Y_2}(l_0|_{Y_2})$ are non-zero elements of $\mathbf{N}_Y(L)$, $\mathbf{N}_{Y_1}(L|_{Y_1})$, and $\mathbf{N}_{Y_2}(L|_{Y_2})$ respectively. So the rule

$$\mathbf{N}_Y(l_0) \mapsto \mathbf{N}_{Y_1}(l_0|_{Y_1}) \otimes \mathbf{N}_{Y_2}(l_0|_{Y_2})$$

defines uniquely an isomorphism

$$\phi\colon \mathbf{N}_Y(L) \xrightarrow{\sim} \mathbf{N}_{Y_1}(L|_{Y_1}) \otimes \mathbf{N}_{Y_2}(L|_{Y_2}),$$

and this isomorphism has the desired property for at least one global section, namely l_0. Now for an arbitrary l, we can write $l = a \cdot l_0$ for some $a = (a_1, a_2)$ in $A_1 \times A_2$. By looking at the K-linear action of $a = (a_1, a_2)$ on $A_1 \times A_2$, it is clear that

$$\mathrm{Norm}_{A_1 \times A_2}(a) = \mathrm{Norm}_{A_1}(a_1) \cdot \mathrm{Norm}_{A_2}(a_2).$$

Therefore, under ϕ,

$$\begin{aligned}
\mathbf{N}_Y(l) &= \mathbf{N}_Y(a \cdot l_0) \\
&= \mathrm{Norm}_{A_1 \times A_2}(a) \cdot \mathbf{N}_Y(l_0) \\
&= \mathrm{Norm}_{A_1}(a_1) \cdot \mathrm{Norm}_{A_2}(a_2) \cdot \mathbf{N}_Y(l_0) \\
&\mapsto \mathrm{Norm}_{A_1}(a_1) \cdot \mathrm{Norm}_{A_2}(a_2) \cdot \mathbf{N}_{Y_1}(l_0|_{Y_1}) \otimes \mathbf{N}_{Y_2}(l_0|_{Y_2}) \\
&= \mathbf{N}_{Y_1}(a_1\, l_0|_{Y_1}) \otimes \mathbf{N}_{Y_2}(a_2\, l_0|_{Y_2}) \\
&= \mathbf{N}_{Y_1}(l|_{Y_1}) \otimes \mathbf{N}_{Y_2}(l|_{Y_2}).
\end{aligned}$$

4. The Definition of the Intersection Pairing

Throughout this section let X be a fixed one-dimensional scheme, projective over a field K. In this section we define the intersection pairing of any pair of line bundles on X. We also define some associated constructions for this pairing.

(4.1) Modelling Data. Let L and M be line bundles of X. We will define the intersection pairing between L and M in two steps. First we will describe how to construct a model of the pairing. Then, in a later section, we will show that any two models of the intersection pairing are canonically isomorphic.

To construct a model of the intersection pairing between L and M we need the following *modelling data*:

1. Line bundles L_1, L_2, M_1, and M_2 on X together with isomorphisms
$$L = \frac{L_1}{L_2} \qquad \text{and} \qquad M = \frac{M_1}{M_2}$$
which we will regarded as identifications;
2. A pair of integers ρ and ρ' (mod 2) satisfying the equation
$$\rho + \rho' = \deg L \cdot \deg M \qquad (\text{mod } 2).$$

For example, we could choose $L_1 = L$, $L_2 = \mathcal{O}_X$, $M_1 = M$, $M_2 = \mathcal{O}_X$, $\rho = 0$, and $\rho' = \deg L \cdot \deg M$.

(4.2) The Intersection Pairing. Let L and M be line bundles on X. Given a choice of modelling data
$$\mathcal{D} = (L_1,\, L_2,\, M_1,\, M_2,\, \rho,\, \rho'),$$
we define the *intersection pairing* of L and M with respect to $\mathcal{D}$, denoted $\langle L, M\rangle_{\mathcal{D}}$ or simply $\langle L, M\rangle$, to be the following one dimensional K-vector space:
$$\langle L, M\rangle_{\mathcal{D}} \overset{\text{def}}{=} \frac{\mathbf{D}(L_1 \otimes M_1) \otimes \mathbf{D}(L_2 \otimes M_2)}{\mathbf{D}(L_1 \otimes M_2) \otimes \mathbf{D}(L_2 \otimes M_1)}.$$

Although the values of ρ and ρ' in $\mathcal{D}$ do not affect the actual definition of the underlying vector space $\langle L, M\rangle_{\mathcal{D}}$, they do affect the associated constructions of Definition 4.6. We will show later that for different choices of $\mathcal{D}$ the resulting intersection pairings are canonically isomorphic, and that the canonical isomorphism is well behaved with respect to the associated constructions of Definition 4.6. Therefore, the modelling data is important only for providing an explicit model of the pairing. If we are not concerned with the explicit model, we will simply write $\langle L, M\rangle$ for the intersection pairing.

(4.3) Pairing of Meromorphic Sections. Let L and M be line bundles on X, and let
$$\mathcal{D} = (L_1,\, L_2,\, M_1,\, M_2,\, \rho,\, \rho')$$
be modelling data for the pair (L, M). Since X is projective of dimension 1, it is always possible to find meromorphic section, l of L and m of M, such that their associated Cartier divisors (l) and (m) have disjoint supports. We will now proceed to construct, for any such l and m, an isomorphism of K-vector spaces:
$$\Phi_{l,m} \colon \langle L, M\rangle_{\mathcal{D}} \to K.$$

We will then define the element $\langle l, m\rangle$ to be the element of $\langle L, M\rangle_{\mathcal{D}}$ which maps to 1 under this isomorphism.

Given l and m meromorphic sections of L and M respectively such that their associated Cartier divisors (l) and (m) have disjoint support, choose effective Cartier divisors D_1, D_2, E_1, and E_2 such that $(l) = D_1 - D_2$ and $(m) = E_1 - E_2$ and such that the support of D_i is disjoint from the support of E_j for each $i, j \in \{1,2\}$.

These give isomorphisms

$$L \xrightarrow{\sim} \mathcal{O}_X(D_1 - D_2) \qquad \text{where} \qquad l \mapsto \mathbf{1}$$

and

$$M \xrightarrow{\sim} \mathcal{O}_X(E_1 - E_2) \qquad \text{where} \qquad m \mapsto \mathbf{1}.$$

Define the following two line bundles:

$$L_0 \stackrel{\text{def}}{=} L_1 \otimes \mathcal{O}_X(-D_1) \qquad \text{and} \qquad M_0 \stackrel{\text{def}}{=} M_1 \otimes \mathcal{O}_X(-E_1).$$

Using the identification of L_1 with $L_2 \otimes L$ and the above isomorphism from L to $\mathcal{O}_X(D_1 - D_2)$, we can form an isomorphism from

$$L_0 = L_1(-D_1) = L_2 \otimes L \otimes \mathcal{O}_X(-D_1) \qquad \text{to} \qquad L_2 \otimes \mathcal{O}_X(-D_2) = L_2(-D_2).$$

This isomorphism can be described as follows. Let l_2 be a meromorphic section of L_2. Using the identification of $L_2 \otimes L$ with L_1, we can consider $l_2 \otimes l$ as a meromorphic section of L_1. Then the above isomorphism takes $(l_2 \otimes l) \otimes \mathbf{1}$ to $l_2 \otimes \mathbf{1}$.

Likewise, we get an isomorphism

$$M_0 = M_1(-E_1) \xrightarrow{\sim} M_2(-E_2) \qquad \text{in which} \qquad (m_2 \otimes m) \otimes \mathbf{1} \mapsto m_2 \otimes \mathbf{1}$$

for any meromorphic section m_2 of M_2.

For $i, j \in \{1,2\}$, we have the exact sequences:

$$0 \longrightarrow L_0 \longrightarrow L_i \longrightarrow L_i|_{D_i} \longrightarrow 0,$$

and

$$0 \longrightarrow M_0 \longrightarrow M_j \longrightarrow M_j|_{E_j} \longrightarrow 0$$

(where $L_i|_{D_i}$ and $M_j|_{E_j}$ are considered as coherent sheaves on X). We can tensor the terms of these sequences together and arrange them in the following commutative diagram

$$
\begin{array}{ccccccccc}
 & & (\zeta^1_{i,j}) & & (\zeta^2_{i,j}) & & (\zeta^3_{i,j}) & & \\
 & & 0 & & 0 & & 0 & & \\
 & & \downarrow & & \downarrow & & \downarrow & & \\
(\eta^1_{i,j}) & 0 \longrightarrow & L_0 \otimes M_0 & \longrightarrow & L_i \otimes M_0 & \longrightarrow & (L_i \otimes M_0)|_{D_i} & \longrightarrow & 0 \\
 & & \downarrow & & \downarrow & & \downarrow & & \\
(\eta^2_{i,j}) & 0 \longrightarrow & L_0 \otimes M_j & \longrightarrow & L_i \otimes M_j & \longrightarrow & (L_i \otimes M_j)|_{D_i} & \longrightarrow & 0 \\
 & & \downarrow & & \downarrow & & \downarrow & & \\
(\eta^3_{i,j}) & 0 \longrightarrow & (L_0 \otimes M_j)|_{E_j} & \longrightarrow & (L_i \otimes M_j)|_{E_j} & \longrightarrow & 0 & & \\
 & & \downarrow & & \downarrow & & & & \\
 & & 0 & & 0 & & & &
\end{array}
$$

(since D_i and E_j are disjoint, the tensor product of $L_i|_{D_i}$ with $M_j|_{E_j}$ is the zero sheaf). The rows and columns of the above diagram are exact: the first and second rows and columns are exact since tensoring by a line bundle is an exact functor; the third row is exact since the map from L_0 to L_i restricts to an isomorphism on E_j because D_i and E_j are disjoint; similarly, the third column is exact.

We define the isomorphism $\phi_{i,j}$ to be the composition of the isomorphism

$$\mathbf{D}(L_i \otimes M_j) \xrightarrow{\mathrm{D}(\eta^2_{i,j})} \mathbf{D}(L_0 \otimes M_j) \otimes \mathbf{D}\big((L_i \otimes M_j)|_{D_i}\big)$$

with

$$\mathbf{D}(L_0 \otimes M_j) \otimes \mathbf{D}\big((L_i \otimes M_j)|_{D_i}\big) \xrightarrow{\mathrm{D}(\zeta^1_{i,j}) \otimes \mathrm{D}(\zeta^3_{i,j})} \mathbf{D}(L_0 \otimes M_0) \otimes \mathbf{D}\big((L_0 \otimes M_j)|_{E_j}\big) \otimes \mathbf{D}\big((L_i \otimes M_0)|_{D_i}\big).$$

Similarly, define $\phi'_{i,j}$ to be the composition of

$$\mathbf{D}(L_i \otimes M_j) \xrightarrow{\mathrm{D}(\zeta^2_{i,j})} \mathbf{D}(L_i \otimes M_0) \otimes \mathbf{D}\big((L_i \otimes M_j)|_{E_j}\big)$$

with

$$\mathbf{D}(L_i \otimes M_0) \otimes \mathbf{D}\big((L_i \otimes M_j)|_{E_j}\big) \xrightarrow{\mathrm{D}(\eta^1_{i,j}) \otimes \mathrm{D}(\eta^3_{i,j})} \mathbf{D}(L_0 \otimes M_0) \otimes \mathbf{D}\big((L_i \otimes M_0)|_{D_i}\big) \otimes \mathbf{D}\big((L_0 \otimes M_j)|_{E_j}\big).$$

By Theorem 2.5.1, the following commutes

$$
\begin{array}{ccc}
\mathbf{D}(L_i \otimes M_j) & \xrightarrow{\phi_{i,j}} & \mathbf{D}(L_0 \otimes M_0) \otimes \mathbf{D}\big((L_0 \otimes M_j)|_{E_j}\big) \otimes \mathbf{D}\big((L_i \otimes M_0)|_{D_i}\big) \\
\| & & \downarrow I \\
\mathbf{D}(L_i \otimes M_j) & \xrightarrow{\phi'_{i,j}} & \mathbf{D}(L_0 \otimes M_0) \otimes \mathbf{D}\big((L_i \otimes M_0)|_{D_i}\big) \otimes \mathbf{D}\big((L_0 \otimes M_j)|_{E_j}\big)
\end{array}
$$

where I is $(-1)^{\pi_{i,j}}$ times the canonical isomorphism switching the two final terms, and where $\pi_{i,j}$ is an integer which is explicitly described in (2.5.6). The following lemma gives a more useful expression for $\pi_{i,j}$.

(4.4) Lemma. *The above integer* $\pi_{i,j}$ *is equal (mod 2) to*

$$\pi_{i,j} \;=\; \deg(D_i)\,\deg(E_j) + u_i + u'_j + h^0(L_i \otimes M_j)\,(v_i + v'_j)$$

where h^0 *of a line bundle is the dimension of its* 0*th cohomology group, and where the integers* u_i, v_i, u'_j, *and* v'_j *are defined as follows:*

$$u_i = h^0(L_i \otimes M_0)\,\big(h^0(L_0 \otimes M_0) + \deg(D_i) + 1\big), \quad v_i = h^0(L_i \otimes M_0) + \deg(D_i),$$
$$u'_j = h^0(L_0 \otimes M_j)\,\big(h^0(L_0 \otimes M_0) + \deg(E_j) + 1\big), \quad v'_j = h^0(L_0 \otimes M_j) + \deg(E_j).$$

Thus the sum, mod 2, of these $\pi_{i,j}$ *terms is as follows:*

$$\begin{aligned}\sum_{i,j\in\{1,2\}} \pi_{i,j} \;&=\; \sum_{i,j\in\{1,2\}} \Big(\deg(D_i)\,\deg(E_j) + u_i + u'_j + h^0(L_i \otimes M_j)\,(v_i + v'_j)\Big) \\ &=\; \deg(L)\,\deg(M) + \sum_{i,j\in\{1,2\}} h^0(L_i \otimes M_j)\,(v_i + v'_j).\end{aligned}$$

Proof. We use the notation of (2.5.6) where

$$H^k_{1,1} = H^k(L_0 \otimes M_0), \quad H^k_{1,2} = H^k(L_i \otimes M_0), \quad H^k_{1,3} = H^k\big(L_i \otimes M_0|_{D_i}\big), \quad \text{etc..}$$

In (2.5.6) we derived the formula

$$\pi_{i,j} = \sum_{k;\ 1\le p,q\le 3} c^k_{p,q}\, s^k_{p,q} + \sum_k a^k_{1,1} + \sum_k \big(b^k_{1,1} + b^k_{2,2}\big) \qquad \text{(mod 2)}.$$

There are terms in the above formula which are obviously zero; for example, the terms with $k > 1$ or $k < 0$. We can use the properties of the special bases constructed in (2.5.6) to see that most of the remaining terms are zero. In fact, in the end we are left only with three terms:

$$\pi_{i,j} \;=\; c^0_{2,2}\cdot s^0_{2,2} \;+\; c^1_{1,1}\cdot s^1_{1,1} \;+\; b^1_{1,1} \qquad \text{(mod 2)}.$$

We can compute the dimensions of each $H^k_{i,j}$ by using the special bases constructed in (2.5.6) and their mapping properties. For example,

$$\begin{aligned}\dim H^1_{1,1} &= h^1(L_0 \otimes M_0) &&= a^1_{2,2} + c^1_{1,1} + s^1_{1,1} + b^1_{1,1} \\ \dim H^1_{1,2} &= h^1(L_i \otimes M_0) &&= a^1_{2,2} + s^1_{1,1} \\ \dim H^1_{2,1} &= h^1(L_0 \otimes M_j) &&= a^1_{2,2} + c^1_{1,1} \\ \dim H^1_{2,2} &= h^1(L_i \otimes M_j) &&= a^1_{2,2} \\ \dim H^0_{1,1} &= h^0(L_0 \otimes M_0) &&= a^0_{2,2} \\ \dim H^0_{1,2} &= h^0(L_i \otimes M_0) &&= a^0_{2,2} + s^0_{2,2} \\ \dim H^0_{2,1} &= h^0(L_0 \otimes M_j) &&= a^0_{2,2} + c^0_{2,2}\end{aligned}$$

so

$$c^0_{2,2} = h^0(L_0 \otimes M_j) - h^0(L_0 \otimes M_0), \qquad s^0_{2,2} = h^0(L_i \otimes M_0) - h^0(L_0 \otimes M_0),$$
$$c^1_{1,1} = h^1(L_0 \otimes M_j) - h^1(L_i \otimes M_j), \qquad s^1_{1,1} = h^1(L_i \otimes M_0) - h^1(L_i \otimes M_j), \quad \text{and}$$
$$b^1_{1,1} = h^1(L_0 \otimes M_0) - h^1(L_0 \otimes M_j) - h^1(L_i \otimes M_0) + h^1(L_i \otimes M_j).$$

By using the additivity of Euler characteristics on short exact sequences we get (mod 2) the following:

$$\begin{aligned} c^1_{1,1} &= h^0(L_0 \otimes M_j) + h^0(L_i \otimes M_j) + \deg(D_i) \\ s^1_{1,1} &= h^0(L_i \otimes M_0) + h^0(L_i \otimes M_j) + \deg(E_j) \\ b^1_{1,1} &= h^0(L_0 \otimes M_0) + h^0(L_0 \otimes M_j) + h^0(L_i \otimes M_0) + h^0(L_i \otimes M_j). \end{aligned}$$

We obtain the lemma by substituting these expressions into the formula for $\pi_{i,j}$, and then simplifying using mod 2 arithmetic.

(4.5) We now continue the discussion and notation of (4.3). Consider the isomorphism

$$\frac{\phi_{1,1} \otimes \phi_{2,2}}{\phi_{1,2} \otimes \phi_{2,1}} \qquad \text{which maps} \qquad \langle L, M \rangle_{\mathcal{D}} = \frac{\mathbf{D}(L_1 \otimes M_1) \otimes \mathbf{D}(L_2 \otimes M_2)}{\mathbf{D}(L_1 \otimes M_2) \otimes \mathbf{D}(L_2 \otimes M_1)}$$

to the space

$$\frac{\mathbf{D}(L_0 \otimes M_0) \otimes \mathbf{D}(L_0 \otimes M_1|_{E_1}) \otimes \mathbf{D}(L_1 \otimes M_0|_{D_1}) \otimes \mathbf{D}(L_0 \otimes M_0) \otimes \mathbf{D}(L_0 \otimes M_2|_{E_2}) \otimes \mathbf{D}(L_2 \otimes M_0|_{D_2})}{\mathbf{D}(L_0 \otimes M_0) \otimes \mathbf{D}(L_0 \otimes M_2|_{E_2}) \otimes \mathbf{D}(L_1 \otimes M_0|_{D_1}) \otimes \mathbf{D}(L_0 \otimes M_0) \otimes \mathbf{D}(L_0 \otimes M_1|_{E_1}) \otimes \mathbf{D}(L_2 \otimes M_0|_{D_2})}.$$

Note that each term of the "numerator" of this image space also occurs in the "denominator" and vice versa. Therefore, there is the canonical "cancellation" map between this space and the field K.

Let ϕ be the K-vector space isomorphism from $\langle L, M \rangle_{\mathcal{D}}$ to K obtained by composing

$$\frac{\phi_{1,1} \otimes \phi_{2,2}}{\phi_{1,2} \otimes \phi_{2,1}}$$

with the natural "cancellation" map.

Similarly, let ϕ' be the K-vector space isomorphism from $\langle L, M \rangle_{\mathcal{D}}$ to K obtained by composing

$$\frac{\phi'_{1,1} \otimes \phi'_{2,2}}{\phi'_{1,2} \otimes \phi'_{2,1}}$$

with the natural "cancellation" map.

Lemma 4.4 above implies that $\phi = (-1)^{\pi} \phi'$ where

$$\begin{aligned} \pi &= \deg(L)\deg(M) \\ &\quad + \sum_{i,j \in \{1,2\}} h^0(L_i \otimes M_j)\Big(h^0(L_i \otimes M_0) + \deg(D_i) + h^0(L_0 \otimes M_j) + \deg(E_j)\Big). \end{aligned}$$

Now define the following two integers

$$\begin{aligned} \tau &= \rho + \sum_{i,j \in \{1,2\}} h^0(L_i \otimes M_j)\Big(h^0(L_0 \otimes M_j) + \deg(D_i)\Big) \\ \tau' &= \rho' + \sum_{i,j \in \{1,2\}} h^0(L_i \otimes M_j)\Big(h^0(L_i \otimes M_0) + \deg(E_j)\Big), \end{aligned}$$

where ρ and ρ' are the integers from the modelling data $\mathcal{D}$ satisfying

$$\rho + \rho' = \deg(L)\deg(M) \qquad (\text{mod } 2).$$

Since $\pi = \tau + \tau' \pmod 2$, we conclude that

$$(-1)^{\tau}\phi = (-1)^{\tau'}\phi'.$$

We are now ready for the promised definition.

(4.6) Definition. Let L and M be line bundles on X, and let $\mathcal{D}$ be a choice of modelling data for L and M. For any pair of meromorphic sections, l of L and m of M, whose associated Cartier divisors, (l) and (m), have disjoint supports, we define an isomorphism of K-vector spaces,

$$\Phi_{l,m}\colon \langle L, M\rangle_{\mathcal{D}} \to K,$$

as follows: first choose effective Cartier divisors D_1, D_2, E_1, E_2 in such a way that $(l) = D_1 - D_2$ and $(m) = E_1 - E_2$, and such that the support of D_i is disjoint from the support of E_j for each $i, j = 1, 2$. Given such choices, we constructed, in (4.3) to (4.5), two isomorphisms, ϕ and ϕ', from $\langle L, M\rangle_{\mathcal{D}}$ to K. Then in (4.5) we defined integers, τ and τ', such that $(-1)^{\tau}\phi = (-1)^{\tau'}\phi'$. Define

$$\Phi_{l,m} \stackrel{\text{def}}{=} (-1)^{\tau}\phi = (-1)^{\tau'}\phi'.$$

From the way it is defined, it appears that $\Phi_{l,m}$ depends, not only on l and m, but also on the choice of D_1, D_2, E_1, and E_2. We will show later in this section (Proposition 4.8) that, in fact, *$\Phi_{l,m}$ is independent of the choice of D_1, D_2, E_1, and E_2.*

For any pair of meromorphic sections, l of L and m of M, whose associated Cartier divisors, (l) and (m), have disjoint supports, we define the *intersection pairing* of l and m, written $\langle l, m\rangle$, to be the inverse image of $1 \in K$ under $\Phi_{l,m}$. Thus, $\langle l, m\rangle$ is a non-zero element of $\langle L, M\rangle_{\mathcal{D}}$.

(4.7) Let L, M, $\mathcal{D}$, l, m, D_1, D_2, E_1, E_2, etc. be as in (4.3 – 4.6). We will also use the notation and constructions introduced in these subsections. Consider the isomorphism

$$\frac{\mathbf{D}(\eta^2_{1,1}) \otimes \mathbf{D}(\eta^2_{2,2})}{\mathbf{D}(\eta^2_{1,2}) \otimes \mathbf{D}(\eta^2_{2,1})} \qquad \text{which maps} \qquad \langle L, M\rangle_{\mathcal{D}} = \frac{\mathbf{D}(L_1 \otimes M_1) \otimes \mathbf{D}(L_2 \otimes M_2)}{\mathbf{D}(L_1 \otimes M_2) \otimes \mathbf{D}(L_2 \otimes M_1)}$$

to

$$\frac{\Big(\mathbf{D}(L_0 \otimes M_1) \otimes \mathbf{D}\big(L_1 \otimes M_1|_{D_1}\big)\Big) \otimes \Big(\mathbf{D}(L_0 \otimes M_2) \otimes \mathbf{D}\big(L_2 \otimes M_2|_{D_2}\big)\Big)}{\Big(\mathbf{D}(L_0 \otimes M_2) \otimes \mathbf{D}\big(L_1 \otimes M_2|_{D_1}\big)\Big) \otimes \Big(\mathbf{D}(L_0 \otimes M_1) \otimes \mathbf{D}\big(L_2 \otimes M_1|_{D_2}\big)\Big)}.$$

Also consider the natural "cancellation" map

$$\frac{\mathbf{D}(L_0 \otimes M_1) \otimes \mathbf{D}(L_1 \otimes M_1|_{D_1}) \otimes \mathbf{D}(L_0 \otimes M_2) \otimes \mathbf{D}(L_2 \otimes M_2|_{D_2})}{\mathbf{D}(L_0 \otimes M_2) \otimes \mathbf{D}(L_1 \otimes M_2|_{D_1}) \otimes \mathbf{D}(L_0 \otimes M_1) \otimes \mathbf{D}(L_2 \otimes M_1|_{D_2})} \xrightarrow{\sim} \frac{\mathbf{D}(L_1 \otimes M_1|_{D_1}) \otimes \mathbf{D}(L_2 \otimes M_2|_{D_2})}{\mathbf{D}(L_1 \otimes M_2|_{D_1}) \otimes \mathbf{D}(L_2 \otimes M_1|_{D_2})}.$$

We define the following isomorphism:

$$\Psi_{(l,D_1,D_2)}\colon \langle L, M\rangle_{\mathcal{D}} \xrightarrow{\sim} \frac{\mathbf{D}(L_1 \otimes M_1|_{D_1}) \otimes \mathbf{D}(L_2 \otimes M_2|_{D_2})}{\mathbf{D}(L_1 \otimes M_2|_{D_1}) \otimes \mathbf{D}(L_2 \otimes M_1|_{D_2})}$$

to be $(-1)^{\tau}$ times the composition of the above two isomorphisms. Since each map $\mathbf{D}(\eta^2_{i,j})$ is independent of E_1,E_2 and M_0, and the same is true of the integer τ (defined in (4.5)), we conclude that *$\Psi_{(l,D_1,D_2)}$ is independent of E_1,E_2 and M_0.*

Similarly, we define

$$\Psi'_{(m,E_1,E_2)}\colon \langle L,M\rangle_{\mathcal{D}} \xrightarrow{\sim} \frac{\mathbf{D}(L_1\otimes M_1|_{E_1})\otimes\mathbf{D}(L_2\otimes M_2|_{E_2})}{\mathbf{D}(L_1\otimes M_2|_{E_2})\otimes\mathbf{D}(L_2\otimes M_1|_{E_1})}$$

to be $(-1)^{\tau'}$ times

$$\frac{\mathbf{D}(\zeta^2_{1,1})\otimes\mathbf{D}(\zeta^2_{2,2})}{\mathbf{D}(\zeta^2_{1,2})\otimes\mathbf{D}(\zeta^2_{2,1})}$$

composed with the natural "cancellation map". We also conclude that $\Psi'_{(m,E_1,E_2)}$ *is independent of* D_1,D_2 *and* L_0.

Now define $\Theta_{(D_1,D_2;\,m)}$ to be the composition of the isomorphism

$$\frac{\mathbf{D}(\zeta^3_{1,1})\otimes\mathbf{D}(\zeta^3_{2,2})}{\mathbf{D}(\zeta^3_{1,2})\otimes\mathbf{D}(\zeta^3_{2,1})}\colon \frac{\mathbf{D}(L_1\otimes M_1|_{D_1})\otimes\mathbf{D}(L_2\otimes M_2|_{D_2})}{\mathbf{D}(L_1\otimes M_2|_{D_1})\otimes\mathbf{D}(L_2\otimes M_1|_{D_2})} \longrightarrow \frac{\mathbf{D}(L_1\otimes M_0|_{D_1})\otimes\mathbf{D}(L_2\otimes M_0|_{D_2})}{\mathbf{D}(L_1\otimes M_0|_{D_1})\otimes\mathbf{D}(L_2\otimes M_0|_{D_2})}$$

followed by the natural "cancellation" isomorphism

$$\frac{\mathbf{D}(L_1\otimes M_0|_{D_1})\otimes\mathbf{D}(L_2\otimes M_0|_{D_2})}{\mathbf{D}(L_1\otimes M_0|_{D_1})\otimes\mathbf{D}(L_2\otimes M_0|_{D_2})} \to K.$$

We observe that the isomorphism $\Phi_{l,m}$, defined in (4.6), can be factored as follows:

$$\Phi_{l,m} = \Theta_{(D_1,D_2;\,m)} \circ \Psi_{(l,D_1,D_2)}.$$

Similarly, define $\Theta'_{(E_1,E_2;\,l)}$ to be the composition of the isomorphism

$$\frac{\mathbf{D}(\eta^3_{1,1})\otimes\mathbf{D}(\eta^3_{2,2})}{\mathbf{D}(\eta^3_{1,2})\otimes\mathbf{D}(\eta^3_{2,1})}\colon \frac{\mathbf{D}(L_1\otimes M_1|_{E_1})\otimes\mathbf{D}(L_2\otimes M_2|_{E_2})}{\mathbf{D}(L_1\otimes M_2|_{E_2})\otimes\mathbf{D}(L_2\otimes M_1|_{E_1})} \longrightarrow \frac{\mathbf{D}(L_0\otimes M_1|_{E_1})\otimes\mathbf{D}(L_0\otimes M_2|_{E_2})}{\mathbf{D}(L_0\otimes M_2|_{E_2})\otimes\mathbf{D}(L_0\otimes M_1|_{E_1})}$$

followed by the natural "cancellation" isomorphism

$$\frac{\mathbf{D}(L_0\otimes M_1|_{E_1})\otimes\mathbf{D}(L_0\otimes M_2|_{E_2})}{\mathbf{D}(L_0\otimes M_2|_{E_2})\otimes\mathbf{D}(L_0\otimes M_1|_{E_1})} \to K.$$

We observe that the isomorphism $\Phi_{l,m}$, defined in (4.6), can be factored as follows:

$$\Phi_{l,m} = \Theta'_{(E_1,E_2;\,l)} \circ \Psi'_{(m,E_1,E_2)}.$$

The notation suggests that $\Theta_{(D_1,D_2;m)}$ is independent of E_1,E_2, and M_0, and that $\Theta'_{(E_1,E_2;l)}$ is independent of D_1 ,D_2, and L_0. This is what we will now show.

For $i,j\in\{1,2\}$, let $\alpha_{i,j}$ be the isomorphism from $(L_i\otimes M_0)|_{D_i}$ to $(L_i\otimes M_j)|_{D_i}$ which occurs in the exact sequence $\zeta^3_{i,j}$ (defined in (4.3)). (For technical reasons, we make the distinction between the exact sequence $\zeta^3_{i,j}$ and the corresponding isomorphism $\alpha_{i,j}$). Let l_1, l_2, and m_2 be meromorphic sections of L_1, L_2, and M_2 respectively, which are defined and nowhere vanishing on D_1 and D_2. Under the identification of M_1 with $M_2\otimes M$, we can regard $m_1 = m_2\otimes m$ as a meromorphic section of M_1 which is defined and nowhere vanishing on D_1 and D_2.

Recall that $M_0 = M_1(-E_1)$, and that

$$M_0 = M_1(-E_1) \to M_1 \qquad \text{sends} \quad m_1\otimes\mathbf{1}\mapsto m_1.$$

So $\alpha_{i,1}$ is characterized by

$$l_i\otimes m_1\otimes\mathbf{1}\,|_{D_i} \mapsto l_i\otimes m_1\,|_{D_i}.$$

In addition, under

$$M_0 = M_1(-E_1) \longrightarrow M_2 \qquad \text{we have} \quad m_1 \otimes \mathbf{1} = (m_2 \otimes m) \otimes \mathbf{1} \mapsto m_2$$

since this map is the composition of the isomorphism $M_0 \xrightarrow{\sim} M_2(-E_2)$, defined in (4.3), followed by the natural injection. So $\alpha_{i,2}$ is characterized by

$$l_i \otimes m_1 \otimes \mathbf{1}\,|_{D_i} \mapsto l_i \otimes m_2\,|_{D_i}.$$

Therefore, the composition $\beta_i \stackrel{\text{def}}{=} \alpha_{i,2} \circ \alpha_{i,1}^{-1}$, which maps $L_i \otimes M_1|_{D_i}$ to $L_i \otimes M_2|_{D_i}$, is characterized by

$$l_i \otimes m_1\,|_{D_i} \mapsto l_i \otimes m_2\,|_{D_i}.$$

Clearly, β_i is independent of E_1, E_2, and M_0. Applying the determinant of cohomology functor, we get that $\mathbf{D}(\beta_i)$ is the composition of $\mathbf{D}(\alpha_{i,1})^{-1} = \mathbf{D}(\zeta^3_{i,1})$ followed by $\mathbf{D}(\alpha_{i,2}) = \mathbf{D}(\zeta^3_{i,2})^{-1}$. Thus, the following commutes

$$\begin{array}{ccc}
\dfrac{\mathbf{D}(L_1 \otimes M_1|_{D_1}) \otimes \mathbf{D}(L_2 \otimes M_2|_{D_2})}{\mathbf{D}(L_1 \otimes M_2|_{D_1}) \otimes \mathbf{D}(L_2 \otimes M_1|_{D_2})} & \xrightarrow{\frac{\mathbf{D}(\zeta^3_{1,1}) \otimes \mathbf{D}(\zeta^3_{2,2})}{\mathbf{D}(\zeta^3_{1,2}) \otimes \mathbf{D}(\zeta^3_{2,1})}} & \dfrac{\mathbf{D}(L_1 \otimes M_0|_{D_1}) \otimes \mathbf{D}(L_2 \otimes M_0|_{D_2})}{\mathbf{D}(L_1 \otimes M_0|_{D_1}) \otimes \mathbf{D}(L_2 \otimes M_0|_{D_2})} \\
\Big\downarrow \frac{\mathbf{D}(\beta_1) \otimes 1}{1 \otimes \mathbf{D}(\beta_2)} & & \Big\| \\
\dfrac{\mathbf{D}(L_1 \otimes M_2|_{D_1}) \otimes \mathbf{D}(L_2 \otimes M_2|_{D_2})}{\mathbf{D}(L_1 \otimes M_2|_{D_1}) \otimes \mathbf{D}(L_2 \otimes M_2|_{D_2})} & \xrightarrow{\frac{\mathbf{D}(\zeta^3_{1,2}) \otimes \mathbf{D}(\zeta^3_{2,2})}{\mathbf{D}(\zeta^3_{1,2}) \otimes \mathbf{D}(\zeta^3_{2,2})}} & \dfrac{\mathbf{D}(L_1 \otimes M_0|_{D_1}) \otimes \mathbf{D}(L_2 \otimes M_0|_{D_2})}{\mathbf{D}(L_1 \otimes M_0|_{D_1}) \otimes \mathbf{D}(L_2 \otimes M_0|_{D_2})}
\end{array}$$

And of course, the following "cancellation" isomorphism commutes:

$$\begin{array}{ccc}
\dfrac{\mathbf{D}(L_1 \otimes M_2|_{D_1}) \otimes \mathbf{D}(L_2 \otimes M_2|_{D_2})}{\mathbf{D}(L_1 \otimes M_2|_{D_1}) \otimes \mathbf{D}(L_2 \otimes M_2|_{D_2})} & \xrightarrow{\frac{\mathbf{D}(\zeta^3_{1,2}) \otimes \mathbf{D}(\zeta^3_{2,2})}{\mathbf{D}(\zeta^3_{1,2}) \otimes \mathbf{D}(\zeta^3_{2,2})}} & \dfrac{\mathbf{D}(L_1 \otimes M_0|_{D_1}) \otimes \mathbf{D}(L_2 \otimes M_0|_{D_2})}{\mathbf{D}(L_1 \otimes M_0|_{D_1}) \otimes \mathbf{D}(L_2 \otimes M_0|_{D_2})} \\
\Big\downarrow \text{cancel} & & \Big\downarrow \text{cancel} \\
K & = = = = & K
\end{array}$$

So $\Theta_{(D_1,D_2;\,m)}$ can be factored as the isomorphism

$$\frac{\mathbf{D}(\beta_1) \otimes 1}{1 \otimes \mathbf{D}(\beta_2)} \colon \frac{\mathbf{D}(L_1 \otimes M_1|_{D_1}) \otimes \mathbf{D}(L_2 \otimes M_2|_{D_2})}{\mathbf{D}(L_1 \otimes M_2|_{D_1}) \otimes \mathbf{D}(L_2 \otimes M_1|_{D_2})} \longrightarrow \frac{\mathbf{D}(L_1 \otimes M_2|_{D_1}) \otimes \mathbf{D}(L_2 \otimes M_2|_{D_2})}{\mathbf{D}(L_1 \otimes M_2|_{D_1}) \otimes \mathbf{D}(L_2 \otimes M_2|_{D_2})}$$

followed by the natural "cancellation" isomorphism. Therefore $\Theta_{(D_1,D_2;\,m)}$ is independent of E_1, E_2, and M_0.

A similar argument shows that $\Theta'_{(E_1,E_2;\,l)}$ is independent of D_1, D_2, and L_0.

(4.8) Proposition. *Let L and M be line bundles on X, and let $\mathcal{D}$ be a choice of modelling data for L and M. Given a pair of meromorphic sections, l of L and m of M, whose associated Cartier divisors (l) and (m) have disjoint supports, the associated isomorphism $\Phi_{l,m}$, defined in (4.6), is well-defined in the sense that it is independent of the choice of D_1, D_2, and E_1, E_2 used to construct it. Similarly, the element $\langle l, m \rangle$ is also independent of the choice of D_1, D_2, and E_1, E_2.*

Proof. In (4.7) we showed that the isomorphism $\Phi_{l,m}$ can be factored as the composition of two isomorphisms, $\Psi_{(l,D_1,D_2)}$ and $\Theta_{(D_1,D_2;\,m)}$, which were shown to be independent of E_1, E_2, and M_0.

Similarly, we showed that $\Phi_{l,m}$ can be factored as the composition of $\Psi'_{(m,E_1,E_2)}$ and $\Theta'_{(E_1,E_2;\,l)}$, which were shown to be independent of D_1, D_2, and L_0.

Finally, recall that the element $\langle l, m\rangle$ was defined as the inverse image of $1 \in K$ under $\Phi_{l,m}$.

5. The Intersection Pairing and Norms

In this section we develop the relationship between the intersection pairing and the norm functor. As in the last section, X will be a fixed one-dimensional scheme projective over a field K.

(5.1) Notation. Let D be a zero-dimensional subscheme of X, and M a line bundle on X. We will use the following notation, adapting the notation of Section 3 to the current situation:

- $\mathbf{N}_D(M)$ will denote the one-dimensional K-vector space $\mathbf{N}_D(M|_D)$.
- For m a meromorphic section of M defined on all of D, $\mathbf{N}_D(m)$ will denote the vector $\mathbf{N}_D(m|_D)$ of the one-dimensional K-vector space $\mathbf{N}_D(M)$.
- If f is an invertible meromorphic function of X defined on all of D, $\mathrm{Norm}_D(f)$ will denote the element $\mathrm{Norm}_D(f|_D)$ of K.

We also mention that if m is a meromorphic section of M defined and nowhere vanishing on D then $\mathbf{N}_D(m)$ is a non-zero vector of $\mathbf{N}_D(M)$.

(5.2) Proposition.
1. Let M be a line bundle on X, and let D_1 and D_2 be non-zero effective Cartier divisors on X. Let $\mathcal{D}$ be modelling data for $\mathcal{O}_X(D_1 - D_2)$ and M. There is a canonical isomorphism

$$\langle\, \mathcal{O}_X(D_1 - D_2),\, M \,\rangle_{\mathcal{D}} \longrightarrow \frac{\mathbf{N}_{D_1}(M)}{\mathbf{N}_{D_2}(M)}$$

characterized by

$$\langle \mathbf{1}, m\rangle \;\mapsto\; \frac{\mathbf{N}_{D_1}(m)}{\mathbf{N}_{D_2}(m)}$$

where $\mathbf{1}$ is the canonical meromorphic section of $\mathcal{O}_X(D_1 - D_2)$, and m is any meromorphic section of M defined and non-vanishing on both D_1 and D_2.
2. Let L be a line bundle on X, and let E_1 and E_2 be non-zero effective Cartier divisors on X. Let $\mathcal{D}$ be modelling data for L and $\mathcal{O}_X(E_1 - E_2)$. There is a canonical isomorphism

$$\langle\, L,\, \mathcal{O}_X(E_1 - E_2) \,\rangle_{\mathcal{D}} \longrightarrow \frac{\mathbf{N}_{E_1}(L)}{\mathbf{N}_{E_2}(L)}$$

characterized by

$$\langle l, \mathbf{1}\rangle \mapsto \frac{\mathbf{N}_{E_1}(l)}{\mathbf{N}_{E_2}(l)}$$

where $\mathbf{1}$ is the canonical meromorphic section of $\mathcal{O}_X(E_1 - E_2)$, and l is any meromorphic section of L defined and non-vanishing on both E_1 and E_2.

Proof. We will only prove the first statement; of course the proof of the second statement proceeds along the same lines. Let L be $\mathcal{O}_X(D_1 - D_2)$, and let

$$\mathcal{D} = (L_1, L_2, M_1, M_2, \rho, \rho')$$

be the chosen modelling data for L and M.

Choose meromorphic sections, m, m_2, l_1, and l_2 of M, M_2, L_1, and L_2 respectively, such that each is defined and nowhere vanishing on both D_1 and D_2. Let $\mathbf{1}$ be the canonical section of $L = \mathcal{O}_X(D_1 - D_2)$. Under the identification of M_1 with $M_2 \otimes M$, we can regard $m_1 = m_2 \otimes m$ as a meromorphic section of M_1 which is also defined and nowhere vanishing on both D_1 and D_2.

For $i, j \in \{1, 2\}$, let $\mu_{i,j}$ be the restriction of $l_i \otimes m_j$ to D_i, so $\mu_{i,j}$ is a nowhere vanishing section of $L_i \otimes M_j|_{D_i}$.

We will now construct an isomorphism

$$\langle L, M\rangle_{\mathcal{D}} \longrightarrow \frac{\mathbf{N}_{D_1}(M)}{\mathbf{N}_{D_2}(M)}$$

as the composition of several intermediary isomorphisms. At each stage, we will keep track of the image of $\langle \mathbf{1}, m\rangle$.

Stage 1. The first isomorphism in the composition is

$$\Psi_{(\mathbf{1}, D_1, D_2)} : \langle L, M\rangle_{\mathcal{D}} \longrightarrow \frac{\mathbf{D}(L_1 \otimes M_1|_{D_1}) \otimes \mathbf{D}(L_2 \otimes M_2|_{D_2})}{\mathbf{D}(L_1 \otimes M_2|_{D_1}) \otimes \mathbf{D}(L_2 \otimes M_1|_{D_2})}$$

defined in (4.7). Since, by definition,

$$\langle \mathbf{1}, m\rangle = \Phi^{-1}_{\mathbf{1},m}(1),$$

and since

$$\Phi_{\mathbf{1},m} = \Theta_{(D_1,D_2;\, m)} \circ \Psi_{(\mathbf{1},D_1,D_2)},$$

the image of $\langle \mathbf{1}, m\rangle$ under $\Psi_{(\mathbf{1},D_1,D_2)}$ is $\Theta^{-1}_{(D_1,D_2;\, m)}(1)$.

We will explicitly compute this image. To this end, let A_1 and A_2 be the rings such that $D_1 = \operatorname{Spec} A_1$ and $D_2 = \operatorname{Spec} A_2$. Let $a_1^1, \ldots, a_1^p$ be a K-basis of A_1, and let $a_2^1, \ldots, a_2^q$ be a K-basis of A_2. Consider the element

$$\frac{\left(a_1^1\,\mu_{1,1} \wedge \ldots \wedge a_1^p\,\mu_{1,1}\right) \otimes \left(a_2^1\,\mu_{2,2} \wedge \ldots \wedge a_2^q\,\mu_{2,2}\right)}{\left(a_1^1\,\mu_{1,2} \wedge \ldots \wedge a_1^p\,\mu_{1,2}\right) \otimes \left(a_2^1\,\mu_{2,1} \wedge \ldots \wedge a_2^q\,\mu_{2,1}\right)} \in \frac{\mathbf{D}(L_1 \otimes M_1|_{D_1}) \otimes \mathbf{D}(L_2 \otimes M_2|_{D_2})}{\mathbf{D}(L_1 \otimes M_2|_{D_1}) \otimes \mathbf{D}(L_2 \otimes M_1|_{D_2})}.$$

In (4.7) we showed that $\Theta_{(D_1,D_2;\, m)}$ was the composition of

$$\frac{\mathbf{D}(\beta_1) \otimes 1}{1 \otimes \mathbf{D}(\beta_2)} : \frac{\mathbf{D}(L_1 \otimes M_1|_{D_1}) \otimes \mathbf{D}(L_2 \otimes M_2|_{D_2})}{\mathbf{D}(L_1 \otimes M_2|_{D_1}) \otimes \mathbf{D}(L_2 \otimes M_1|_{D_2})} \longrightarrow \frac{\mathbf{D}(L_1 \otimes M_2|_{D_1}) \otimes \mathbf{D}(L_2 \otimes M_2|_{D_2})}{\mathbf{D}(L_1 \otimes M_2|_{D_1}) \otimes \mathbf{D}(L_2 \otimes M_2|_{D_2})}$$

followed by the natural "cancellation" isomorphism, where here 1 is the identity map, and, for $i = 1, 2$, the isomorphism β_i mapping $(L_i \otimes M_1)|_{D_i}$ to $(L_i \otimes M_2)|_{D_i}$ is characterized by sending $\mu_{i,1} = l_i \otimes (m_2 \otimes m)|_{D_i}$ to $\mu_{i,2} = l_i \otimes m_2\,|_{D_i}$. Therefore,

$$\frac{\mathbf{D}(\beta_1) \otimes 1}{1 \otimes \mathbf{D}(\beta_2)} \quad \text{sends} \quad \frac{\left(a_1^1\,\mu_{1,1} \wedge \ldots \wedge a_1^p\,\mu_{1,1}\right) \otimes \left(a_2^1\,\mu_{2,2} \wedge \ldots \wedge a_2^q\,\mu_{2,2}\right)}{\left(a_1^1\,\mu_{1,2} \wedge \ldots \wedge a_1^p\,\mu_{1,2}\right) \otimes \left(a_2^1\,\mu_{2,1} \wedge \ldots \wedge a_2^q\,\mu_{2,1}\right)}$$

$$\text{to} \quad \frac{\left(a_1^1\,\mu_{1,2} \wedge \ldots \wedge a_1^p\,\mu_{1,2}\right) \otimes \left(a_2^1\,\mu_{2,2} \wedge \ldots \wedge a_2^q\,\mu_{2,2}\right)}{\left(a_1^1\,\mu_{1,2} \wedge \ldots \wedge a_1^p\,\mu_{1,2}\right) \otimes \left(a_2^1\,\mu_{2,2} \wedge \ldots \wedge a_2^q\,\mu_{2,2}\right)}.$$

Which is then sent to $1 \in K$ under the "cancellation" map. Therefore,

$$\frac{\left(a_1^1\,\mu_{1,1} \wedge \ldots \wedge a_1^p\,\mu_{1,1}\right) \otimes \left(a_2^1\,\mu_{2,2} \wedge \ldots \wedge a_2^q\,\mu_{2,2}\right)}{\left(a_1^1\,\mu_{1,2} \wedge \ldots \wedge a_1^p\,\mu_{1,2}\right) \otimes \left(a_2^1\,\mu_{2,1} \wedge \ldots \wedge a_2^q\,\mu_{2,1}\right)}$$

is the image of $\langle \mathbf{1}, m\rangle$ under $\Psi_{(\mathbf{1},D_1,D_2)}$.

Stage 2. The second isomorphism is the natural isomorphism from

$$\frac{\mathbf{D}(L_1 \otimes M_1|_{D_1}) \otimes \mathbf{D}(L_2 \otimes M_2|_{D_2})}{\mathbf{D}(L_1 \otimes M_2|_{D_1}) \otimes \mathbf{D}(L_2 \otimes M_1|_{D_2})} \quad \text{to} \quad \frac{\left(\frac{\mathbf{D}(L_1 \otimes M_1|_{D_1})}{\mathbf{D}(\mathcal{O}_{D_1})}\right) \otimes \left(\frac{\mathbf{D}(L_2 \otimes M_2|_{D_2})}{\mathbf{D}(\mathcal{O}_{D_2})}\right)}{\left(\frac{\mathbf{D}(L_1 \otimes M_2|_{D_1})}{\mathbf{D}(\mathcal{O}_{D_1})}\right) \otimes \left(\frac{\mathbf{D}(L_2 \otimes M_1|_{D_2})}{\mathbf{D}(\mathcal{O}_{D_2})}\right)}.$$

Using the definition of (3.2), we can write this second space as

$$\frac{\mathbf{N}_{D_1}(L_1 \otimes M_1) \otimes \mathbf{N}_{D_2}(L_2 \otimes M_2)}{\mathbf{N}_{D_1}(L_1 \otimes M_2) \otimes \mathbf{N}_{D_2}(L_2 \otimes M_1)}.$$

This map sends the element

$$\frac{\left(a_1^1\,\mu_{1,1} \wedge \ldots \wedge a_1^p\,\mu_{1,1}\right) \otimes \left(a_2^1\,\mu_{2,2} \wedge \ldots \wedge a_2^q\,\mu_{2,2}\right)}{\left(a_1^1\,\mu_{1,2} \wedge \ldots \wedge a_1^p\,\mu_{1,2}\right) \otimes \left(a_2^1\,\mu_{2,1} \wedge \ldots \wedge a_2^q\,\mu_{2,1}\right)}$$

to

$$\frac{\left(\frac{a_1^1\,\mu_{1,1} \wedge \ldots \wedge a_1^p\,\mu_{1,1}}{a_1^1 \wedge \ldots \wedge a_1^p}\right) \otimes \left(\frac{a_2^1\,\mu_{2,2} \wedge \ldots \wedge a_2^q\,\mu_{2,2}}{a_2^1 \wedge \ldots \wedge a_2^q}\right)}{\left(\frac{a_1^1\,\mu_{1,2} \wedge \ldots \wedge a_1^p\,\mu_{1,2}}{a_1^1 \wedge \ldots \wedge a_1^p}\right) \otimes \left(\frac{a_2^1\,\mu_{2,1} \wedge \ldots \wedge a_2^q\,\mu_{2,1}}{a_2^1 \wedge \ldots \wedge a_2^q}\right)} = \frac{\mathbf{N}_{D_1}(\mu_{1,1}) \otimes \mathbf{N}_{D_2}(\mu_{2,2})}{\mathbf{N}_{D_1}(\mu_{1,2}) \otimes \mathbf{N}_{D_2}(\mu_{2,1})}.$$

Recall that, for $i, j \in \{1, 2\}$, the element $\mu_{i,j}$ was defined to be the restriction of $l_i \otimes m_j$ to D_i, and recall that $m_1 = m_2 \otimes m$. So

$$\frac{\mathbf{N}_{D_1}(\mu_{1,1}) \otimes \mathbf{N}_{D_2}(\mu_{2,2})}{\mathbf{N}_{D_1}(\mu_{1,2}) \otimes \mathbf{N}_{D_2}(\mu_{2,1})} = \frac{\mathbf{N}_{D_1}\big(l_1 \otimes (m_2 \otimes m)\big) \otimes \mathbf{N}_{D_2}\big(l_2 \otimes m_2\big)}{\mathbf{N}_{D_1}\big(l_1 \otimes m_2\big) \otimes \mathbf{N}_{D_2}\big(l_2 \otimes (m_2 \otimes m)\big)}.$$

Stage 3. The third and final isomorphism is the canonical isomorphism from

$$\frac{\mathbf{N}_{D_1}(L_1 \otimes M_1) \otimes \mathbf{N}_{D_2}(L_2 \otimes M_2)}{\mathbf{N}_{D_1}(L_1 \otimes M_2) \otimes \mathbf{N}_{D_2}(L_2 \otimes M_1)} \quad \text{to} \quad \frac{\mathbf{N}_{D_1}\left(\frac{L_1 \otimes M_1}{L_1 \otimes M_2}\right)}{\mathbf{N}_{D_2}\left(\frac{L_2 \otimes M_1}{L_2 \otimes M_2}\right)} = \frac{\mathbf{N}_{D_1}(M)}{\mathbf{N}_{D_2}(M)}$$

whose existence is proved in Proposition 3.5. This map sends

$$\frac{\mathbf{N}_{D_1}\big(l_1 \otimes (m_2 \otimes m)\big) \otimes \mathbf{N}_{D_2}(l_2 \otimes m_2)}{\mathbf{N}_{D_1}(l_1 \otimes m_2) \otimes \mathbf{N}_{D_2}\big(l_2 \otimes (m_2 \otimes m)\big)} \quad \text{to} \quad \frac{\mathbf{N}_{D_1}\left(\frac{l_1 \otimes (m_2 \otimes m)}{l_1 \otimes m_2}\right)}{\mathbf{N}_{D_2}\left(\frac{l_2 \otimes (m_2 \otimes m)}{l_2 \otimes m_2}\right)} = \frac{\mathbf{N}_{D_1}(m)}{\mathbf{N}_{D_2}(m)}.$$

The composition of these three isomorphisms gives the map we want.

(5.3) Corollary. *Let D be a Cartier divisor on X, and suppose that*

$$D = D_1 - D_2 \qquad \text{and} \qquad D = D_1' - D_2'$$

are two expression of D in terms of non-trivial effective Cartier divisors. Then, for any line bundle M on X, there is a canonical isomorphism

$$\frac{\mathbf{N}_{D_1}(M)}{\mathbf{N}_{D_2}(M)} \xrightarrow{\sim} \frac{\mathbf{N}_{D_1'}(M)}{\mathbf{N}_{D_2'}(M)}$$

characterized by

$$\frac{\mathbf{N}_{D_1}(m)}{\mathbf{N}_{D_2}(m)} \mapsto \frac{\mathbf{N}_{D_1'}(m)}{\mathbf{N}_{D_2'}(m)}$$

for any meromorphic section m of M which is defined and non-vanishing on D_1, D_2, D_1', and D_2'.

Proof. Let $L = \mathcal{O}_X(D)$. By Proposition 5.2, we have canonical isomorphisms

$$\langle L, M\rangle \xrightarrow{\sim} \frac{\mathbf{N}_{D_1}(M)}{\mathbf{N}_{D_2}(M)} \qquad \text{and} \qquad \langle L, M\rangle \xrightarrow{\sim} \frac{\mathbf{N}_{D_1'}(M)}{\mathbf{N}_{D_2'}(M)}.$$

What we want is the inverse of the first, composed with the second.

(5.4) Corollary. *Let D be a non-trivial effective Cartier divisor on X, and suppose that $D = D_1 - D_2$ where D_1 and D_2 are also non-trivial effective Cartier divisors. For any line bundle M on X there is a canonical isomorphism*

$$\mathbf{N}_D(M) \xrightarrow{\sim} \frac{\mathbf{N}_{D_1}(M)}{\mathbf{N}_{D_2}(M)}$$

characterized by

$$\mathbf{N}_D(m) \mapsto \frac{\mathbf{N}_{D_1}(m)}{\mathbf{N}_{D_2}(m)}$$

for any meromorphic section m of M which is defined and nowhere vanishing on D, D_1, and D_2.

Proof. Let D_0 be a non-zero effective Cartier divisor on X which is disjoint from D. By Proposition 3.6 there is a canonical isomorphism

$$\mathbf{N}_D(M) \xrightarrow{\sim} \frac{\mathbf{N}_{D+D_0}(M)}{\mathbf{N}_{D_0}(M)} \qquad \text{under which} \qquad \mathbf{N}_D(m) \mapsto \frac{\mathbf{N}_{D+D_0}(m)}{\mathbf{N}_{D_0}(m)}$$

for any meromorphic section m of M defined and non-vanishing on D and D_0. By Corollary 5.3 there is also a canonical isomorphism

$$\frac{\mathbf{N}_{D+D_0}(M)}{\mathbf{N}_{D_0}(M)} \xrightarrow{\sim} \frac{\mathbf{N}_{D_1}(M)}{\mathbf{N}_{D_2}(M)} \qquad \text{under which} \qquad \frac{\mathbf{N}_{D+D_0}(m)}{\mathbf{N}_{D_0}(m)} \mapsto \frac{\mathbf{N}_{D_1}(m)}{\mathbf{N}_{D_2}(m)}$$

for any meromorphic section m of M defined and non-vanishing on D, D_1, D_2, and D_0.

The isomorphism we want is the composition of these two:

$$\mathbf{N}_D(M) \xrightarrow{\sim} \frac{\mathbf{N}_{D_1}(M)}{\mathbf{N}_{D_2}(M)}, \qquad \text{under which} \qquad \mathbf{N}_D(m) \mapsto \frac{\mathbf{N}_{D_1}(m)}{\mathbf{N}_{D_2}(m)}$$

for any meromorphic section m of M defined and non-vanishing on D, D_1, D_2, and D_0.

If we choose, instead of D_0, another non-zero effective divisor D_0' disjoint from D, we get another isomorphism. But if we look at the image of $\mathbf{N}_D(m)$, where m is chosen to be defined and non-vanishing on D, D_1, D_2, D_0, *and* D_0', we see that the two maps agree. Hence, this map is independent of the choice of D_0. Thus given any meromorphic section m of M defined and non-vanishing on D, D_1, and D_2. We can choose D_0 disjoint from the support of the divisor of m, and we conclude that

$$\mathbf{N}_D(m) \mapsto \frac{\mathbf{N}_{D_1}(m)}{\mathbf{N}_{D_2}(m)}.$$

6. The Norm Functor for Divisors

Throughout this section let X be a fixed one-dimensional scheme projective over a field K. In Section 3 we defined a norm functor on any zero-dimensional scheme, which automatically gives a norm functor for any non-zero, effective Cartier divisor on X. Corollary 5.3 suggests that we can extend this definition to arbitrary Cartier divisors. In this section we define such an extension and give its basic properties. We apply this theory to the intersection pairing, proving a basic equality for the intersection pairing of sections. We then conclude by showing that Weil reciprocity follows from this theory.

(6.1) Definition. Let D be a Cartier divisor of X. We can always write D as $D_1 - D_2$ where D_1 and D_2 are non-zero effective Cartier divisors. For any line bundle M on X define the norm of M at D, written $N_D(M)$, by the formula

$$N_D(M) \overset{\text{def}}{=} \frac{\mathbf{N}_{D_1}(M)}{\mathbf{N}_{D_2}(M)}.$$

Corollary 5.3 shows that $N_D(M)$ is well-defined up to a canonical isomorphism. Corollary 5.4 shows that, if D is a non-zero effective Cartier, then, up to a canonical isomorphism, this definition of $N_D(M)$ agrees with the old definition of (3.2).

For any meromorphic section m of M, defined and nowhere vanishing on the support of D, define $N_D(m)$ to be the element

$$N_D(m) \overset{\text{def}}{=} \frac{\mathbf{N}_{D_1}(m)}{\mathbf{N}_{D_2}(m)} \quad \text{in} \quad \frac{\mathbf{N}_{D_1}(M)}{\mathbf{N}_{D_2}(M)} = N_D(M)$$

where $D = D_1 - D_2$, and where D_1 and D_2 are effective Cartier divisors chosen such that m is defined and non-zero at every point of D_1 and D_2. The element $N_D(m)$ is well-defined, i.e., it is well-defined for any model of $N_D(m)$, and is well-behaved under the canonical isomorphisms relating the various models.

For any isomorphism θ between line bundles of X, we define $N_D(\theta)$ to be the obvious isomorphism. It is clear that N_D is a functor from the category of line bundles of X (whose arrows are isomorphisms) to the category of one-dimensional K-vector spaces.

These new definitions allow us to give an alternate form for Proposition 5.2.

(6.2) Proposition (An alternate form of Proposition 5.2).
1. Let M be a line bundle on X, and let D be a Cartier divisors on X. Let $\mathcal{D}$ be modelling data for $\mathcal{O}_X(D)$ and M. There is a canonical isomorphism

$$\langle\, \mathcal{O}_X(D),\, M\, \rangle_{\mathcal{D}} \ \stackrel{\sim}{\rightarrow}\ \mathbf{N}_D(M)$$

characterized by

$$\langle \mathbf{1}, m\rangle \ \mapsto\ \mathbf{N}_D(m)$$

where $\mathbf{1}$ is the canonical meromorphic section of $\mathcal{O}_X(D)$, and m is any *meromorphic section of M defined and non-zero at every point of the support of D.*
2. Let L be a line bundle on X, and let E be a Cartier divisors on X. Let $\mathcal{D}$ be modelling data for L and $\mathcal{O}_X(E)$. There is a canonical isomorphism

$$\langle\, L,\, \mathcal{O}_X(E)\, \rangle_{\mathcal{D}} \ \stackrel{\sim}{\rightarrow}\ \mathbf{N}_E(L)$$

characterized by

$$\langle l, \mathbf{1}\rangle \mapsto \mathbf{N}_E(l)$$

where $\mathbf{1}$ is the canonical meromorphic section of $\mathcal{O}_X(E)$, and l is any *meromorphic section of L defined and non-zero at every point of the support of E.*

(6.3) Definition. Let D be a Cartier divisor of X, and let f be an invertible meromorphic function of X defined and non-zero at every point of the support of D. Then $\mathbf{N}_D(f)$ and $\mathbf{N}_D(1)$ are both non-zero elements of $\mathbf{N}_D(\mathcal{O}_X)$. We define $f(D)$, *f evaluated at D*, to be the element of K^* such that

$$\mathbf{N}_D(f) = f(D)\ \mathbf{N}_D(1).$$

The following proposition gives an alternate characterization of $f(D)$.

(6.4) Proposition.
1. Let D be a non-trivial, effective Cartier divisor, and let f be an invertible meromorphic function of X defined and non-zero at every point of D. Then

$$f(D) \ =\ \mathrm{Norm}_D(f).$$

2. Let $D = D_1 - D_2$, where D_1 and D_2 are non-trivial effective Cartier divisors, and let f be an invertible meromorphic function of X defined and non-zero at every point of D_1 and D_2. Then

$$f(\mathcal{D}) \ =\ \frac{\mathrm{Norm}_{D_1}(f)}{\mathrm{Norm}_{D_2}(f)}.$$

3. Let D be a Cartier divisor of X, and let f and g be invertible meromorphic functions of X defined and non-zero at every point of the support of D. Then

$$f(D)\, g(D) \ =\ fg\,(D).$$

4. Let D_1 and D_2 be Cartier divisors of X, and let f be an invertible meromorphic function of X defined and non-zero at every point of the supports of D_1 and D_2. Then

$$f(D_1 + D_2) \ =\ f(D_1)\, f(D_2).$$

Proof.
1. By Proposition 3.4 and using the notation of (5.1), we have that, as elements of $N_D(\mathcal{O}_X)$,

$$\mathbf{N}_D(f) = \mathbf{N}_D(f\,1) = \operatorname{Norm}_D(f)\cdot\mathbf{N}_D(1).$$

So by definition $f(D) = \operatorname{Norm}_D(f)$.
2. By definition,

$$\mathbf{N}_D(\mathcal{O}_X) = \frac{\mathbf{N}_{D_1}(\mathcal{O}_X)}{\mathbf{N}_{D_2}(\mathcal{O}_X)}.$$

Under this identification, and using Proposition 3.4,

$$\mathbf{N}_D(f) = \mathbf{N}_D(f\,1) = \frac{\mathbf{N}_{D_1}(f\,1)}{\mathbf{N}_{D_2}(f\,1)} = \frac{\operatorname{Norm}_{D_1}(f)\,\mathbf{N}_{D_1}(1)}{\operatorname{Norm}_{D_2}(f)\,\mathbf{N}_{D_2}(1)} = \frac{\operatorname{Norm}_{D_1}(f)}{\operatorname{Norm}_{D_2}(f)}\mathbf{N}_D(1).$$

So

$$f(D) = \frac{\operatorname{Norm}_{D_1}(f)}{\operatorname{Norm}_{D_2}(f)}.$$

3. Let $D = D_1 - D_2$ where D_1 and D_2 are non-zero, effective Cartier divisors of X. Then, using part 2 of the current proposition,

$$fg(D) = \frac{\operatorname{Norm}_{D_1}(f\,g)}{\operatorname{Norm}_{D_2}(f\,g)} = \frac{\operatorname{Norm}_{D_1}(f)\operatorname{Norm}_{D_1}(g)}{\operatorname{Norm}_{D_2}(f)\operatorname{Norm}_{D_2}(g)} = f(D)\,g(D)$$

4. Let $D_1 = D_{1,1} - D_{1,2}$ and $D_2 = D_{2,1} - D_{2,2}$ where $D_{1,1}$, $D_{1,2}$, $D_{2,1}$, and $D_{2,2}$ are effective, non-trivial Cartier divisor, chosen so that f is defined and non-zero at every point of $D_{1,1}$, $D_{1,2}$, $D_{2,1}$, and $D_{2,2}$. By parts 1 and 2 of the current proposition,

$$f(D_{1,1}) = f(D_{1,1} + D_{2,1} - D_{2,1}) = \frac{\operatorname{Norm}_{(D_{1,1}+D_{2,1})}(f)}{\operatorname{Norm}_{D_{2,1}}(f)} = \frac{f(D_{1,1} + D_{2,1})}{f(D_{2,1})}.$$

So $f(D_{1,1} + D_{2,1}) = f(D_{1,1})\,f(D_{2,1})$. Likewise, $f(D_{1,2} + D_{2,2}) = f(D_{1,2})\,f(D_{2,2})$.
By parts 1 and 2 of the current proposition,

$$\begin{aligned}
f(D_1 + D_2) &= f\big((D_{1,1} + D_{2,1}) - (D_{1,2} + D_{2,2})\big)\\
&= \frac{\operatorname{Norm}_{(D_{1,1}+D_{2,1})}(f)}{\operatorname{Norm}_{(D_{1,2}+D_{2,2})}(f)}\\
&= \frac{f(D_{1,1} + D_{2,1})}{f(D_{1,2} + D_{2,2})}\\
&= \frac{f(D_{1,1})\,f(D_{2,1})}{f(D_{1,2})\,f(D_{2,2})}.
\end{aligned}$$

Parts 1 and 2 of the current proposition also imply that

$$f(D_1) = f(D_{1,1} - D_{1,2}) = \frac{\operatorname{Norm}_{D_{1,1}}(f)}{\operatorname{Norm}_{D_{1,2}}(f)} = \frac{f(D_{1,1})}{f(D_{1,2})}$$

and

$$f(D_2) = f(D_{2,1} - D_{2,2}) = \frac{\operatorname{Norm}_{D_{2,1}}(f)}{\operatorname{Norm}_{D_{2,2}}(f)} = \frac{f(D_{2,1})}{f(D_{2,2})}.$$

Therefore,

$$f(D_1 + D_2) = f(D_1)\, f(D_2).$$

(6.5) Proposition. *Let M_1 and M_2 be line bundles on X, and let D be a Cartier divisor on X. There is a canonical isomorphism*

$$\mathbf{N}_D(M_1 \otimes M_2) \xrightarrow{\sim} \mathbf{N}_D(M_1) \otimes \mathbf{N}_D(M_2)$$

characterized by the requirement that, for any choice of meromorphic sections, m_1 of M_1 and m_2 of M_2, defined and non-zero at all the points of the support of D, we have

$$\mathbf{N}_D(m_1 \otimes m_2) \mapsto \mathbf{N}_D(m_1) \otimes \mathbf{N}_D(m_2).$$

Proof. Let $D = D_1 - D_2$ where D_1 and D_2 are non-trivial, effective Cartier divisors. We can choose D_1 and D_2 so that the support of D is the union of the supports of D_1 and D_2. By Proposition 3.5, there is a canonical isomorphism

$$\mathbf{N}_{D_1}(M_1 \otimes M_2) \xrightarrow{\sim} \mathbf{N}_{D_1}(M_1) \otimes \mathbf{N}_{D_1}(M_2)$$

such that

$$\mathbf{N}_{D_1}(m_1 \otimes m_2) \xrightarrow{\sim} \mathbf{N}_{D_1}(m_1) \otimes \mathbf{N}_{D_1}(m_2)$$

for any pair of meromorphic sections, m_1 of M_1 and m_2 of M_2, defined and non-zero at all the points of D_1. Likewise, there is a canonical isomorphism

$$\mathbf{N}_{D_2}(M_1 \otimes M_2) \xrightarrow{\sim} \mathbf{N}_{D_2}(M_1) \otimes \mathbf{N}_{D_2}(M_2)$$

such that

$$\mathbf{N}_{D_2}(m_1 \otimes m_2) \xrightarrow{\sim} \mathbf{N}_{D_2}(m_1) \otimes \mathbf{N}_{D_2}(m_2)$$

for any pair of meromorphic sections, m_1 of M_1 and m_2 of M_2, defined and non-zero at all the points of D_2.

Putting these two isomorphisms together, we get an isomorphism

$$\mathbf{N}_D(M_1 \otimes M_2) = \frac{\mathbf{N}_{D_1}(M_1 \otimes M_2)}{\mathbf{N}_{D_2}(M_1 \otimes M_2)} \xrightarrow{\sim} \frac{\mathbf{N}_{D_1}(M_1) \otimes \mathbf{N}_{D_1}(M_2)}{\mathbf{N}_{D_2}(M_1) \otimes \mathbf{N}_{D_2}(M_2)} = \mathbf{N}_D(M_1) \otimes \mathbf{N}_D(M_2)$$

such that

$$\mathbf{N}_D(m_1 \otimes m_2) = \frac{\mathbf{N}_{D_1}(m_1 \otimes m_2)}{\mathbf{N}_{D_2}(m_1 \otimes m_2)} \mapsto \frac{\mathbf{N}_{D_1}(m_1) \otimes \mathbf{N}_{D_1}(m_2)}{\mathbf{N}_{D_2}(m_1) \otimes \mathbf{N}_{D_2}(m_2)} = \mathbf{N}_D(m_1) \otimes \mathbf{N}_D(m_2)$$

for any pair of meromorphic sections, m_1 of M_1 and m_2 of M_2, defined and non-zero at all the points of D_1 and D_2.

(6.6) Proposition. *Let D be a Cartier divisor of X, and let M be a line bundle on X. Let m be a meromorphic section of M defined and non-zero at every point of the support of D. Finally, let f be an invertible meromorphic function of X, also defined and non-zero at every point of the support of D. Then in $\mathbf{N}_D(M)$ we have the following equation*

$$\mathbf{N}_D(f\,m) \;=\; f(D)\,\mathbf{N}_D(m).$$

Proof. Under the canonical isomorphism

$$\mathbf{N}_D(\mathcal{O}_X) \otimes \mathbf{N}_D(M) \xrightarrow{\sim} \mathbf{N}_D(\mathcal{O}_X \otimes M) \xrightarrow{\sim} \mathbf{N}_D(M),$$

we have that

$$\mathbf{N}_D(1) \otimes \mathbf{N}_D(m) \mapsto \mathbf{N}_D(1 \otimes m) \mapsto \mathbf{N}_D(m),$$

and that

$$\mathbf{N}_D(f) \otimes \mathbf{N}_D(m) \mapsto \mathbf{N}_D(f \otimes m) \mapsto \mathbf{N}_D(f\, m).$$

But $\mathbf{N}_D(f) = f(D)\ \mathbf{N}_D(1)$, so

$$\mathbf{N}_D(f) \otimes \mathbf{N}_D(m) = f(D)\ \mathbf{N}_D(1) \otimes \mathbf{N}_D(m).$$

Taking the images in $\mathbf{N}_D(M)$ of both sides of this equation gives the equation

$$\mathbf{N}_D(f\, m) = f(D)\ \mathbf{N}_D(m).$$

(6.7) Proposition. *Let D_1 and D_2 be Cartier divisors of X, and let M be a line bundle on X. There is a canonical isomorphism*

$$\mathbf{N}_{D_1+D_2}(M) \xrightarrow{\sim} \mathbf{N}_{D_1}(M) \otimes \mathbf{N}_{D_2}(M)$$

characterized by

$$\mathbf{N}_{D_1+D_2}(m) \mapsto \mathbf{N}_{D_1}(m) \otimes \mathbf{N}_{D_2}(m)$$

for any meromorphic section m of M which is defined and non-zero at every point of the supports of D_1 and D_2.

Proof. Let m_0 be a meromorphic section of M, defined and non-zero at every point of the supports of D_1 and D_2. The requirement

$$\mathbf{N}_{D_1+D_2}(m_0) \mapsto \mathbf{N}_{D_1}(m_0) \otimes \mathbf{N}_{D_2}(m_0)$$

defines an isomorphism

$$\mathbf{N}_{D_1+D_2}(M) \xrightarrow{\sim} \mathbf{N}_{D_1}(M) \otimes \mathbf{N}_{D_2}(M).$$

Let m be any other meromorphic section of M, defined and non-zero at every point of the supports of D_1 and D_2. We write $m = f\, m_0$ where f is an invertible meromorphic function. By using Proposition 6.4 and 6.6, we have that, under this isomorphism,

$$\begin{aligned}
\mathbf{N}_{D_1+D_2}(m) &= \mathbf{N}_{D_1+D_2}(f\, m_0) \\
&= f(D_1 + D_2)\ \mathbf{N}_{D_1+D_2}(m_0) \\
&\mapsto f(D_1 + D_2)\ \mathbf{N}_{D_1}(m_0) \otimes \mathbf{N}_{D_2}(m_0) \\
&= f(D_1)\, f(D_2)\ \mathbf{N}_{D_1}(m_0) \otimes \mathbf{N}_{D_2}(m_0) \\
&= \mathbf{N}_{D_1}(f\, m_0) \otimes \mathbf{N}_{D_2}(f\, m_0) \\
&= \mathbf{N}_{D_1}(m) \otimes \mathbf{N}_{D_2}(m).
\end{aligned}$$

(6.8) Proposition.

Let L and M be line bundles on X, and let $\mathcal{D}$ be modelling data for the pair L and M. Let l and m be meromorphic sections of L and M respectively, such that their associated Cartier divisors, (l) and (m), have disjoint supports. We have the following equalities relating elements of the K-vector space $\langle L, M\rangle_{\mathcal{D}}$:

1. Let g be an invertible meromorphic function on X which is defined and non-zero at every point of the support of (l). Hence $(g\,m)$ and (l) have disjoint supports. For all such g, we have the relation

$$\langle l,\, g\,m\rangle = g\big((l)\big)\,\langle l, m\rangle.$$

2. Let f be an invertible meromorphic function on X which is defined and non-zero at every point of the support of (m). Hence $(f\,l)$ and (m) have disjoint supports. For all such f, we have the relation

$$\langle f\,l,\, m\rangle = f\big((m)\big)\,\langle l, m\rangle.$$

Proof. We will prove only the first assertion; the proof of the second follows exactly the same pattern.

Write $(l) = D_1 - D_2$ where D_1 and D_2 are non-trivial, effective Cartier divisor with support disjoint from the supports of (m) and $(g\,m)$. By Proposition 6.2, we have a canonical isomorphism

$$\langle L, M\rangle_{\mathcal{D}} \xrightarrow{\sim} \mathbf{N}_{(l)}(M)$$

under which

$$\langle l, m\rangle \mapsto \mathbf{N}_{(l)}(m) \qquad \text{and} \qquad \langle l, g\,m\rangle \mapsto \mathbf{N}_{(l)}(g\,m)$$

But, by Proposition 6.6,

$$\mathbf{N}_{(l)}(g\,m) = g\big((l)\big)\;\mathbf{N}_{(l)}(m).$$

Comparing the pre-images of both sides of this equation gives

$$\langle l,\, g\,m\rangle = g\big((l)\big)\,\langle l, m\rangle.$$

(6.9) Corollary (Weil Reciprocity). *Let f and g be invertible meromorphic functions on X whose divisors (f) and (g) have disjoint support. Then*

$$f\big((g)\big) = g\big((f)\big).$$

Proof. Consider the element $\langle f, g\rangle$ of the vector space $\langle \mathcal{O}_X, \mathcal{O}_X\rangle$. By Proposition 6.8, we have that

$$\langle f, g\rangle = \langle f\,1, g\,1\rangle = f\big((g)\big)\,\langle 1, g\,1\rangle = f\big((g)\big)\,g(0)\,\langle 1, 1\rangle$$

where 0 is the zero divisor. Now, proposition 6.4 (part 4) implies that $g(0) = 1$. Therefore,

$$\langle f, g\rangle = f\big((g)\big)\,\langle 1, 1\rangle.$$

Likewise,

$$\langle f, g\rangle = \langle f\,1, g\,1\rangle = g\big((f)\big)\,\langle f\,1, 1\rangle = g\big((f)\big)\,f(0)\,\langle 1, 1\rangle = g\big((f)\big)\,\langle 1, 1\rangle.$$

Since $\langle 1, 1\rangle$ is a non-zero element of $\langle \mathcal{O}_X, \mathcal{O}_X\rangle$,

$$f\big((g)\big) = g\big((f)\big).$$

7. Other Properties of the Intersection Pairing

Throughout this section let X be a fixed one-dimensional scheme projective over a field K. In this section we continue the study of the intersection pairing. We show the existence of a series of canonical isomorphisms which will (i) establish the independence of the intersection pairing (up to canonical isomorphism) from the choice of the modelling data, (ii) establish the bilinearity and symmetry of the intersection pairing, and (iii) describe the behavior of the intersection pairing with respect to birational maps.

Now we will state a simple lemma which will be used repeatedly in this section.

(7.1) Lemma. *Let L and M be two line bundles of X, and let l_1 and m_1 be meromorphic sections of L and M respectively whose associated Cartier divisors have disjoint supports. Suppose we are given another pair of meromorphic sections, l_2 and m_2 of L and M respectively, whose associated Cartier divisors have disjoint supports. Then we can find a meromorphic section l' of L such that the following chain of statements are true:*

0. The Cartier divisors associated with l_1 and m_1 have disjoint supports.
1. The Cartier divisors associated with l' and m_1 have disjoint supports.
2. The Cartier divisors associated with l' and m_2 have disjoint supports.
3. The Cartier divisors associated with l_2 and m_2 have disjoint supports.

Note that the pair occurring at each step differs by one element from the pair occurring at any adjacent step.

Proof. Since X is projective, we can find a meromorphic section of L whose associated Cartier divisor has support disjoint from any given finite set of points. In this case, simply choose a meromorphic section l' whose associated Cartier divisor has support disjoint from the supports of (m_1) and (m_2).

(7.2) Proposition. *Let L and M be line bundles on X. Suppose we are given two sets of modelling data, $\mathcal{D}$ and $\mathcal{D}'$, for L and M. Then there is a canonical isomorphism*

$$\langle L, M\rangle_{\mathcal{D}} \xrightarrow{\sim} \langle L, M\rangle_{\mathcal{D}'}$$

characterized by the requirement that

$$\langle l, m\rangle \mapsto \langle l, m\rangle$$

for all pairs of meromorphic sections, l of L and m of M, whose associated Cartier divisors have disjoint supports.

Proof. Let l_0 and m_0 be meromorphic sections of L and M respectively whose associated Cartier divisors have disjoint supports. The rule

$$\langle l_0, m_0\rangle \mapsto \langle l_0, m_0\rangle$$

defines an isomorphism, call it ϕ, between the one-dimensional K-vector spaces $\langle L, M\rangle_{\mathcal{D}}$ and $\langle L, M\rangle_{\mathcal{D}'}$.

We want to show, for *all* pairs of meromorphic sections l of L and m of M whose associated Cartier divisors have disjoint supports, that, under ϕ,

$$\langle l, m\rangle \mapsto \langle l, m\rangle.$$

By Lemma 7.1, it is enough to show that

(1) $$\langle l_1, m\rangle \mapsto \langle l_1, m\rangle \quad \text{implies} \quad \langle l_2, m\rangle \mapsto \langle l_2, m\rangle$$

for all meromorphic sections l_1, l_2 of L and m of M where each (l_i) is disjoint from (m), and

(2) $$\langle l, m_1\rangle \mapsto \langle l, m_1\rangle \quad \text{implies} \quad \langle l, m_2\rangle \mapsto \langle l, m_2\rangle$$

for all meromorphic sections l of L and m_1, m_2 of M where (l) is disjoint from each (m_j).

We will show the first implication. The demonstration of the second is, of course, just like the first. Let l_1, l_2, and m be as above, and let f be such that $l_2 = f\,l_1$. If, under ϕ,

$$\langle l_1, m\rangle \mapsto \langle l_1, m\rangle,$$

then Proposition 6.8 implies that, under ϕ,

$$\begin{aligned}
\langle l_2, m\rangle \in \langle L, M\rangle_{\mathcal{D}} &= \langle f\,l_1, m\rangle \\
&= f((m))\,\langle l_1, m\rangle \\
&\mapsto f((m))\,\langle l_1, m\rangle \in \langle L, M\rangle_{\mathcal{D}'} \\
&= \langle f\,l_1, m\rangle \\
&= \langle l_2, m\rangle \in \langle L, M\rangle_{\mathcal{D}'}.
\end{aligned}$$

Note The above proposition implies that, up to canonical isomorphism, the definition of the intersection pairing of line bundles is independent of the modelling data chosen to construct it. Therefore, from this point on, we will suppress all mention of the modelling data unless we are interested in the specific model.

(7.3) Proposition. *The intersection pairing is symmetric; i.e., given L and M line bundles on X, there is a canonical isomorphism*

$$\langle L, M\rangle \xrightarrow{\sim} \langle M, L\rangle$$

characterized by the requirement that

$$\langle l, m\rangle \mapsto \langle m, l\,\rangle$$

for all pairs of meromorphic sections, l of L and m of M, whose associated Cartier divisors have disjoint supports.

Proof. Let l_0 and m_0 be meromorphic sections of L and M respectively whose associated Cartier divisors have disjoint supports. Since $\langle l_0, m_0\rangle$ is a non-zero element of the one-dimensional K-vector space $\langle L, M\rangle$, and $\langle m_0, l_0\rangle$ is a non-zero element of $\langle M, L\rangle$, the rule

$$\langle l_0, m_0\rangle \mapsto \langle m_0, l_0\rangle$$

defines an isomorphism, call it ϕ, between $\langle L, M\rangle$ and $\langle M, L\rangle$.

We want to show, for *all* pairs of meromorphic sections l of L and m of M whose associated Cartier divisors have disjoint supports, that, under ϕ,

$$\langle l, m\rangle \mapsto \langle m, l\,\rangle.$$

By Lemma 7.1, it is enough to show that

$$\text{(1)} \qquad \langle l_1, m\rangle \mapsto \langle m, l_1\rangle \qquad \text{implies} \qquad \langle l_2, m\rangle \mapsto \langle m, l_2\rangle$$

for all meromorphic sections l_1, l_2 of L and m of M where each (l_i) is disjoint from (m), and

$$\text{(2)} \qquad \langle l, m_1\rangle \mapsto \langle m_1, l\rangle \qquad \text{implies} \qquad \langle l, m_2\rangle \mapsto \langle m_2, l\rangle$$

for all meromorphic sections l of L and m_1, m_2 of M where (l) is disjoint from each (m_j).

We will show the first implication. The demonstration of the second is, of course, just like the first. Let l_1, l_2, and m be as above, and let f be such that $l_2 = f\, l_1$. If, under ϕ,

$$\langle l_1, m\rangle \mapsto \langle m, l_1\rangle,$$

then Proposition 6.8 implies that, under ϕ,

$$\begin{aligned} \langle l_2, m\rangle &= \langle f\, l_1, m\rangle \\ &= f\big((m)\big)\, \langle l_1, m\rangle \\ &\mapsto f\big((m)\big)\, \langle m, l_1\rangle \\ &= \langle m, f\, l_1\rangle \\ &= \langle m, l_2\rangle. \end{aligned}$$

(7.4) Proposition. *The Intersection pairing is bilinear; i.e., we have the following.*

1. *Given L, L', and M line bundles on X, there is a canonical isomorphism*

$$\langle L, M\rangle \otimes \langle L', M\rangle \xrightarrow{\sim} \langle L \otimes L', M\rangle$$

characterized by the requirement that

$$\langle l, m\rangle \otimes \langle l', m\rangle \mapsto \langle l \otimes l', m\rangle$$

for all triples of meromorphic sections, l, l' and m of L, L' and M respectively, such that the supports of the Cartier divisors (l) and (l') are both disjoint from the support of (m).

2. *Given L, M, and M' line bundles on X, there is a canonical isomorphism*

$$\langle L, M\rangle \otimes \langle L, M'\rangle \xrightarrow{\sim} \langle L, M \otimes M'\rangle$$

characterized by the requirement that

$$\langle l, m\rangle \otimes \langle l, m'\rangle \mapsto \langle l, m \otimes m'\rangle$$

for all triples of meromorphic sections, l, m and m' of L, M and M' respectively, such that the supports of the Cartier divisors (m) and (m') are both disjoint from the support of (l).

Proof. We will prove the first assertion; the proof of the second assertion follows exactly the same pattern.

Let l_0, l'_0 and m_0 be meromorphic sections of L, L' and M respectively such that the supports of the Cartier divisors (l_0) and (l'_0) are both disjoint from the support of (m_0). Note that $\langle l_0, m_0\rangle$ is a non-zero element of the one-dimensional K-vector space $\langle L, M\rangle$, $\langle l'_0, m_0\rangle$ is a non-zero element of $\langle L', M\rangle$, and $\langle l_0 \otimes l'_0, m_0\rangle$ is a non-zero element of $\langle L \otimes L', M\rangle$. So the requirement that

$$\langle l_0, m_0\rangle \otimes \langle l'_0, m_0\rangle \mapsto \langle l_0 \otimes l'_0, m_0\rangle$$

defines an isomorphism, call it ϕ, between $\langle L, M\rangle \otimes \langle L', M\rangle$ and $\langle L \otimes L', M\rangle$.

Suppose that l, l' and m are meromorphic sections of L, L' and M respectively, such that the supports of the Cartier divisors (l) and (l') are both disjoint from the support of (m), and such that, under ϕ,

$$\langle l, m\rangle \otimes \langle l', m\rangle \mapsto \langle l \otimes l', m\rangle.$$

Let l_1 and l'_1 be arbitrary meromorphic section of L and L' respectively, such that the supports of the Cartier divisors (l_1) and (l'_1) are both disjoint from the support of (m). Let f and f' be the invertible meromorphic functions on X such that $l_1 = f\, l$ and $l'_1 = f'\, l'$. Then, using Propositions 6.8 and 6.4, we have that, under ϕ,

$$\begin{aligned}
\langle l_1, m\rangle \otimes \langle l'_1, m\rangle &= \langle f\, l, m\rangle \otimes \langle f'\, l', m\rangle \\
&= f\big((m)\big)\, f'\big((m)\big)\, \langle l, m\rangle \otimes \langle l', m\rangle \\
&\mapsto f\big((m)\big)\, f'\big((m)\big)\, \langle l \otimes l', m\rangle \\
&= f\, f'\big((m)\big)\, \langle l \otimes l', m\rangle \\
&= \langle f\, f'\, l \otimes l', m\rangle \\
&= \langle (f\, l) \otimes (f'\, l'), m\rangle \\
&= \langle l_1 \otimes l'_1, m\rangle.
\end{aligned}$$

Now, let m_1 be another meromorphic section of M whose associated Cartier divisor has support disjoint from those of both l and l'. Let g be the invertible meromorphic function on X such that $m_1 = g\, m$. Then, using Propositions 6.8 and 6.4, we have that, under ϕ,

$$\begin{aligned}
\langle l, m_1\rangle \otimes \langle l', m_1\rangle &= \langle l, g\, m\rangle \otimes \langle l', g\, m\rangle \\
&= g\big((l)\big)\, g\big((l')\big)\, \langle l, m\rangle \otimes \langle l', m\rangle \\
&\mapsto g\big((l)\big)\, g\big((l')\big)\, \langle l \otimes l', m\rangle \\
&= g\big((l) + (l')\big)\, \langle l \otimes l', m\rangle \\
&= \langle l \otimes l', g\, m\rangle \\
&= \langle l \otimes l', m_1\rangle.
\end{aligned}$$

Now let l, l' and m be arbitrary meromorphic sections of L, L' and M respectively, such that the supports of the Cartier divisors (l) and (l') are both disjoint from the support of (m). Then the above discussion, together with a slight generalization of Lemma 7.1, implies that, under ϕ,

$$\langle l, m\rangle \otimes \langle l', m\rangle \mapsto \langle l \otimes l', m\rangle.$$

(7.5) Proposition. *Let X and X' be one-dimensional schemes projective over K, and let ρ be a morphism from X' to X defined over K. Suppose that there is an open subscheme U of X such that the restriction of ρ to U' is an isomorphism, where U' is the inverse image of U under ρ. Also suppose that the complement of U in X consists of only a finite number of closed points.*

Let L and M be line bundles on X, and let L' be the pull-back of L and M' the pull-back of M to X' under ρ. There is a canonical isomorphism

$$\langle L, M\rangle \xrightarrow{\sim} \langle L', M'\rangle$$

characterized by the requirement that

$$\langle l, m\rangle \mapsto \langle l', m'\rangle$$

for all pairs of meromorphic sections, l of L and m of M,whose associated Cartier divisors have disjoint supports contained in U. Here l' is the meromorphic section of L' obtained by pulling back l through ρ, and m' is the meromorphic section of M' obtained by pulling back m through ρ

Proof. Let l_0 and m_0 be meromorphic sections of L and M respectively, such that their associated Cartier divisors, (l_0) and (m_0), are disjoint and contained in U. Let l_0' be the pull-back of l_0 to L', and let m_0' be the pull-back of m_0 to M'. Since l_0 and m_0 are defined and non-zero at ever point of X outside U, the sections l_0' and m_0' are defined and non-zero at every associated point of X' which is outside U'. Since l_0 and m_0 are defined and non-zero at every associated point of U, and since U is isomorphic to U', the sections l_0' and m_0' are defined and non-zero at every associated point of U', i.e., every associated point of X' contained in U'. These two facts imply that l_0' and m_0' are meromorphic sections on X'. Also note that their associated Cartier divisors, (l_0') and (m_0'), are contained in U' and have disjoint support.

Therefore, $\langle l_0, m_0\rangle$ is a non-zero vector of $\langle L, M\rangle$, and $\langle l_0', m_0'\rangle$ is a non-zero vector of $\langle L', M'\rangle$. Since $\langle L, M\rangle$ and $\langle L', M'\rangle$ are one-dimensional K-vector spaces, the rule

$$\langle l_0, m_0\rangle \mapsto \langle l_0', m_0'\rangle$$

specifies an isomorphism, call it ϕ, between $\langle L, M\rangle$ and $\langle L', M'\rangle$.

Now let l and m be any meromorphic sections of L and M, respectively, whose associated Cartier divisors, (l) and (m), have disjoint supports contained in U.

Let m_1 be a meromorphic section of M whose associated Cartier divisor (m_1) has support contained in U and disjoint from the supports of (l_0) and (l). As argued above in the case of l_0' and m_0', the pull-backs l', m', and m_1' of l, m, and m_1 via ρ are meromorphic sections of L', M', and M' respectively all with supports contained in U'. Furthermore, (l') and (m') have disjoint supports, and (m_1') has support disjoint from the supports of both (l') and (l_0').

Let g be the invertible meromorphic function on X such that $m_1 = g\, m_0$. So g is an invertible meromorphic function of X whose associated Cartier divisor (g) has support which is contained in U, and g is defined and non-zero at every point of the support of (l_0). Let g' be the pull-back of g to X', hence $m_1' = g'\, m_0'$, g' is an invertible meromorphic function of X' whose associated Cartier divisor (g') has support which is contained in U', and g' is defined and non-zero at every point of the support of (l'_0). Proposition 6.8 implies that, under ϕ,

$$\langle l_0, m_1\rangle = \langle l_0, g\, m_0\rangle = g\big((l_0)\big)\,\langle l_0, m_0\rangle \mapsto g\big((l_0)\big)\,\langle l'_0, m'_0\rangle.$$

By Proposition 6.4 (part 2), it is clear that $g\big((l_0)\big)$ depends only on the behavior of g near the support of (l_0), which is contained in U. Since U is isomorphic to U', it follows that $g\big((l_0)\big) = g'\big((l'_0)\big)$. Therefore,

$$\langle l_0, m_1\rangle \mapsto g\big((l_0)\big)\,\langle l'_0, m'_0\rangle = g'\big((l'_0)\big)\,\langle l'_0, m'_0\rangle = \langle l'_0, g'\, m'_0\rangle = \langle l'_0, m'_1\rangle.$$

We have shown that ϕ maps $\langle l_0, m_1\rangle$ to $\langle l'_0, m'_1\rangle$. We can adapt the above argument to show that this in turn implies that

$$\langle l, m_1\rangle \mapsto \langle l', m'_1\rangle.$$

Finally, by again adapting the above argument, we can show that this in turn implies that

$$\langle l, m\rangle \mapsto \langle l', m'\rangle.$$

Exercise. Let X be a one-dimensional scheme projective over K, and let L and M be line bundles of X. Also, let $\mathcal{D} = (L_1,\, L_2,\, M_1,\, M_2,\, \rho,\, \rho')$, be a choice of modelling data for the pair (L, M). Then, of course, $\mathcal{D}' = (M_1,\, M_2,\, L_1,\, L_2,\, \rho',\, \rho)$ is a set of modelling data for (M, L). Consider the natural isomorphism

$$\frac{\mathbf{D}(L_1\otimes M_1)\otimes \mathbf{D}(L_2\otimes M_2)}{\mathbf{D}(L_1\otimes M_2)\otimes \mathbf{D}(L_2\otimes M_1)} \overset{\sim}{\to} \frac{\mathbf{D}(M_1\otimes L_1)\otimes \mathbf{D}(M_2\otimes L_2)}{\mathbf{D}(M_1\otimes L_2)\otimes \mathbf{D}(M_2\otimes L_1)}.$$

Show that this is exactly the isomorphism

$$\langle L, M\rangle_{\mathcal{D}} \overset{\sim}{\to} \langle M, L\rangle_{\mathcal{D}'}$$

of Proposition 7.3.

Exercise. Let X be a one-dimensional scheme projective over K, and let L and M be line bundles of X. Define the *default modelling data* for the pair (L, M) to be $\mathcal{D} = (L,\, \mathcal{O}_X,\, M,\, \mathcal{O}_X,\, \rho,\, 0)$, where $\rho = \deg L \cdot \deg M$. Let $\mathcal{D}' = (M,\, \mathcal{O}_X,\, L,\, \mathcal{O}_X,\, \rho,\, 0)$ be the default modelling data of the pair (M, L) where ρ is as above. Consider the natural isomorphism

$$\frac{\mathbf{D}(L_1\otimes M_1)\otimes \mathbf{D}(L_2\otimes M_2)}{\mathbf{D}(L_1\otimes M_2)\otimes \mathbf{D}(L_2\otimes M_1)} \overset{\sim}{\to} \frac{\mathbf{D}(M_1\otimes L_1)\otimes \mathbf{D}(M_2\otimes L_2)}{\mathbf{D}(M_1\otimes L_2)\otimes \mathbf{D}(M_2\otimes L_1)}.$$

Show that this is $(-1)^{\deg L\cdot \deg M}$ times the isomorphism

$$\langle L, M\rangle_{\mathcal{D}} \overset{\sim}{\to} \langle M, L\rangle_{\mathcal{D}'}$$

of Proposition 7.3.

8. Extensions of the Base Field

Throughout this section let K be a field, and let K' a field extension of K. In this section we will study the behavior of the norm functor and the intersection pairing under field extension from K to K'.

(8.1) Notation. The following notation will be specific to this section:

• For any scheme Y over K, Y' will denote the scheme obtained from Y by change of base field from K to K', i.e.,

$$Y' = Y \times_{\operatorname{Spec} K} \operatorname{Spec} K'.$$

• If Y is a scheme over K and L is a line bundle on Y, then L' will denote the pull-back of L to Y' via the natural map $Y' \to Y$. Since $Y' \to Y$ is flat, any meromorphic section l of L pulls back via this map to give a meromorphic section, denoted l', of L'.

In particular, if f an invertible meromorphic function of Y, we will denote by f' the pull-back of f to Y'. Suppose we have a line bundle L on Y, a meromorphic section l of L, and an invertible meromorphic function f of Y. Let $l_1 = fl$. Then $l'_1 = f'l'$ as meromorphic sections of L'.

• Let Y be a scheme over K and D a Cartier divisor of Y. We define the pull-back D' of D to Y' as follows. Let y' be a point of Y', and let $y \in Y$ be its image under the natural map $Y' \to Y$. At y the divisor D is represented by an element f/g of the total quotient ring of $\mathcal{O}_{Y,y}$, where f and g are elements of $\mathcal{O}_{Y,y}$ which are not zero divisors. Since $\mathcal{O}_{Y',y'}$ is flat over $\mathcal{O}_{Y,y}$, the pull-backs f' and g' are elements of $\mathcal{O}_{Y',y'}$ which are not zero divisors, and so f'/g' is an element of the total quotient ring of $\mathcal{O}_{Y',y'}$. This procedure defines a divisor D' locally, and these local definitions patch together to form a global divisor on all of Y'.

Given a line bundle L and a meromorphic section l of L, if D is the Cartier divisor associated with l, then D' is the Cartier divisor associated with l'.

• If D is a non-trivial effective Cartier divisor then the notation D' appears at first sight to be ambiguous: it might refer to the pull-back of the Cartier divisor D, or, considering D as a scheme, it might refer to the scheme obtained by changing base field. However, it is easy to check that these definitions are compatible; i.e., the scheme associated to the effective Cartier divisor D' is exactly the scheme D'.

(8.2) Proposition. *Let $Y = \operatorname{Spec} A$ be a scheme, finite over the field K. Under field extension, we have*

$$Y' = \operatorname{Spec} A' \qquad \textit{where} \qquad A' \stackrel{\text{def}}{=} A \otimes_K K'.$$

For any element a of A, and a' its image in A',

$$\operatorname{Norm}_{Y'}(a') = \operatorname{Norm}_Y(a)$$

where the left hand side is taken with respect to the ground field K', and the right hand side is taken with respect to K but is considered as an element of the extension K'.

Proof. Let $b_1, \ldots, b_n$ be a basis of A over K. Then $b'_1, \ldots, b'_n$ is a basis of A' over K'. The matrix representing the action of a on $b_1, \ldots, b_n$ is the same as that of a' on $b'_1, \ldots, b'_n$. Thus their determinants are the same.

(8.3) Proposition. *Let X be a one-dimensional scheme projective over the field K, let D be a Cartier divisor on X, and let f be an invertible meromorphic function on X defined and non-zero at every point of the support of D. Then*

$$f(D) = f'(D').$$

Proof. Choose non-trivial, effective Cartier divisors D_1 and D_2 in such a way that $D = D_1 - D_2$ and f is defined and non-zero at every point of D_1 and D_2. As Cartier divisors $D' = D_1' - D_2'$. So, by Proposition 6.4 part 2 and Proposition 8.2,

$$f'(D') = \frac{\mathrm{Norm}_{D_1'}(f')}{\mathrm{Norm}_{D_2'}(f')} = \frac{\mathrm{Norm}_{D_1}(f)}{\mathrm{Norm}_{D_2}(f)} = f(D).$$

(8.4) Proposition. *Let X be a one-dimensional scheme projective over the field K. Given L and M line bundles on X, there is a canonical isomorphism*

$$\langle L, M\rangle \otimes K' \xrightarrow{\sim} \langle L', M'\rangle$$

characterized by the requirement that

$$\langle l, m\rangle \otimes 1 \mapsto \langle l', m'\rangle$$

for all pairs of meromorphic sections l of L and m of M whose associated Cartier divisors (l) and (m) have disjoint supports.

Proof. Let l_0 and m_0 be meromorphic sections of L and M respectively whose associated Cartier divisors, (l_0) and (m_0), have disjoint supports. Thus, the Cartier divisors associated with l_0' and m_0' must also have disjoint supports. Since $\langle l_0, m_0\rangle$ is a non-zero element of the one-dimensional K-vector space $\langle L, M\rangle$, and $\langle l_0', m_0'\rangle$ is a non-zero element of the one-dimensional K'-vector space $\langle L', M'\rangle$, the rule

$$\langle l_0, m_0\rangle \otimes 1 \mapsto \langle l_0', m_0'\rangle$$

defines an isomorphism, call it ϕ, between $\langle L, M\rangle \otimes K'$ and $\langle L', M'\rangle$.

Now let l and m be any pair of meromorphic sections of L and M respectively, whose associated Cartier divisors, (l) and (m), have disjoint supports.

Choose a meromorphic section m_1 of M defined and non-zero at every point of the supports of (l_0) and (l). Let g be the invertible meromorphic function on X such that $m_1 = g\, m_0$. Note that g is defined and non-zero at every point of the support of (l_0). Let g' be the pull-back of g to X', hence $m_1' = g'\, m_0'$, and g' is defined and non-zero at every point of the support of (l_0'). Propositions 6.8 and 8.3 imply that, under ϕ,

$$\begin{aligned}
\langle l_0, m_1\rangle \otimes 1 &= \langle l_0, g\, m_0\rangle \otimes 1 \\
&= g\big((l_0)\big)\, \langle l_0, m_0\rangle \otimes 1 \\
&\mapsto g\big((l_0)\big)\, \langle l_0', m_0'\rangle \\
&= g'\big((l_0)'\big)\, \langle l_0', m_0'\rangle \\
&= g'\big((l_0')\big)\, \langle l_0', m_0'\rangle \\
&= \langle l_0', g'\, m_0'\rangle \\
&= \langle l_0', m_1'\rangle
\end{aligned}$$

A similar argument shows that, since $\langle l_0, m_1\rangle \otimes 1$ maps to $\langle l_0', m_1'\rangle$, the vector $\langle l, m_1\rangle \otimes 1$ must map to $\langle l', m_1'\rangle$. Again, a similar argument shows that this in turn implies that, under ϕ, the element $\langle l, m\rangle \otimes 1$ is sent to $\langle l', m'\rangle$.

(8.5) Proposition. *Let X be a one-dimensional scheme projective over the field K, let M be a line bundle on X, and let D be a Cartier divisor of X. There is a canonical isomorphism*

$$\mathbf{N}_D(M) \otimes K' \xrightarrow{\sim} \mathbf{N}_{D'}(M')$$

characterized by the requirement that

$$\mathbf{N}_D(m) \otimes 1 \mapsto \mathbf{N}_{D'}(m')$$

for all meromorphic sections m of M defined and non-zero at every point of the support of D.

Proof. Let m be a meromorphic section of M defined and non-zero at every point of the support of D. Apply Proposition 8.4 to the line bundle $L = \mathcal{O}_X(D)$. Note that $\mathcal{O}_X(D)' = \mathcal{O}_{X'}(D')$. So we have a canonical isomorphism

$$\langle\, \mathcal{O}_X(D),\, M\,\rangle \otimes K' \xrightarrow{\sim} \langle\, \mathcal{O}_{X'}(D'),\, M'\,\rangle$$

under which

$$\langle \mathbf{1}_D, m\rangle \otimes 1 \mapsto \langle \mathbf{1}_{D'}, m'\rangle$$

where $\mathbf{1}_D$ and $\mathbf{1}_{D'}$ are the canonical meromorphic sections of $\mathcal{O}_X(D)$ and $\mathcal{O}_{X'}(D')$ respectively.

By Proposition 5.2, as restated in (6.2), we have a canonical isomorphism

$$\langle\, \mathcal{O}_X(D),\, M\,\rangle \xrightarrow{\sim} \mathbf{N}_D(M)$$

under which

$$\langle \mathbf{1}_D, m\rangle \mapsto \mathbf{N}_D(m),$$

and a canonical isomorphism

$$\langle\, \mathcal{O}_{X'}(D'),\, M'\,\rangle \xrightarrow{\sim} \mathbf{N}_{D'}(M')$$

under which

$$\langle \mathbf{1}_{D'}, m'\rangle \mapsto \mathbf{N}_{D'}(m').$$

Putting these three together, we get the desired isomorphism. (Take the inverse of the second, tensored with K', followed by the first and then by the third.)

Exercise. Let X be a one-dimensional scheme projective over K, and let L be a line bundle on X. Note that $\mathbf{D}(L) \otimes_K K'$ is naturally isomorphic to $\mathbf{D}(L')$.

Now let L and M be line bundles of X, and let $\mathcal{D} = (L_1, L_2, M_1, M_2, \rho, \rho')$ be a choice of modelling data for the pair L, M. Then, of course,

$$\mathcal{D}' = (L_1', L_2', M_1', M_2', \rho, \rho')$$

is a set of modelling data for the pair L', M'. Consider the natural isomorphism

$$\frac{\mathbf{D}(L_1 \otimes M_1) \otimes \mathbf{D}(L_2 \otimes M_2)}{\mathbf{D}(L_1 \otimes M_2) \otimes \mathbf{D}(L_2 \otimes M_1)} \otimes_K K' \xrightarrow{\sim} \frac{\mathbf{D}(L_1' \otimes M_1') \otimes \mathbf{D}(L_2' \otimes M_2')}{\mathbf{D}(L_1' \otimes M_2') \otimes \mathbf{D}(L_2' \otimes M_1')}.$$

Show that this is exactly the isomorphism

$$\langle L, M\rangle_{\mathcal{D}} \otimes_K K' \xrightarrow{\sim} \langle L', M'\rangle_{\mathcal{D}'}$$

of Proposition 8.4.

Chapter 2

The Intersection Pairing for Families of One-dimensional Schemes

1. Introduction.

In Chapter 1 we developed an intersection pairing for pairs of line bundles on a one-dimensional scheme projective over a field. In this chapter we will consider families of one-dimensional schemes, flat and projective over an integral base. For any pair of line bundles on such a family, we will define their intersection pairing which will be a line bundle on the base. After going through the main properties of this pairing, we will develop the determinant of cohomology which assigns to each line bundle on such a family a line bundle on the base. Finally we will relate the intersection pairing and the determinant of cohomology.

We will use the following notation and conventions throughout:

• We will always assume that the schemes we are dealing with are separated and noetherian.

• Let $\pi\colon X \to S$ be a projective morphism. If U is an open subscheme of S, then X_U will denote $\pi^{-1}(U) = X \times_S U$. So X_U is projective and flat over U, and is an open subscheme of X.

• Let X be a scheme over an integral scheme S, and let K be the function field of S. Then X_K will denote the fiber over the generic point of S, i.e., $X_K = X \times_S \operatorname{Spec} K$.

• If S is a scheme and if s is a point of S, then $\mathbf{F}(s)$ will denote the function field of the closure of s. If s is a closed point, this field is also called the residue field of s.

• Let X be a scheme over S. If $s \in S$ is a point, then X_s will denote $X \times_S \operatorname{Spec} \mathbf{F}(s)$, the fiber of X over s.

2. Horizontal Divisors

(2.1) Definitions. Let $\pi\colon X \to S$ be a projective, flat morphism whose fibers are one-dimensional. In this situation we will make use of the following definitions:

• An *effective horizontal divisor D of X* is a non-zero effective Cartier divisor of X which is, as a scheme, finite and flat over S. Note that, since the fibers of X over S are all one-dimensional, any non-zero effective Cartier divisor D which is flat over S is automatically quasi-finite over S; since such a divisor D is projective over S, Zariski's Main Theorem implies that D is actually finite over S. Therefore, a non-zero effective Cartier divisor of X is an effective horizontal divisor if and only if it is, as a scheme, flat over S.

• A *horizontal divisor $D = D_1 - D_2$ of X* is a Cartier divisor D of X, together with a pair of effective horizontal divisors, D_1 and D_2, of X, such that $D = D_1 - D_2$.

• For $D = D_1 - D_2$ a horizontal divisor of X, the *support* of D is defined to be set of points of D_1 together with the set of points of D_2.

(2.2) Lemma. *Let $\pi\colon X \to S$ be a projective, flat morphism, let D be a non-zero effective Cartier divisor of X, and let $s \in S$ be a point of S such that D intersects the fiber X_s. Then there is an open subscheme U of S containing s such that the restriction of D to X_U is flat over U if and only if D does not contain any associated points of the fiber X_s.*

Proof. Let x be a point of the intersection of the fiber X_s with D, and let $f \in \mathcal{O}_{X,x}$ be a local generator of the defining ideal of D at x. One version of the local flatness criterion (see [M], Corollary to Theorem 22.5, pg. 177) states that the following are equivalent:

1. f is not a zero divisor of $\mathcal{O}_{X,x}$, and $\mathcal{O}_{X,x}/(f) = \mathcal{O}_{D,x}$ is flat over $\mathcal{O}_{S,s}$.
2. The image of f in $\mathcal{O}_{X,x} \otimes \mathbf{F}(s) = \mathcal{O}_{X_s,x}$ is not a zero divisor and $\mathcal{O}_{X,x}$ is flat over $\mathcal{O}_{S,s}$.

By assumption, $\mathcal{O}_{X,x}$ is flat over $\mathcal{O}_{S,s}$, and, since f defines a Cartier divisor, f is not a zero divisor of $\mathcal{O}_{X,x}$. Therefore, $\mathcal{O}_{D,x}$ is flat over $\mathcal{O}_{S,s}$ if and only if the image of f in $\mathcal{O}_{X_s,x}$ is not a zero divisor. But the image of f in $\mathcal{O}_{X_s,x}$ is a zero divisor if and only if f is contained in an associated prime of $\mathcal{O}_{X_s,x}$, which is true if and only if D contains an associated point of the fiber X_s whose closure contains x.

So, $\mathcal{O}_{D,x}$ is flat over $\mathcal{O}_{S,s}$ for all points x in the intersection of X_s and D if and only if D does not contain any associated points of the fiber X_s.

The set $\{x \in D \mid \pi \text{ is not flat at } x\}$ is a closed subset of X (see [M], Theorem 24.3, pg. 187). Since π is a projective morphism, the image of this set in S is a closed set. So $\mathcal{O}_{D,x}$ is flat over $\mathcal{O}_{S,s}$ for all x in the intersection of X_s and D if and only if there is an open subscheme U of S containing s such that $D|_{X_U}$ is flat over U.

(2.3) Lemma. *Let $\pi\colon X \to S$ be a projective, flat morphism whose fibers are all one-dimensional. Let $\mathcal{O}(1)$ be a line bundle on X which is very ample with respect to π, let L be a line bundle on X, let $s \in S$ be a closed point, and let Z be a closed subset of X intersecting the fiber X_s in at most a finite number of points. Then there is an integer N such that, for all $n \geq N$, there exists an open subscheme U*

of S containing s, an effective horizontal divisor D on X_U whose support is disjoint from the restriction of Z to X_U, and an isomorphism

$$L(n)\,|\,X_U \;\tilde{\to}\; \mathcal{O}_{X_U}(D)$$

where $L(n)$ is defined to be $L \otimes \mathcal{O}(1)^{\otimes n}$.

Proof. Let $U_0 = \operatorname{Spec} R$ be an affine open subscheme of S containing s. The closure of any associated point of X maps to a closed set of S. If such an image does not contain s, we can assume U_0 was chosen not to intersect this image. In other words, we can assume that every associated point of X_{U_0} has closure intersecting the fiber X_s.

The above question is local in S, so we can, without loss of generality, assume that $S = U_0 = \operatorname{Spec} R$. Let $x_1, \ldots, x_q$ be a finite set of closed points of X_s which contains every point of the intersection of Z with X_s. Also assume that the set $x_1, \ldots, x_q$ contains a closed point of X_s from the closure of each associated point of X, and from the closure of each associated point of the fiber X_s.

Since $\mathcal{O}(1)$ is very ample, we can factor the projective morphism $\pi: X \to S$ as follows:

$$X \;\overset{i}{\hookrightarrow}\; \mathbf{P}^t_R \;\to\; S = \operatorname{Spec} R,$$

where i is a closed embedding into $P = \mathbf{P}^t_R$, projective t-space over R, such that $\mathcal{O}(1)$ is canonically isomorphic to the restriction of $\mathcal{O}_P(1)$, the canonical very ample sheaf on P, to X.

Identify X with its image in $P = \mathbf{P}^t_R$. We can find homogeneous polynomials $F_1, \ldots, F_q$ of $R[T_0, \ldots, T_t]$, such that, for each F_i, the corresponding hypersurface of $\mathbf{P}^t_R$ does not contain x_i, but contains x_j for all $j \neq i$. By replacing some of the F_i's by powers of themselves, we can assume that all the F_i's have the same degree N_1. We can regard $F_1, \ldots, F_q$ as global sections of $\mathcal{O}(N_1) \overset{\text{def}}{=} \mathcal{O}(1)^{\otimes N_1}$.

Since $\mathcal{O}(1)$ is very ample, there is an integer N_0 such that if $n \geq N_0$ then $L(n) \overset{\text{def}}{=} L \otimes \mathcal{O}(1)^{\otimes n}$ is generated by global sections (see [H], II, Theorem 5.17).

Let $N = N_0 + N_1$. Let n be an integer such that $n \geq N$, and let $n_0 = n - N_1$. Since $n_0 \geq N_0$, we can find global sections $\lambda_1, \ldots, \lambda_q$ of $L(n_0)$, such that λ_i does not vanish at x_i. Then $\lambda = \sum_{i=0}^{q} \lambda_i \otimes F_i$ is a global section of $L(n_0 + N_1) = L(n)$ which is non-zero at $x_1, \ldots, x_q$.

Since the set $\{x_1, \ldots, x_q\}$ contains a point from the closure of each associated point of X, the zero set of the section λ does not contain any associated point of X. Therefore, λ is meromorphic (in the sense of (1.2) of Chapter 1), so λ determines an effective Cartier divisor D whose support does not contain the points $x_1, \ldots, x_q$. This gives an isomorphism $L(n) \tilde{\to} \mathcal{O}_X(D)$ under which $\lambda \mapsto \mathbf{1}_D$.

Since the set $x_1, \ldots, x_q$, contains a closed point from the closure of each associated point of X_s, the effective Cartier divisor D cannot contain any associated points of X_s. So by Lemma 2.2, there is an open subscheme U_1 of S containing s such that the restriction of D to X_{U_1} is flat over U_1.

The set Z intersects $D\,|X_{U_1}$ in a closed set of X_{U_1} not intersecting the fiber X_s. Since π takes closed sets to closed sets, we can find an open subscheme U of U_1 containing s such that the restriction of D to X_U does not intersect Z.

(2.4) Proposition. *Let $\pi: X \to S$ be a projective, flat morphism whose fibers are one-dimensional. Let L be a line bundle on X, $s \in S$ a closed point, and Z a closed subset of X intersecting the fiber X_s in at most a finite number of points. Then there is an open subscheme U of S containing s such that*

$$L|_{X_U} \xrightarrow{\sim} \mathcal{O}_{X_U}(D)$$

for some horizontal divisor D whose support is disjoint from the restriction of Z to X_U.

Proof. Let $\mathcal{O}(1)$ be a line bundle of X, very ample with respect to the projective morphism $\pi: X \to S$. It follows from Lemma 2.3 that there is an integer n and an open subscheme U of S containing s such that

$$L \otimes \mathcal{O}(n) \,|\, X_U \xrightarrow{\sim} \mathcal{O}_{X_U}(D_1) \qquad \text{and} \qquad \mathcal{O}(n) \,|\, X_U \xrightarrow{\sim} \mathcal{O}_{X_U}(D_2)$$

where D_1 and D_2 are effective horizontal divisors on X_U disjoint from the restriction of Z to X_U. Therefore,

$$L \,|\, X_U \xrightarrow{\sim} (L \otimes \mathcal{O}(n) \,|\, X_U) \otimes (\mathcal{O}(n) \,|\, X_U)^{-1} \xrightarrow{\sim} \mathcal{O}_{X_U}(D_1 - D_2).$$

(2.5) Proposition. *Let $\pi: X \to S$ be a projective, flat morphism whose fibers are all one-dimensional. Let D_1 and D_2 be two effective horizontal divisors on X. Then the divisor $D_1 + D_2$ is also an effective horizontal divisor on X.*

Proof. By Lemma 2.2, both D_1 and D_2 do not contain any associated points of any fiber of π. So the same is true of $D_1 + D_2$. Therefore, by Lemma 2.2 again, $D_1 + D_2$ is flat over S.

(2.6) Corollary. *Let $\pi: X \to S$ be a projective, flat morphism whose fibers are all one-dimensional. Let $D = D_1 - D_2$ and $D' = D'_1 - D'_2$ be two horizontal divisors on X. Then their sum*

$$D + D' = (D_1 + D'_1) - (D_2 + D'_2)$$

is also a horizontal divisor on X.

3. The Norm Functor

In this section we extend the constructions and results of Chapter 1, Section 6, to the case of a family of one-dimensional schemes, flat and projective over an integral base scheme.

(3.1) Definition. Let Y be a scheme, finite and flat over an integral base scheme S, and let K be the function field of S. As usual, Y_K will denote the generic fiber of Y over S.

For any regular function f on Y, we define the norm of f by the equation

$$\mathrm{Norm}_{Y/S}(f) \stackrel{\mathrm{def}}{=} \mathrm{Norm}_{Y_K}(f|_{Y_K})$$

where the right-hand side is as in Section 3 of Chapter 1. This is an element of K, and the multiplicative group K^* can be canonically identified with the group of invertible meromorphic functions of S. Thus we can regard $\mathrm{Norm}_{Y/S}(f)$ as a function defined on some (maximal) open subscheme of S. The following proposition shows that it is in fact defined at every point of S.

(3.2) Proposition. *Let Y be a scheme, finite and flat over an integral base scheme S, and let f be a regular function on all of Y. Then $\mathrm{Norm}_{Y/S}(f)$ is a regular function on all of S. In addition, if f is non-zero at every point of Y, then $\mathrm{Norm}_{Y/S}(f)$ is non-zero at every point of S.*

Proof. Since Y is finite and flat over S, we can find, around each point $s \in S$, an affine open subscheme $U = \mathrm{Spec}\, R$ of S such that $Y_U = \mathrm{Spec}\, A$ where the ring A is a finite, free R-module. By restricting to Y_U, we can consider f as an element of A.

The generic fiber Y_K is $\mathrm{Spec}(A \otimes_R K)$. The element $\mathrm{Norm}_{Y/S}(f)$ is defined to be $\mathrm{Norm}_{A\otimes_R K}(f \otimes 1)$, i.e., the determinant of the K-linear action of $f \otimes 1$ on $A \otimes_R K$. Since A is a free R-module and since R is a sub-ring of K, this is the same as the determinant of the R-linear action of f on A. Therefore, $\mathrm{Norm}_{Y/S}(f)$ is an element of R, and so is defined at $s \in S$.

If f is also non-zero at every point of Y, then $1/f$ is regular at every point of Y. Hence, by what we just showed, $\mathrm{Norm}_{Y/S}(1/f)$ is regular at every point of S. However,

$$\mathrm{Norm}_{Y/S}(1/f)\, \mathrm{Norm}_{Y/S}(f) = \mathrm{Norm}_{Y/S}(1) = 1.$$

Therefore, $\mathrm{Norm}_{Y/S}(f)$ is an invertible function of S.

(3.3) Definition. Let X be flat and projective over an integral scheme S, with fibers which are all one-dimensional schemes, and let K be the function field of S. Let D be an effective horizontal divisor of X, i.e., as a scheme, D is finite and flat over S. For any invertible meromorphic function f of X defined at every point of D, Proposition 3.2 demonstrates that $\mathrm{Norm}_{D/S}(f|_D)$ is defined at every point of S. The expression $\mathrm{Norm}_{D/S}(f)$ will denote $\mathrm{Norm}_{D/S}(f|_D)$.

Now let D be a horizontal divisor on X, and f an invertible meromorphic function of X defined and non-zero at every point of the support of D. Let f_K be the restriction of f to the generic fiber X_K, and let D_K be the restriction of the Cartier divisor D to X_K. Then $f_K\,(D_K)$, as defined in Chapter 1, Section 6, is a non-zero element of K. Define $f(D)$ to be the invertible meromorphic function of S associated to $f_K\,(D_K)$. As we will see in the following proposition, $f(D)$ is actually defined and non-zero at every point of S.

(3.4) Proposition. *Let X be flat and projective over an integral scheme S, with fibers which are all one-dimensional schemes.*
1. Let D be a effective horizontal divisor of X, and let f be an invertible meromorphic function of X defined and non-zero at every point of D. Then

$$f(D) = \mathrm{Norm}_{D/S}(f).$$

2. Let $D = D_1 - D_2$ be a horizontal divisor of X, i.e., D_1 and D_2 are effective horizontal divisors of X, and let f be an invertible meromorphic function of X defined and non-zero at every point of D_1 and D_2. Then

$$f(D) = \frac{\mathrm{Norm}_{D_1/S}(f)}{\mathrm{Norm}_{D_2/S}(f)}.$$

Therefore, Proposition 3.2 implies that $f(D)$ is defined and non-zero at every point of S.
3. Let D be a horizontal divisor of X, and let f and g be invertible meromorphic functions of X defined and non-zero at every point of the support of D. Then

$$f(D)\,g(D) = fg\,(D).$$

4. Let D and E be horizontal divisors of X, and let f be an invertible meromorphic function of X defined and non-zero at every point of the support of D and E. Then

$$f(D+E) = f(D)\,f(E).$$

Proof. This is a restatement of Proposition 6.4 of Chapter 1 in terms of the notation of this section.

(3.5) The following lemma will be our basic tool for defining and comparing line bundles in this and the following sections.

(3.5) Lemma. *Let $\mathcal{G}$ be a quasi-coherent sheaf on a scheme S.*
1. Let $\mathcal{U}$ be a family of open sets which cover S, and, for each member U of $\mathcal{U}$, let $\mathcal{F}_U$ be a quasi-coherent subsheaf of $\mathcal{G}|_U$. Then there is a quasi-coherent subsheaf $\mathcal{F}$ of $\mathcal{G}$ such that $\mathcal{F}|_U = \mathcal{F}_U$ for each $U \in \mathcal{U}$ if and only if, for any pair $U, V \in \mathcal{U}$ and any point x in their intersection, the stalks $\mathcal{F}_U$ and $\mathcal{F}_V$ at x are equal as submodules of $\mathcal{G}_x$.
2. Let $\mathcal{F}$ and $\mathcal{F}'$ be two quasi-coherent subsheaves of $\mathcal{G}$. Then $\mathcal{F} = \mathcal{F}'$ if and only if, for every $x \in S$, the stalks $\mathcal{F}_x$ and $\mathcal{F}'_x$ are equal as submodules of $\mathcal{G}$.
3. Let $\mathcal{G}$ and $\mathcal{G}'$ be quasi-coherent sheaves of S and let $\rho\colon \mathcal{G} \xrightarrow{\sim} \mathcal{G}'$ be an isomorphism between them. Let $\mathcal{F}$ and $\mathcal{F}'$ be quasi-coherent subsheaves of $\mathcal{G}$ and $\mathcal{G}'$ respectively. Then ρ restricts to an isomorphism between $\mathcal{F}$ and $\mathcal{F}'$ if and only if, for each point x of S, the induced isomorphism of stalks from $\mathcal{G}_x$ to $\mathcal{G}'_x$ restricts to an isomorphism between their submodules $\mathcal{F}_x$ and $\mathcal{F}'_x$.

Proof. First show that, for any quasi-coherent subsheaf $\mathcal{F}$ of $\mathcal{G}$ and for any open subscheme U of S, the global sections of $\mathcal{F}$ on U are exactly the global sections of $\mathcal{G}$ on U whose image, for each $x \in U$, in the stalk $\mathcal{G}_x$ is in $\mathcal{F}_x$. This suggests a definition for $\mathcal{F}$ in part 1. In part 2, this gives a description of $\mathcal{F}$ and $\mathcal{F}'$ which agree. Part 3 is essentially the same as part 2. We leave the details to the reader.

(3.6) Definition of the Norm: Part 1. Let X be flat and projective over an integral scheme S, with fibers which are all one-dimensional schemes, and let K be the function field of S. Let $D = D_1 - D_2$ be a horizontal divisor of X, i.e., D_1 and D_2 are non-zero effective Cartier divisors which are finite and flat as schemes over S. Finally, let M be a line bundle on X.

Let D_K be the Cartier divisor of the generic fiber X_K associated to D, and let M_K be the restriction of M to X_K. Let $\mathcal{C}$ be the constant sheaf on S whose stalk at every point is the one-dimensional K-vector space $\mathbf{N}_{D_K}(M_K)$ defined in (6.1) of Chapter 1.

We consider now only the case where there is a meromorphic section m of M which is defined and non-zero at every point of D_1 and D_2. Then $\mathbf{N}_{D_K}(m|_{X_K})$ is a non-zero vector of $\mathbf{N}_{D_K}(M_K)$. Define $\mathbf{N}_{D/S}(m)$ to be the global section of $\mathcal{C}$ associated with $\mathbf{N}_{D_K}(m|_{X_K})$, and define $\mathbf{N}_{D/S}(M)$ to be the $\mathcal{O}_S$-submodule of $\mathcal{C}$ generated by $\mathbf{N}_{D/S}(m)$. Clearly, this is a line bundle on S. The following lemma, in light of Lemma 3.5, part 2, shows that $\mathbf{N}_{D/S}(M)$ is independent of the choice of m.

(3.6.1) Lemma. *We defined $\mathbf{N}_{D/S}(M)$ to be an $\mathcal{O}_S$-submodule of the constant sheaf whose stalks are equal to $\mathbf{N}_{D_K}(M_K)$. So, at any point $s \in S$, the stalk of $\mathbf{N}_{D/S}(M)$ at s is a $\mathcal{O}_{S,s}$-submodule of $\mathbf{N}_{D_K}(M_K)$. This submodule can be described as follows: Let U be an open subscheme of S containing s, and let m' be a meromorphic section of $M|_{X_U}$ which is defined and non-zero at every point of D_1 and D_2 contained in X_U. The stalk of $\mathbf{N}_{D/S}(M)$ at $s \in S$ is the $\mathcal{O}_{S,s}$-submodule of $\mathbf{N}_{D_K}(M_K)$ generated by $\mathbf{N}_{D_K}(m'|_{X_K})$.*

Proof. Let m be as chosen in Definition 3.6. Then, straight from the definition, the stalk of $\mathbf{N}_{D/S}(M)$ at $s \in S$ the $\mathcal{O}_{S,s}$-submodule of $\mathbf{N}_{D_K}(M_K)$ generated by $\mathbf{N}_{D_K}(m|_{X_K})$.

Let U be an open subscheme of S containing s, and let m' be a meromorphic section of $M|_{X_U}$ which is defined and non-zero at every point of D_1 and D_2 contained in X_U. Let D', D'_1, and D'_2 be the restrictions of D, D_1, and D_2 respectively to X_U.

Let f be the invertible meromorphic function of X_U such that $m' = f \cdot m|_{X_U}$. Note that f is defined and non-zero at every point of D'_1 and D'_2. By Proposition 6.6 of Chapter 1 and Definition 3.3 above,

$$\mathbf{N}_{D_K}(m'|_{X_K}) \;=\; f(D')\,\mathbf{N}_{D_K}(m|_{X_K}).$$

Now according to part 2 of Proposition 3.4, $f(D')$ is defined and non-zero at every point of U. Therefore, the image of $f(D')$ in $\mathcal{O}_{S,s}$ is a unit. So $\mathbf{N}_{D_K}(m|_{X_K})$ and $\mathbf{N}_{D_K}(m'|_{X_K})$ generate the same $\mathcal{O}_{S,s}$-submodule of $\mathbf{N}_{D_K}(M_K)$.

(3.7) Definition of the Norm: Part 2. As in the first part of the definition, let X be flat and projective over an integral scheme S, with fibers which are all one-dimensional schemes, and let K be the function field of S. Let D be a horizontal divisor of X, and let M be a line bundle on X. Let M_K be the restriction of M to the generic fiber X_K, and let D_K be the Cartier divisor of X_K associated to D. Let $\mathcal{C}$ be the constant sheaf on S whose stalk at every point is the one-dimensional K-vector space $\mathbf{N}_{D_K}(M_K)$ defined in (6.1) of Chapter 1.

Let $\mathcal{U}$ be a cover of S by open subschemes such that, for each $U \in \mathcal{U}$, there is a meromorphic section m_U of $M|_{X_U}$ which is defined and non-zero at every point of the support of $D_U = D\,|X_U$. Such a cover exists by Proposition 2.4 (taking Z to be the support of D).

For each $U \in \mathcal{U}$, Definition 3.6 defines the line bundle $\mathbf{N}_{D_U/U}(M|_{X_U})$ as a certain subsheaf of the constant sheaf $\mathcal{C}\,|_U$. Using the description of the stalks given in Lemma 3.6.1, part 1 of Lemma 3.5 implies that this defines a line bundle $N_{D/S}(M)$ on all of S such that

$$N_{D/S}(M)|\,U = N_{D_U/U}(M|_{X_U})$$

for each $U \in \mathcal{U}$. Clearly, the description of the stalks given in Lemma 3.6.1, also applies to this definition of $N_{D/S}(M)$. This description of the stalks together with part 2 of Lemma 3.5 implies that this line bundle is independent of the choice of the cover $\mathcal{U}$ and that, when Definition 3.6 applies, the present definition agrees with this earlier definition.

(3.8) Proposition. *Let X be flat and projective over an integral scheme S, with fibers which are all one-dimensional schemes, and let K be the function field of S. Let D be a horizontal divisor of X, and let M be a line bundle on X.*
1. The restriction of the line bundle $\mathbf{N}_{D/S}(M)$ to the generic point of S is naturally isomorphic to the K-vector space $\mathbf{N}_{D_K}(M\,|_{X_K})$.
2. Let S' be an open subscheme of S, and let D' be the restriction of D to S'. Then there is a natural isomorphism

$$\mathbf{N}_{D/S}(M)\,|\,S' \overset{\sim}{\rightarrow} \mathbf{N}_{D'/S'}(M\,|X_{S'}).$$

3. If there is a meromorphic section m of M which is defined and non-zero at every point of the support of D, then $\mathbf{N}_{D/S}(M)$ is a trivial line-bundle on S with nowhere-zero global section $\mathbf{N}_{D/S}(m)$.

If, in addition, f is an invertible meromorphic function of X, defined and non-zero at every point of the support of D, then $\mathbf{N}_{D/S}(f\,m)$ is another nowhere-zero global section of $\mathbf{N}_{D/S}(M)$, and

$$\mathbf{N}_{D/S}(f\,m) \;=\; f(D)\,\mathbf{N}_{D/S}(m).$$

Proof. Let D_K be the restriction of D to the generic fiber X_K.
1. As in Lemma 3.6.1, the stalks of $\mathbf{N}_{D/S}(M)$ are submodules of $\mathbf{N}_{D_K}(M\,|_{X_K})$. Clearly the stalk at the generic point is the whole of $\mathbf{N}_{D_K}(M\,|_{X_K})$. This identification gives the desired isomorphism.
2. Both line bundles are defined to be $\mathcal{O}_{S'}$-submodules of the same constant sheaf: the constant sheaf whose stalk at every $s \in S'$ is $\mathbf{N}_{D_K}(M|_{X_K})$. Lemma 3.6.1, in light of part 2 of Lemma 3.5, shows that they both have the same stalks and so are equal.
3. The first claim is an immediate consequence of Definition 3.6. Let η be the generic point of S. Let f_K be the restriction of f to the generic fiber X_K. Under the natural isomorphism, mentioned in part 1 above, from $\mathbf{N}_{D/S}(M)|\,\eta$ to $\mathbf{N}_{D_K}(M|_{X_K})$, the elements $\mathbf{N}_{D/S}(fm)|\,\eta$ and $f(D)\,\mathbf{N}_{D/S}(m)|\,\eta$ are sent to $\mathbf{N}_{D_K}(f_K m|_{X_K})$ and $f_K(D_K)\,\mathbf{N}_{D_K}(m|_{X_K})$ respectively; this follows from the definition of $\mathbf{N}_{D/S}(m)$ in Definition 3.6. By proposition 6.6 of Chapter 1,

$$\mathbf{N}_{D_K}(f_K\,m|_{X_K}) = f_K(D_K)\,\mathbf{N}_{D_K}(m|_{X_K}).$$

Therefore,

$$\mathbf{N}_{D/S}(f\,m)|\,\eta \;=\; f(D)\;\mathbf{N}_{D/S}(m)|\,\eta.$$

This implies in turn that $\mathbf{N}_{D/S}(f\,m)$ and $f(D)\;\mathbf{N}_{D/S}(m)$ are themselves equal.

(3.9) Proposition. *Let X be flat and projective over an integral scheme S, with fibers that are all one-dimensional schemes, and let K be the function field of S. Let D be a horizontal divisor of X, and let M_1 and M_2 be line bundles on X.*

Consider the canonical isomorphism of Proposition 6.5 of Chapter 1,

$$\mathbf{N}_{D_K}(M_1 \otimes M_2|_{X_K}) \xrightarrow{\sim} \mathbf{N}_{D_K}(M_1|_{X_K}) \otimes \mathbf{N}_{D_K}(M_2|_{X_K}),$$

where D_K is the restriction of D to the generic fiber X_K. This isomorphism induces a canonical isomorphism

$$\mathbf{N}_{D/S}(M_1 \otimes M_2) \xrightarrow{\sim} \mathbf{N}_{D/S}(M_1) \otimes \mathbf{N}_{D/S}(M_2).$$

Proof. We defined $\mathbf{N}_{D/S}(M_1)$, $\mathbf{N}_{D/S}(M_2)$, and $\mathbf{N}_{D/S}(M_1 \otimes M_2)$ as subsheaves of the constant sheaves associated to the vector spaces $\mathbf{N}_{D_K}(M_1|_{X_K})$, $\mathbf{N}_{D_K}(M_2|_{X_K})$, and $\mathbf{N}_{D_K}(M_1 \otimes M_2|_{X_K})$ respectively. Therefore, (since all the sheaves involved are flat over S), the sheaf $\mathbf{N}_{D/S}(M_1) \otimes \mathbf{N}_{D/S}(M_2)$ is a subsheaf of the constant sheaf associated with the vector space $\mathbf{N}_{D_K}(M_1|_{X_K}) \otimes \mathbf{N}_{D_K}(M_2|_{X_K})$. The canonical isomorphism of Proposition 6.5 of Chapter 1, induces an isomorphism between the corresponding constant sheaves of S. By Lemma 3.5 part 3, to show that this isomorphism restricts to an isomorphism between their respective subsheaves, $\mathbf{N}_{D/S}(M_1 \otimes M_2)$ and $\mathbf{N}_{D/S}(M_1) \otimes \mathbf{N}_{D/S}(M_2)$, we only need to show that it restricts to an isomorphism at the level of stalks.

Let $s \in S$ be a point, let U be an open subscheme of S containing s, and let m_1 and m_2 be meromorphic sections of $M_1|_{X_U}$ and $M_2|_{X_U}$ respectively which are defined and non-zero at every point of the support of the restriction of D to X_U. Such U, m_1, and m_2 exist by Proposition 2.4.

According to Lemma 3.6.1, the stalk of $\mathbf{N}_{D/S}(M_1)$ at s is the $\mathcal{O}_{S,s}$-submodule of $\mathbf{N}_{D_K}(M_1|_{X_K})$ generated by $\mathbf{N}_{D_K}(m_1|_{X_K})$. Likewise, the stalk of $\mathbf{N}_{D/S}(M_2)$ at s is generated by $\mathbf{N}_{D_K}(m_2|_{X_K})$ and the stalk of $\mathbf{N}_{D/S}(M_1 \otimes M_2)$ at s is generated by $\mathbf{N}_{D_K}(m_1 \otimes m_2|_{X_K})$. Therefore, the stalk of $\mathbf{N}_{D/S}(M_1) \otimes \mathbf{N}_{D/S}(M_2)$ is generated by $\mathbf{N}_{D_K}(m_1|_{X_K}) \otimes \mathbf{N}_{D_K}(m_2|_{X_K})$. Proposition 6.5 of Chapter 1 implies that $\mathbf{N}_{D_K}(m_1 \otimes m_2|_{X_K})$ is sent to $\mathbf{N}_{D_K}(m_1|_{X_K}) \otimes \mathbf{N}_{D_K}(m_2|_{X_K})$. Since these elements are generators of the respective stalks, the isomorphism restricts to an isomorphism from the stalk of $\mathbf{N}_{D/S}(M_1 \otimes M_2)$ at s to the stalk of $\mathbf{N}_{D/S}(M_1) \otimes \mathbf{N}_{D/S}(M_2)$ at s.

(3.10) Proposition. *Let X be flat and projective over an integral scheme S, with fibers which are all one-dimensional schemes, and let K be the function field of S. Let D and E be horizontal divisors of X, and let M be a line bundle on X.*

Consider the canonical isomorphism of Chapter 1, Proposition 6.7:

$$\mathbf{N}_{D_K+E_K}(M|_{X_K}) \xrightarrow{\sim} \mathbf{N}_{D_K}(M|_{X_K}) \otimes \mathbf{N}_{E_K}(M|_{X_K})$$

where D_K and E_K are the restrictions of D and E, respectively, to the generic fiber X_K. This isomorphism induces a canonical isomorphism

$$\mathbf{N}_{D+E/S}(M) \xrightarrow{\sim} \mathbf{N}_{D/S}(M) \otimes \mathbf{N}_{E/S}(M).$$

Proof. We defined $\mathbf{N}_{D/S}(M)$, $\mathbf{N}_{E/S}(M)$, and $\mathbf{N}_{D+E/S}(M)$ as subsheaves of the constant sheaves associated with the vector spaces $\mathbf{N}_{D_K}(M|_{X_K})$, $\mathbf{N}_{E_K}(M|_{X_K})$, and $\mathbf{N}_{D_K+E_K}(M|_{X_K})$ respectively. So, since all the sheaves in question are flat over S, the sheaf $\mathbf{N}_{D/S}(M) \otimes \mathbf{N}_{E/S}(M)$ is a subsheaf of the constant sheaf associated with $\mathbf{N}_{D_K}(M|_{X_K}) \otimes \mathbf{N}_{E_K}(M|_{X_K})$. The canonical isomorphism of Proposition 6.7 in Chapter 1 induces an isomorphism between the corresponding constant sheaves. By Lemma 3.5 part 3, to show that this isomorphism restricts to an isomorphism between their respective subsheaves, $\mathbf{N}_{D+E/S}(M)$ and $\mathbf{N}_{D/S}(M) \otimes \mathbf{N}_{E/S}(M)$, we only need to show that it restricts to an isomorphism at the level of stalks.

Let $s \in S$ be a point, let U be an open subscheme of S containing s, and let m be a meromorphic section of $M|_{X_U}$ which is defined and non-zero at every point of the supports of the restrictions of D and E to X_U. Such U and m exist by Proposition 2.4.

According to Lemma 3.6.1, the stalk of $\mathbf{N}_{D/S}(M)$ at s is the $\mathcal{O}_{S,s}$-submodule of $\mathbf{N}_{D_K}(M|_{X_K})$ generated by $\mathbf{N}_{D_K}(m|_{X_K})$. Likewise, the stalk of $\mathbf{N}_{E/S}(M)$ at s is generated by $\mathbf{N}_{E_K}(m|_{X_K})$ and the stalk of $\mathbf{N}_{D+E/S}(M)$ at s is generated by $\mathbf{N}_{D_K+E_K}(m|_{X_K})$. Therefore, the stalk of $\mathbf{N}_{D/S}(M) \otimes \mathbf{N}_{E/S}(M)$ is generated by $\mathbf{N}_{D_K}(m|_{X_K}) \otimes \mathbf{N}_{E_K}(m|_{X_K})$. Under the above isomorphism, Proposition 6.7 of Chapter 1 implies that $\mathbf{N}_{D_K+E_K}(m|_{X_K})$ is sent to $\mathbf{N}_{D_K}(m|_{X_K}) \otimes \mathbf{N}_{E_K}(m|_{X_K})$. Since these elements are generators of the respective stalks, the isomorphism restricts to an isomorphism from the stalk of $\mathbf{N}_{D+E/S}(M)$ at s to the stalk of $\mathbf{N}_{D/S}(M) \otimes \mathbf{N}_{E/S}(M)$ at s.

4. The Intersection Pairing

(4.1) Definition of the Intersection Pairing: Part 1. Let X be flat and projective over an integral scheme S, with fibers which are all one-dimensional schemes, and let K be the function field of S. Let L and M be a line bundle on X, and let L_K and M_K be their respective restrictions to the generic fiber X_K. Finally, let $\mathcal{C}$ be the constant sheaf on S whose stalks are all equal to the one-dimensional K-vector space $\langle L_K, M_K \rangle$ defined in Chapter 1.

We consider now only the case where there are meromorphic sections, l of L and m of M, such that their associated Cartier divisors can be written as horizontal divisors, (l) as D and (m) as E, where D and E have disjoint supports. Then define the *intersection pairing of the meromorphic sections l and m*, written $\langle l, m \rangle$, to be the global section of the constant sheaf $\mathcal{C}$ determined by the non-zero vector $\langle\, l|_{X_K}, m|_{X_K} \,\rangle$ of $\langle L_K, M_K \rangle$. Define the *intersection pairing of L with M*, written $\langle L, M \rangle$, to be to be the $\mathcal{O}_S$-submodule of $\mathcal{C}$ generated by the global section $\langle l, m \rangle$. Clearly, this is a line bundle on S. The following lemma, in light of Lemma 3.5 part 2, shows that $\langle L, M \rangle$ is independent of the choice of l and m.

(4.1.1) Lemma. *The line bundle $\langle L, M \rangle$ was defined as an $\mathcal{O}_S$-submodule of the constant sheaf whose stalks are equal to $\langle L_K, M_K \rangle$. So, at any point $s \in S$, the stalk of $\langle L, M \rangle$ at s is an $\mathcal{O}_{S,s}$-submodule of $\langle L_K, M_K \rangle$. This submodule can be described as follows: Let U be an open subscheme of S containing s, and let l_1 and m_1 be meromorphic sections of $L|_{X_U}$ and $M|_{X_U}$, respectively, whose associated Cartier divisors can be written as horizontal divisors over U with disjoint supports. Then, for any such l_1 and m_1, the stalk of $\langle L, M \rangle$ at $s \in S$ is the $\mathcal{O}_{S,s}$-submodule of $\langle L_K, M_K \rangle$ generated by $\langle\, l_1|_{X_K}, m_1|_{X_K} \,\rangle$.*

Proof. Let l, m, D, and E be as chosen in Definition 4.1. Then, straight from the definition, the stalk of $\langle L, M\rangle$ at $s \in S$ is the $\mathcal{O}_{S,s}$-submodule of $\langle L_K, M_K\rangle$ generated by $\langle\, l|_{X_K}, m|_{X_K}\,\rangle$.

Let U be an open subscheme of S containing s, and let l_1 and m_1 be meromorphic sections of $L|_{X_U}$ and $M|_{X_U}$, respectively, whose associated Cartier divisors can be written as horizontal divisors, (l_1) as D_1 and (m_1) as E_1, such that D_1 and E_1 have disjoint supports. Then, by Proposition 2.4, there is an open subscheme U' of U containing s, and a meromorphic section m_2 of $M|_{X_{U'}}$, such that the Cartier divisor associated with m_2 can be written as a horizontal divisor E_2, and such that the support of E_2 is disjoint from the supports of $D|_{X_{U'}}$ and $D_1|_{X_{U'}}$.

Let D' be the restriction of D to $X_{U'}$. Let g be the meromorphic function of $X_{U'}$ such that $m_2 = g \cdot m|_{X_{U'}}$. Note that g is defined and non-zero at every point of the support of D'. By Proposition 6.8 of Chapter 1 and Definition 3.3,

$$\langle\, l|_{X_K}, m_2\,|_{X_K}\,\rangle = \langle\, l|_{X_K}, g\,m\,|_{X_K}\,\rangle = g(D')\,\langle\, l|_{X_K}, m|_{X_K}\,\rangle.$$

But as mentioned in Definition 3.3, $g(D')$ is defined and non-zero at every point of U'. Therefore, the image of $g(D')$ in $\mathcal{O}_{S,s}$ is a unit. So $\langle\, l|_{X_K}, m|_{X_K}\,\rangle$ and $\langle\, l|_{X_K}, m_2|_{X_K}\,\rangle$ generate the same $\mathcal{O}_{S,s}$-submodule of $\langle\, L_K, M_K\,\rangle$.

A similar argument, applied to the pair $\langle\, l|_{X_K}, m_2\,|_{X_K}\,\rangle$ and $\langle\, l_1\,|_{X_K}, m_2\,|_{X_K}\,\rangle$, and then to the pair $\langle\, l_1\,|_{X_K}, m_2\,|_{X_K}\,\rangle$ and $\langle\, l_1\,|_{X_K}, m_1\,|_{X_K}\,\rangle$, shows that not only $\langle\, l|_{X_K}, m|_{X_K}\,\rangle$ and $\langle\, l|_{X_K}, m_2\,|_{X_K}\,\rangle$, but also $\langle\, l_1\,|_{X_K}, m_2\,|_{X_K}\,\rangle$, and additionally $\langle\, l_1\,|_{X_K}, m_1\,|_{X_K}\,\rangle$ are generators of the same $\mathcal{O}_{S,s}$-submodule of $\langle\, L_K, M_K\,\rangle$.

(4.2) Definition of the Intersection Pairing: Part 2. As in the first part of the definition, let X be flat and projective over an integral scheme S, with fibers which are all one-dimensional schemes, and let K be the function field of S. Let L and M be a line bundle on X, and let $\mathcal{C}$ be the constant sheaf on S whose stalks are all equal to $\langle\, L\,|_{X_K}, M\,|_{X_K}\,\rangle$.

Let $\mathcal{U}$ be a cover of S by open subschemes such that, for each $U \in \mathcal{U}$, there are meromorphic section, l_U and m_U of $L|_{X_U}$ and $M|_{X_U}$ respectively, whose associated Cartier divisors, (l_U) and (m_U), can be written as horizontal divisors with disjoint supports. Such a cover exists by Proposition 2.4.

So, for each $U \in \mathcal{U}$, Definition 4.1 gives the intersection pairing $\langle\, L|_{X_U}, M|_{X_U}\,\rangle$ which is a line bundle on U and is a subsheaf of the constant sheaf $\mathcal{C}\,|_U$. Using the description of the stalks given in Lemma 4.1.1, part 1 of Lemma 3.5 implies that there is a line bundle $\langle L, M\rangle$ on all of S such that

$$\langle L, M\rangle|_U = \langle\, L|_{X_U}, M|_{X_U}\,\rangle$$

for each $U \in \mathcal{U}$. We call this line bundle the *intersection pairing of L and M*. Clearly, the description of the stalks given in Lemma 4.1.1 also applies to this definition of $\langle L, M\rangle$. This description of the stalks, in light of part 2 of Lemma 3.5, implies that this line bundle is independent of the choice of the cover $\mathcal{U}$ and that, when Definition 4.1 applies, the present definition agrees with this earlier definition.

(4.3) Proposition. *Let X be flat and projective over an integral scheme S, with fibers which are all one-dimensional schemes, and let K be the function field of S. Let L and M be line bundles on X.*

1. The restriction of the line bundle $\langle L, M\rangle$ to the generic point of S is naturally isomorphic to the K-vector space $\langle\, L|_{X_K}, M|_{X_K}\,\rangle$.
2. Let U be an open subscheme of S. There is a natural isomorphism

$$\langle L, M\rangle|_U \xrightarrow{\sim} \langle\, L|_{X_U}, M|_{X_U}\,\rangle.$$

3. Suppose there are meromorphic section, l of L and m of M, whose associated Cartier divisors can be written as horizontal divisors, (l) as D and (m) as E, with disjoint supports. Then $\langle L, M\rangle$ is a trivial line-bundle on S with nowhere-vanishing global section $\langle l, m\rangle$.

If, in addition, f is an invertible meromorphic function of X, defined and non-zero at every point of the support of E, then $\langle fl, m\rangle$ is another nowhere-vanishing global section of $\langle L, M\rangle$, and the two sections are related by the formula

$$\langle fl, m\rangle = f(E)\,\langle l, m\rangle.$$

Likewise, if g is an invertible meromorphic function of X, defined and non-zero at every point of the support of D, then $\langle l, gm\rangle$ is another nowhere-vanishing global section of $\langle L, M\rangle$, and the two sections are related by the formula

$$\langle l, gm\rangle = g(D)\,\langle l, m\rangle.$$

Proof.
1. As in Lemma 4.1.1, the stalks of $\langle L, M\rangle$ can be considered as submodules of $\langle\, L|_{X_K}, M|_{X_K}\,\rangle$. Clearly the stalk at the generic point is all of $\langle\, L|_{X_K}, M|_{X_K}\,\rangle$. This identification gives the desired isomorphism.
2. Both line bundles are defined to be $\mathcal{O}_U$-submodules of the same constant sheaf: the constant sheaf whose stalk at every $s \in U$ is $\langle\, L|_{X_K}, M|_{X_K}\,\rangle$. Lemma 4.1.1 shows that they both have the same stalks. So part 2 of Lemma 3.5 implies that they are in fact the same $\mathcal{O}_U$-submodule.

3. The first claim is an immediate consequence of Definition 4.1. Let η be the generic point of S, let f_K be the restriction of f to the generic fiber X_K, and let E_K be the restriction of E to X_K. Under the natural isomorphism, mentioned in part 1 above, from $\langle L, M\rangle|\,\eta$ to $\langle\, L|_{X_K}, M|_{X_K}\,\rangle$, the elements $\langle fl, m\rangle\,|\eta$ and $f(E)\langle l, m\rangle\,|\eta$ are sent to $\langle\, f_K l\,|_{X_K}, m|_{X_K}\,\rangle$ and $f_K(E_K)\,\langle\, l|_{X_K}, m|_{X_K}\,\rangle$ respectively. This follows from the definition of $\langle l, m\rangle$ in (4.1). By Proposition 6.8 of Chapter 1,

$$\langle\, f_K\, l\,|_{X_K}, m|_{X_K}\,\rangle \;=\; f_K(E_K)\,\langle\, l|_{X_K}, m|_{X_K}\,\rangle.$$

Therefore,

$$\langle fl, m\rangle\,|\eta \;=\; f(E)\,\langle l, m\rangle\,|\eta.$$

This implies in turn that $\langle fl, m\rangle$ and $f(E)\,\langle l, m\rangle$ are themselves equal. The other statement follows from a similar argument.

(4.4) Proposition. *Let X be flat and projective over an integral scheme S, with fibers that are all one-dimensional schemes, and let K be the function field of S.*
1. Let D be a horizontal divisor of X, and let M be a line bundle on X. Consider the canonical isomorphism of Chapter 1, Proposition 6.2,

$$\langle\, \mathcal{O}_{X_K}(D_K), M|_{X_K}\,\rangle \xrightarrow{\sim} \mathbf{N}_{D_K}(M|_{X_K}),$$

where D_K is the restriction of D to the generic fiber X_K. This isomorphism induces a canonical isomorphism

$$\langle\, \mathcal{O}_X(D),\, M\,\rangle \xrightarrow{\sim} \mathbf{N}_{D/S}(M).$$

2. Let E be a horizontal divisor of X, and let L be a line bundle on X. Consider the canonical isomorphism of Proposition 6.2 in Chapter 1,

$$\langle\, L|_{X_K},\, \mathcal{O}_{X_K}(E_K)\,\rangle \xrightarrow{\sim} \mathbf{N}_{E_K}(L|_{X_K}),$$

where E_K is the restriction of E to the generic fiber X_K. This isomorphism induces a canonical isomorphism

$$\langle\, L,\, \mathcal{O}_X(E)\,\rangle \xrightarrow{\sim} \mathbf{N}_{E/S}(L).$$

Proof. We will only prove part 1, the proof of part 2 is similar. By definition, $\langle\, \mathcal{O}_X(D),\, M\,\rangle$ is a subsheaf of the constant sheaf associated to $\langle\, \mathcal{O}_{X_K}(D_K),\, M|_{X_K}\,\rangle$. Likewise, $\mathbf{N}_{D/S}(M)$ is a subsheaf of the constant sheaf associated to $\mathbf{N}_{D_K}(M|_{X_K})$. The canonical isomorphism, mentioned in the statement of the theorem, can be regarded as an isomorphism between these two constant sheaves. By Lemma 3.5 part 3, to show that this isomorphism restricts to an isomorphism between their respective subsheaves, $\langle\, \mathcal{O}_X(D),\, M\,\rangle$ and $\mathbf{N}_{D/S}(M)$, we only need to show that it restricts to an isomorphism at the level of stalks.

Let $s \in S$ be a point, let U be an open subscheme of S containing s, and let m be a meromorphic section of $M|_{X_U}$ which is defined and non-zero at every point of the support of $D|_{X_U}$. Such U and m exist by Proposition 2.4.

By Lemma 4.1.1, the stalk of $\langle\, \mathcal{O}_X(D),\, M\,\rangle$ at s is the $\mathcal{O}_{S,s}$-submodule of $\langle\, \mathcal{O}_{X_K}(D_K),\, M|_{X_K}\,\rangle$ generated by $\langle\, \mathbf{1}|_{X_K},\, m|_{X_K}\,\rangle$, where $\mathbf{1}$ is the canonical meromorphic section of $\mathcal{O}_X(D)$. By Lemma 3.6.1, the stalk of $\mathbf{N}_{D/S}(M)$ at s is the $\mathcal{O}_{S,s}$-submodule of $\mathbf{N}_{D_K}(M|_{X_K})$ generated by $\mathbf{N}_{D_K}(m|_{X_K})$. By Proposition 6.2 of Chapter 1, the isomorphism in question sends $\langle\, \mathbf{1}|_{X_K},\, m|_{X_K}\,\rangle$ to $\mathbf{N}_{D_K}(m|_{X_K})$. Since these elements are generators of the respective stalks, the isomorphism restricts to an isomorphism from the stalk of $\langle\, \mathcal{O}_X(D),\, M\,\rangle$ at s to the stalk of $\mathbf{N}_{D/S}(M)$ at s.

(4.5) Proposition. *Let X be flat and projective over an integral scheme S, with fibers that are all one-dimensional schemes, and let K be the function field of S. Let L and M be line bundles on X. Consider the canonical isomorphism of Chapter 1, Proposition 7.3:*

$$\langle\, L|_{X_K},\, M|_{X_K}\,\rangle \xrightarrow{\sim} \langle\, M|_{X_K},\, L|_{X_K}\,\rangle.$$

This isomorphism induces a canonical isomorphism

$$\langle L, M\rangle \xrightarrow{\sim} \langle M, L\rangle.$$

Proof. By definition, $\langle L, M\rangle$ is a subsheaf of the constant sheaf associated to $\langle\, L|_{X_K},\, M|_{X_K}\,\rangle$, and $\langle M, L\rangle$ is a subsheaf of the constant sheaf associated to $\langle\, M|_{X_K},\, L|_{X_K}\,\rangle$. The canonical isomorphism, mentioned in the statement of the theorem, can be regarded as an isomorphism between these two constant sheaves. By Lemma 3.5 part 3, to show that this isomorphism restricts to an isomorphism between their respective subsheaves, $\langle L, M\rangle$ and $\langle M, L\rangle$, we only need to show that it restricts to an isomorphism at the level of stalks.

Let $s \in S$ be a point, let U be an open subscheme of S containing s, and let l and m be meromorphic sections of $L|_{X_U}$ and $M|_{X_U}$, respectively, whose associated Cartier divisors, (l) and (m), can be written as horizontal divisors with disjoint support. Such U, l, and m exist by Proposition 2.4.

By Lemma 4.1.1, the stalk of $\langle L, M\rangle$ at s is the $\mathcal{O}_{S,s}$-submodule of $\langle L|_{X_K}, M|_{X_K}\rangle$ generated by $\langle\, l|_{X_K}, m|_{X_K}\,\rangle$, and the stalk of $\langle M, L\rangle$ at s is the $\mathcal{O}_{S,s}$-submodule of $\langle\, M|_{X_K}, L|_{X_K}\,\rangle$ generated by $\langle\, m|_{X_K}, l|_{X_K}\,\rangle$. By Proposition 7.3 of Chapter 1, the isomorphism in question sends $\langle\, l|_{X_K}, m|_{X_K}\,\rangle$ to $\langle\, m|_{X_K}, l|_{X_K}\,\rangle$. Since these elements are generators of the respective stalks, the isomorphism restricts to an isomorphism from the stalk of $\langle L, M\rangle$ at s to the stalk of $\langle M, L\rangle$ at s.

(4.6) Proposition. *Let X be flat and projective over an integral scheme S, with fibers that are all one-dimensional schemes, and let K be the function field of S.*

1. Given L, L', and M line bundles on X. Consider the canonical isomorphism of Chapter 1, Proposition 7.4:

$$\langle\, L|_{X_K}, M|_{X_K}\,\rangle \otimes \langle\, L'|_{X_K}, M|_{X_K}\,\rangle \xrightarrow{\sim} \langle\, L\otimes L'|_{X_K}, M|_{X_K}\,\rangle.$$

This isomorphism induces a canonical isomorphism

$$\langle L, M\rangle \otimes \langle L', M\rangle \xrightarrow{\sim} \langle L\otimes L', M\rangle.$$

2. Given L, M, and M' line bundles on X. Consider the canonical isomorphism of Chapter 1, Proposition 7.4:

$$\langle\, L|_{X_K}, M|_{X_K}\,\rangle \otimes \langle\, L|_{X_K}, M'|_{X_K}\,\rangle \xrightarrow{\sim} \langle\, L|_{X_K}, M\otimes M'|_{X_K}\,\rangle.$$

This isomorphism induces a canonical isomorphism

$$\langle L, M\rangle \otimes \langle L, M'\rangle \xrightarrow{\sim} \langle L, M\otimes M'\rangle.$$

Proof. We only prove part 1, the proof of part 2 is similar. By definition, $\langle L, M\rangle$, $\langle L', M\rangle$, and $\langle L\otimes L', M\rangle$ are subsheaves of the constant sheaves associated to the vector spaces $\langle\, L|_{X_K}, M|_{X_K}\,\rangle$, $\langle\, L'|_{X_K}, M|_{X_K}\,\rangle$, and $\langle\, L\otimes L'|_{X_K}, M|_{X_K}\,\rangle$ respectively. Therefore, since all the sheaves involved are flat over S, $\langle L, M\rangle\otimes\langle L', M\rangle$ is a subsheaf of the constant sheaf associated with $\langle\, L|_{X_K}, M|_{X_K}\,\rangle\otimes\langle\, L'|_{X_K}, M|_{X_K}\,\rangle$. The canonical isomorphism, mentioned in the statement of the theorem, can be regarded as an isomorphism between two constant sheaves on S. By Lemma 3.5 part 3, to show that this isomorphism restricts to an isomorphism between their respective subsheaves, $\langle L, M\rangle\otimes\langle L', M\rangle$ and $\langle L\otimes L', M\rangle$, we only need to show that it restricts to an isomorphism at the level of stalks.

Let $s \in S$ be a point, let U be an open subscheme of S containing s, and let l, l', and m be meromorphic sections of $L|_{X_U}$, $L'|_{X_U}$ and $M|_{X_U}$, respectively, whose Cartier divisors (l), (l') and (m) can be written as horizontal divisors. Also assume that the horizontal divisors associated with (l) and (l') have support disjoint from the horizontal divisor associated with (m). Such U, l, l', and m exist by Proposition 2.4.

By Lemma 4.1.1, the stalk of $\langle L, M\rangle$ at s is the $\mathcal{O}_{S,s}$-submodule of $\langle L|_{X_K}, M|_{X_K}\rangle$ generated by $\langle\, l|_{X_K}, m|_{X_K}\,\rangle$, the stalk of $\langle L', M\rangle$ at s is generated by $\langle\, l'|_{X_K}, m|_{X_K}\,\rangle$, and the stalk of $\langle L\otimes L', M\rangle$ at s is generated by $\langle\, l\otimes l'|_{X_K}, m|_{X_K}\,\rangle$. Therefore, the stalk of $\langle L, M\rangle \otimes \langle L', M\rangle$ at s is generated by $\langle\, l|_{X_K}, m|_{X_K}\,\rangle \otimes \langle\, l'|_{X_K}, m|_{X_K}\,\rangle$. By Proposition 7.4 of Chapter 1, $\langle\, l|_{X_K}, m|_{X_K}\,\rangle \otimes \langle\, l'|_{X_K}, m|_{X_K}\,\rangle$ is sent to the element $\langle\, l\otimes l'|_{X_K}, m|_{X_K}\,\rangle$. Since these elements are generators of the respective stalks, the isomorphism restricts to an isomorphism from the stalk of $\langle L, M\rangle \otimes \langle L', M\rangle$ at s to the stalk of $\langle L\otimes L', M\rangle$ at s.

(4.7) Proposition. *Let X and X' be flat and projective over an integral scheme S such that all the fibers of $X \to S$ and $X' \to S$ are one-dimensional schemes. Let K be the function field of S.*

Suppose there is an S-morphism ρ: $X' \to X$ and an open subscheme W of X such that the restriction of ρ to $W' = \rho^{-1}(W)$ is an isomorphism from W' to W. Also suppose that the complement of W in X intersects each fiber of $X \to S$ in at most a finite number of points.

Given L and M line bundles on X, and L' and M' be their respective pull-backs to X', we have the canonical isomorphism

$$\langle\, L|_{X_K}, M|_{X_K}\,\rangle \xrightarrow{\sim} \langle\, L'|_{X'_K}, M'|_{X'_K}\,\rangle$$

from Proposition 7.5 of Chapter 1. This isomorphism induces a canonical isomorphism

$$\langle L, M\rangle \xrightarrow{\sim} \langle L', M'\rangle.$$

Proof. By definition, $\langle L, M\rangle$ is a subsheaf of the constant sheaf associated to $\langle\, L|_{X_K}, M|_{X_K}\,\rangle$, and $\langle L', M'\rangle$ is a subsheaf of the constant sheaf associated to $\langle\, L'|_{X'_K}, M'|_{X'_K}\,\rangle$. The canonical isomorphism, mentioned in the statement of the theorem, can be regarded as an isomorphism between these two constant sheaves. By Lemma 3.5 part 3, to show that this isomorphism restricts to an isomorphism between their subsheaves, $\langle L, M\rangle$ and $\langle L', M'\rangle$, we only need to show that it restricts to an isomorphism at the level of stalks.

Let s be a point of S. By Proposition 2.4, there is an open subscheme U of S containing s such that $L|_{X_U}$ and $M|_{X_U}$ have meromorphic sections, l and m respectively, such that the associated Cartier divisors can be written as horizontal divisors, (l) as $D = D_1 - D_2$ and (m) as $E = E_1 - E_2$, where D and E have disjoint supports. In addition, since the complement of W in X intersects each fiber in a finite number of points, we can choose U, D and E so that the supports of D and E are contained in $W_U = W \cap X_U$.

Let W'_U be the intersection of W' with X'_U. Since $D = D_1 - D_2$ and $E = E_1 - E_2$ have supports contained in W_U, and W_U is isomorphic to W'_U, these divisors pull back to horizontal divisors $D' = D'_1 - D'_2$ and $E' = E'_1 - E'_2$ of X'_U with supports contained in W'_U. In addition, D' and E' have disjoint supports. Therefore, the meromorphic sections l and m pull back to meromorphic sections l' of $L'|_{X'_U}$ and m' of $M'|_{X'_U}$ respectively, and the Cartier divisors (l') and (m') can be written as horizontal divisors: (l') as D' and (m') as E'.

By Lemma 4.1.1, the stalk of $\langle L, M\rangle$ at s is the $\mathcal{O}_{S,s}$-submodule of $\langle L|_{X_K}, M|_{X_K}\rangle$ generated by $\langle\, l|_{X_K}, m|_{X_K}\,\rangle$, and the stalk of $\langle L', M'\rangle$ at s is the $\mathcal{O}_{S,s}$-submodule of $\langle\, L'|_{X'_K}, M'|_{X'_K}\,\rangle$ generated by $\langle\, l'|_{X'_K}, m'|_{X'_K}\,\rangle$. By Proposition 7.5 of Chapter 1, the isomorphism in question sends $\langle\, l|_{X_K}, m|_{X_K}\,\rangle$ to $\langle\, l'|_{X'_K}, m'|_{X'_K}\,\rangle$. Since these elements are generators of the respective stalks, this isomorphism restricts to an isomorphism from the stalk of $\langle L, M\rangle$ at s to the stalk of $\langle L', M'\rangle$ at s.

5. The Determinant of Cohomology Line Bundle

(5.1) Definition of the Determinant of Cohomology: Part 1. Let X be flat and projective over an integral scheme S, with fibers which are all one-dimensional schemes, and let K be the function field of S. Let L be a line bundle on X. We will now define the determinant of cohomology of L, but first only in a very specific situation. We require that S be an affine scheme, $S = \operatorname{Spec} R$. We also require that there be an effective horizontal divisor D on X such that (i) $H^1(X, L(D)) = 0$, (ii) $H^0(X, L(D))$ is a free R-module, and (iii) $H^0(D, L(D)|_D)$ is a free R-module.

Let L_K and D_K be the restrictions of L and D to the generic fiber X_K. By (2.4) of Chapter 1, the exact sequence

$$0 \longrightarrow L_K \longrightarrow L_K(D_K) \longrightarrow L_K(D_K)|_{D_K} \longrightarrow 0$$

yields an isomorphism

$$\mathbf{D}(L_K(D_K)) \xrightarrow{\sim} \mathbf{D}(L_K) \otimes \mathbf{D}(L_K(D_K)|_{D_K})$$

and hence an isomorphism

$$\frac{\operatorname{Det} H^0(X_K, L_K(D_K))}{\operatorname{Det} H^0(D_K, L_K(D_K)|_{D_K})} = \frac{\mathbf{D}(L_K(D_K))}{\mathbf{D}(L_K(D_K)|_{D_K})} \xrightarrow{\sim} \mathbf{D}(L_K).$$

The equality between the first two terms follows from the fact that all the higher cohomology is trivial: first of all, $H^1(X_K, L_K(D_K)) = 0$ because, by assumption, $H^1(X, L(D)) = 0$, and cohomology commutes with flat base change; in addition, the generic fiber X_K is one-dimensional, so $H^i(X_K, L_K(D_K)) = 0$ for $i > 1$; finally, the scheme D_K is zero-dimensional, so $H^i(D_K, L_K(D_K)|_{D_K}) = 0$ for $i > 0$.

Choose an R-basis $l_1, \ldots, l_n$ of $H^0(X, L(D))$. Hence, their restrictions $\hat{l}_1, \ldots, \hat{l}_n$ to the generic fiber X_K form a K-basis for $H^0(X_K, L_K(D_K))$. Likewise, choose an R-basis $m_1, \ldots, m_p$ for $H^0(D, L(D)|_D)$, and so their restrictions $\hat{m}_1, \ldots, \hat{m}_p$ to D_K form a K-basis of $H^0(D_K, L_K(D_K)|_{D_K})$.

Let δ be the image in $\mathbf{D}(L_K)$ of

$$\frac{\hat{l}_1 \wedge \cdots \wedge \hat{l}_n}{\hat{m}_1 \wedge \cdots \wedge \hat{m}_p}$$

under the above isomorphism; so δ is a non-zero vector of $\mathbf{D}(L_K)$. Let N be the R-submodule of $\mathbf{D}(L_K)$ generated by δ. We define the *determinant of cohomology line bundle of* L, written $\mathbf{D}(L)$, to be the sheaf on $S = \operatorname{Spec} R$ associated to N. Clearly $\mathbf{D}(L)$ is a line bundle on $S = \operatorname{Spec} R$.

Choosing another R-basis for $H^0(X, L(D))$ or $H^0(D, L(D)|_D)$ will change δ by a unit in R; therefore, the module N and hence the line bundle $\mathbf{D}(L)$ is independent of this choice.

Another way to describe $\mathbf{D}(L)$ more in line with Lemma 3.5 is to consider the constant sheaf $\mathcal{C}$ whose stalks are all the one-dimensional K-vector space $\mathbf{D}(L_K)$. Then δ defines a global section of $\mathcal{C}$, and $\mathbf{D}(L)$ is the $\mathcal{O}_S$-submodule of the constant sheaf $\mathcal{C}$ generated by this global section.

We will show below, in Lemma 5.1.2, that the stalks of the sheaf $\mathbf{D}(L)$ are independent of the choice of D. So, by Lemma 3.5 part 2, the sheaf $\mathbf{D}(L)$ itself is independent of the choice of D, and thus depends only on L.

(5.1.1) Lemma. *Let $\pi: X \to S$ be a projective, flat morphism with one-dimensional fibers, where S is an affine scheme. Let L be a line bundle on X such that $H^1(X,L)=0$. Then, around any $s \in S$, there is an affine open neighborhood $U = \operatorname{Spec} A$ such that $H^0(X_U, L|_{X_U})$ is a free A-module of finite rank.*

Proof. For each point $s \in S$, we have $H^2(X_s, L|_{X_s}) = 0$, and so the natural map

$$R^2\pi_*(L) \otimes \mathbf{F}(s) \to H^2(X_s, L|_{X_s})$$

is trivially surjective. Also, by [H], Chapter III, Corollary 11.2, $R^2\pi_*(L) = 0$. These two facts together imply, by the base change theorem (i=2) of [H], Chapter III, 12.11, that the natural map

$$R^1\pi_*(L) \otimes \mathbf{F}(s) \to H^1(X_s, L|_{X_s})$$

is an isomorphism at ever point $s \in S$. Since S is affine, $R^1\pi_*(L)$ is the sheaf associated with the R-module $H^1(X,L)$, which is 0 by assumption, i.e., $R^1\pi_*(L)$ is locally free of rank 0. This fact together with the fact that the above natural map is an isomorphism imply, by the base change change theorem ([H], Chapter III, 12.11 – in the case i=1), that the natural map

$$R^0\pi_*(L) \otimes \mathbf{F}(s) \to H^0(X_s, L|_{X_s})$$

is an isomorphism at ever point $s \in S$. This same theorem (in the case i=0) implies that, since the above map is surjective, $R^0\pi_*(L)$ is locally free.

In particular, around any $s \in S$ there is an affine neighborhood $U = \operatorname{Spec} A$, such that the U-sections of $R^0\pi_*(L)$, which is $H^0(X_U, L|_{X_U})$ since U is affine, form a free A-module. The map $X_U \to U$ is projective, therefore $H^0(X_U, L|_{X_U})$ is finitely generated as an A-module ([H], Chapter III, Theorem 5.2(a)).

(5.1.2) Lemma. *The line bundle $\mathbf{D}(L)$ was described in (5.1) above as an $\mathcal{O}_S$-submodule of the constant sheaf whose stalks are equal to $\mathbf{D}(L_K)$. So the stalk of $\mathbf{D}(L)$ at any point $s \in S$ is an $\mathcal{O}_{S,s}$-submodule of $\mathbf{D}(L_K)$. This submodule can be described as follows:*

Let $U = \operatorname{Spec} A$ be an affine open subscheme of S containing s, let D' be an effective horizontal divisor on X_U, and let L_U be the restriction of L to X_U. Suppose that $H^1(X_U, L_U(D')) = 0$, and that both $H^0(X_U, L_U(D'))$ and $H^0(D', L_U(D')|_{D'})$ are free A-modules. Let D'_K be the restriction of D' to the generic fiber X_K. Let λ be an ordered A-basis of $H^0(X_U, L_U(D'))$, and $\hat{\lambda}$ the ordered K-basis of $H^0(X_K, L_K(D'_K))$ obtained by restricting the members of λ to the generic fiber X_K. Likewise, let μ be an ordered A-basis of $H^0(D', L_U(D')|_{D'})$, and $\hat{\mu}$ the associated ordered K-basis of $H^0(D'_K, L_K(D'_K)|_{D'_K})$.

In a way similar to that described in Definition 5.1 above we get an isomorphism

$$\frac{\operatorname{Det} H^0(X_K, L_K(D'_K))}{\operatorname{Det} H^0(D'_K, L_K(D'_K)|_{D'_K})} = \frac{\mathbf{D}(L_K(D'_K))}{\mathbf{D}(L_K(D'_K)|_{D'_K})} \xrightarrow{\sim} \mathbf{D}(L_K).$$

Let $\langle\lambda\rangle$ be the non-zero vector of $\operatorname{Det} H^0(X_K, L_K(D'_K))$ associated to the basis $\hat{\lambda}$ and let $\langle\mu\rangle$ be the non-zero vector of $\operatorname{Det} H^0(D'_K, L_K(D'_K)|_{D'_K})$ associated to the basis $\hat{\mu}$.

Then the stalk of $\mathbf{D}(L)$ *at* $s \in S$ *is the* $\mathcal{O}_{S,s}$*-submodule of* $\mathbf{D}(L_K)$ *generated by the image in* $\mathbf{D}(L_K)$ *of* $\frac{\langle \lambda \rangle}{\langle \mu \rangle}$ *under the above isomorphism.*

Proof. We assume that we are in the situation of Definition 5.1. Let D be the effective horizontal divisor of X used to define $\mathbf{D}(L)$. For the remainder of the proof, D will denote the restriction of D to the open subscheme X_U. We note that $H^1(X_U, L_U(D + D')) = 0$. To see this, consider the short exact sequence

$$0 \longrightarrow L_U(D') \longrightarrow L_U(D + D') \longrightarrow L_U(D + D')|_D \longrightarrow 0$$

which gives rise to the exact sequence

$$H^1(X_U, L_U(D')) \longrightarrow H^1(X_U, L_U(D + D')) \longrightarrow H^1(D, L_U(D + D')|_D),$$

but the two outer terms are zero: $H^1(X_U, L_U(D')) = 0$ by assumption and additionally $H^1(D, L_U(D + D')|_D) = 0$ by [H], Chapter III, Corollary 11.2. Using Lemma 5.1.1, we can, without loss of generality, assume that U is small enough so that $H^0(X_U, L_U(D + D'))$ is a free A-module of finite rank.

The line bundle $L_U(D + D')|_{D+D'}$ is flat over $D + D'$, and, as a scheme, $D + D'$ is finite and flat over U, so $L_U(D + D')|_{D+D'}$ is flat over U. This implies that the push forward of $L_U(D + D')|_{D+D'}$to U is a locally free coherent sheaf. Therefore, after possibly replacing U by a smaller affine open neighborhood of s, we can assume that $H^0(D + D', L_U(D + D')|_{D+D'})$ is a free A-module of finite rank. Similarly, we can assume that $H^0(D', L_U(D + D')|_{D'})$ is also a free A-module of finite rank.

Consider the following commutative diagram:

$$\begin{array}{ccccccccc}
 & & 0 & & 0 & & & & \\
 & & \downarrow & & \downarrow & & & & \\
0 & \longrightarrow & L_U & \longrightarrow & L_U(D) & \longrightarrow & L_U(D)|_D & \longrightarrow & 0 \\
 & & \downarrow & & \downarrow & & & & \\
0 & \longrightarrow & L_U & \longrightarrow & L_U(D + D') & \longrightarrow & L_U(D + D')|_{D+D'} & \longrightarrow & 0 \\
 & & \downarrow & & \downarrow & & \downarrow & & \\
 & & 0 & \longrightarrow & L_U(D + D')|_{D'} & \longrightarrow & L_U(D + D')|_{D'} & \longrightarrow & 0 \\
 & & & & \downarrow & & \downarrow & & \\
 & & & & 0 & & 0 & &
\end{array}$$

where all the morphisms are the obvious ones. By a "diagram chase" argument, it is straightforward to check that this extends, uniquely, to the following commutative diagram, and that this diagram has exact rows and columns:

$$\begin{array}{ccccccccc}
 & & (\zeta_1) & & (\zeta_2) & & (\zeta_3) & & \\
 & & 0 & & 0 & & 0 & & \\
 & & \downarrow & & \downarrow & & \downarrow & & \\
(\eta_1) & 0 \longrightarrow & L_U & \longrightarrow & L_U(D) & \longrightarrow & L_U(D)|_D & \longrightarrow & 0 \\
 & & \downarrow & & \downarrow & & \downarrow & & \\
(\eta_2) & 0 \longrightarrow & L_U & \longrightarrow & L_U(D+D') & \longrightarrow & L_U(D+D')|_{D+D'} & \longrightarrow & 0 \\
 & & \downarrow & & \downarrow & & \downarrow & & \\
(\eta_3) & & 0 & \longrightarrow & L_U(D+D')|_{D'} & \longrightarrow & L_U(D+D')|_{D'} & \longrightarrow & 0 \\
 & & & & \downarrow & & \downarrow & & \\
 & & & & 0 & & 0 & &
\end{array}$$

Consider also the restriction of this diagram to the generic fiber X_K, and label the rows of this restriction $(\eta_1'), (\eta_2'), (\eta_3')$ and label the columns $(\zeta_1'), (\zeta_2'), (\zeta_3')$. Then, by Theorem 2.5.1 of Chapter 1, the following diagram commutes up to sign:

$$\begin{array}{ccc}
\mathbf{D}\Big(L_K\left(D_K + D'_K\right)\Big) & \xrightarrow{\mathbf{D}(\eta_2')} & \mathbf{D}(L_K) \otimes \mathbf{D}\Big(L_K(D_K + D'_K)|_{D_K+D'_K}\Big) \\
\Big\downarrow {\scriptstyle \mathbf{D}(\zeta_2')} & & \Big\downarrow {\scriptstyle 1\otimes \mathbf{D}(\zeta_3')} \\
\mathbf{D}\Big(L_K(D_K)\Big) \otimes \mathbf{D}\Big(L_K(D_K + D'_K)|_{D'_K}\Big) & \xrightarrow{\mathbf{D}(\eta_1')\otimes 1} & \mathbf{D}(L_K)\otimes\mathbf{D}\Big(L_K(D_K)|_{D_K}\Big)\otimes\mathbf{D}\Big(L_K(D_K + D'_K)|_{D'_K}\Big)
\end{array}$$

Trivially, the following commutes:

$$\begin{array}{ccc}
\mathbf{D}\Big(L_K(D_K + D'_K)|_{D_K+D'_K}\Big) & \xrightarrow{=} & \mathbf{D}\Big(L_K(D_K + D'_K)|_{D_K+D'_K}\Big) \\
\Big\downarrow {\scriptstyle \mathbf{D}(\zeta_3')} & & \Big\downarrow {\scriptstyle \mathbf{D}(\zeta_3')} \\
\mathbf{D}\Big(L_K(D_K)|_{D_K}\Big) \otimes \mathbf{D}\Big(L_K(D_K + D'_K)|_{D'_K}\Big) & \xrightarrow{=} & \mathbf{D}\Big(L_K(D_K)|_{D_K}\Big) \otimes \mathbf{D}\Big(L_K(D_K + D'_K)|_{D'_K}\Big)
\end{array}$$

By "dividing" the terms of the first diagram by the second and "cancelling" like terms, we get the following diagram, which commutes up to sign:

$$\begin{array}{ccc}
\dfrac{\mathbf{D}\Big(L_K(D_K+D'_K)\Big)}{\mathbf{D}\Big(L_K(D_K+D'_K)|_{D_K+D'_K}\Big)} & \xrightarrow{\frac{\mathbf{D}(\eta_2')}{1}} & \mathbf{D}(L_K) \\
\Big\downarrow {\scriptstyle \frac{\mathbf{D}(\zeta_2')}{\mathbf{D}(\zeta_3')}} & & \Big\| \\
\dfrac{\mathbf{D}\Big(L_K(D_K)\Big)}{\mathbf{D}\Big(L_K(D_K)|_{D_K}\Big)} & \xrightarrow{\frac{\mathbf{D}(\eta_1')}{1}} & \mathbf{D}(L_K)
\end{array}$$

where the isomorphisms are not as labelled, but are rather the labelled isomorphisms followed by natural "cancellation" isomorphisms.

Now we set up some notation. If M is a free A-module and $M_K = M \otimes_A K$ is the corresponding K-vector space, then, for any ordered A-basis β of M, the expression $\langle\beta\rangle$ will denote the non-zero vector of $\mathrm{Det}(M_K)$ corresponding to β. So, for example, if λ_D is an ordered A-basis of $H^0(X_U, L_U(D))$, then $\langle\lambda_D\rangle$ is a non-zero vector of

$$\mathbf{D}(L_K(D_K)) = \operatorname{Det} H^0(X_K, L_K(D_K)).$$

Now, since the sequence of A-modules

$$0 \longrightarrow H^0(X_U, L_U(D)) \longrightarrow H^0(X_U, L_U(D+D')) \longrightarrow H^0(D', L_U(D+D')|_{D'}) \longrightarrow 0$$

is exact, we can choose ordered A-bases, λ_D a basis of $H^0(X_U, L_U(D))$, $\lambda_{D+D'}$ a basis of $H^0(X_U, L_U(D+D'))$, and ν a basis of $H^0(D', L_U(D+D')|_{D'})$, such that, under the map

$$\mathbf{D}(L_K(D_K+D'_K)) \xrightarrow{\mathbf{D}(\zeta'_2)} \mathbf{D}(L_K(D_K)) \otimes \mathbf{D}(L_K(D_K+D'_K)|_{D'_K}),$$

the vector $\langle\lambda_{D+D'}\rangle$ is sent to $\langle\lambda_D\rangle \otimes \langle\nu\rangle$.

Similarly, since the sequence of A-modules, derived from the sequence (ζ'_3),

$$0 \to H^0(D, L_U(D)|_D) \to H^0(D+D', L_U(D+D')|_{D+D'}) \to H^0(D', L_U(D+D')|_{D'}) \to 0$$

is exact, we can choose ordered A-bases, μ_D of $H^0(D, L_U(D)|_D)$ and $\mu_{D+D'}$ of $H^0(D+D', L_U(D+D')|_{D+D'})$, such that under the map

$$\mathbf{D}(L_K(D_K+D'_K)|_{D_K+D'_K}) \xrightarrow{\mathbf{D}(\zeta'_3)} \mathbf{D}(L_K(D_K)|_{D_K}) \otimes \mathbf{D}(L_K(D_K+D'_K)|_{D'_K})$$

the vector $\langle\mu_{D+D'}\rangle$ is sent to $\langle\mu_D\rangle \otimes \langle\nu\rangle$.

Therefore, under the map

$$\frac{\mathbf{D}(L_K(D_K+D'_K))}{\mathbf{D}(L_K(D_K+D'_K)|_{D_K+D'_K})} \xrightarrow{\frac{\mathbf{D}(\zeta'_2)}{\mathbf{D}(\zeta'_3)}} \frac{\mathbf{D}(L_K(D_K))}{\mathbf{D}(L_K(D_K)|_{D_K})},$$

we have

$$\frac{\langle\lambda_{D+D'}\rangle}{\langle\mu_{D+D'}\rangle} \mapsto \frac{\langle\lambda_D\rangle}{\langle\mu_D\rangle}.$$

Since the diagram

$$\begin{array}{ccc}
\dfrac{\mathbf{D}(L_K(D_K+D'_K))}{\mathbf{D}(L_K(D_K+D'_K)|_{D_K+D'_K})} & \xrightarrow{\frac{\mathbf{D}(\eta'_2)}{1}} & \mathbf{D}(L_K) \\
\Big\downarrow \frac{\mathbf{D}(\zeta'_2)}{\mathbf{D}(\zeta'_3)} & & \Big\| \\
\dfrac{\mathbf{D}(L_K(D_K))}{\mathbf{D}(L_K(D_K)|_{D_K})} & \xrightarrow{\frac{\mathbf{D}(\eta'_1)}{1}} & \mathbf{D}(L_K)
\end{array}$$

commutes up to sign, the image of $\frac{\langle\lambda_{D+D'}\rangle}{\langle\mu_{D+D'}\rangle}$ in $\mathbf{D}(L_K)$ is equal to ± 1 times the image of $\frac{\langle\lambda_D\rangle}{\langle\mu_D\rangle}$ in $\mathbf{D}(L_K)$.

This is what is true of the specially chosen bases $\lambda_{D+D'}$, $\mu_{D+D'}$, λ_D, and μ_D. Now, if $\lambda_{D+D'}$, $\mu_{D+D'}$, λ_D, and μ_D are arbitrary ordered A-bases of their respective spaces, then all we can say is that the image of $\frac{\langle\lambda_{D+D'}\rangle}{\langle\mu_{D+D'}\rangle}$ in $\mathbf{D}(L_K)$ is equal to an invertible element of A times the image of $\frac{\langle\lambda_D\rangle}{\langle\mu_D\rangle}$ in $\mathbf{D}(L_K)$.

A similar argument shows that if, in addition, $\lambda_{D'}$ and $\mu_{D'}$ are ordered A-bases of $\mathrm{Det} H^0(X_U, L_U(D'))$ and $H^0(D, L_U(D')|_{D'})$ respectively, then the image of $\frac{\langle \lambda_{D+D'} \rangle}{\langle \mu_{D+D'} \rangle}$ under the map

$$\frac{\mathbf{D}\big(L_K(D_K+D'_K)\big)}{\mathbf{D}\big(L(D_K+D'_K)|_{D_K+D'_K}\big)} \xrightarrow{\frac{\mathbf{D}(\eta'_2)}{1}} \mathbf{D}(L_K)$$

is equal to an invertible element of A times the image of $\frac{\langle \lambda_{D'} \rangle}{\langle \mu_{D'} \rangle}$ under the corresponding map

$$\frac{\mathbf{D}\big(L_K(D'_K)\big)}{\mathbf{D}\big(L(D'_K)|_{D'_K}\big)} \longrightarrow \mathbf{D}(L_K).$$

So, under the above maps, the images of

$$\frac{\langle \lambda_D \rangle}{\langle \mu_D \rangle}, \qquad \frac{\langle \lambda_{D+D'} \rangle}{\langle \mu_{D+D'} \rangle}, \qquad \text{and} \qquad \frac{\langle \lambda_{D'} \rangle}{\langle \mu_{D'} \rangle}$$

all generate the same A-submodule, and consequently the same $\mathcal{O}_{S,s}$-submodule, of $\mathbf{D}(L_K)$. Definition 5.1 implies that the stalk of the line bundle $\mathbf{D}(L)$ at s is the $\mathcal{O}_{S,s}$-submodule, of $\mathbf{D}(L_K)$ generated by the image of $\frac{\langle \lambda_D \rangle}{\langle \mu_D \rangle}$. Therefore, it is also the $\mathcal{O}_{S,s}$-submodule of $\mathbf{D}(L_K)$ generated by the image of $\frac{\langle \lambda_{D'} \rangle}{\langle \mu_{D'} \rangle}$, which is exactly what we set out to prove.

(5.1.3) Lemma. *Let $X \to S$ be a projective, flat morphism whose fibers are all of dimension one, and where S is an integral scheme. Let L be a line bundle on S.*

The base scheme S can be covered by affine open subschemes in such a way that, given any member of the cover, $U = \mathrm{Spec}\, A$, there is an effective horizontal divisor D of X_U such that $H^1(X_U, L_U(D)) = 0$, and such that both $H^0(X_U, L_U(D))$ and $H^0(D, L_U(D)|_D)$ are free A-modules of finite rank. Here L_U denotes the restriction of L to X_U.

Proof. Given any point $s \in S$, let $U = \mathrm{Spec}\, A$ be an affine open subscheme of S containing s, let L_U be the restriction of L to X_U, and let M be a line bundle which is very ample with respect to the projective morphism $X_U \to U$. After possibly replacing M by a tensor power of itself we can also assume that, for all $i > 0$ and $n \geq 1$, $H^i(X_U, L_U \otimes M^{\otimes n}) = 0$ (see [H], Theorem 5.2(b) of Chapter III).

After possibly replacing M by a power of itself and replacing U by a smaller affine open neighborhood of s, Lemma 2.3 allows us to assume that M is isomorphic to $\mathcal{O}_{X_U}(D)$ for some effective horizontal divisor D of X_U. Now, since

$$H^1(X_U, L_U \otimes M) = H^1(X_U, L_U(D)) = 0,$$

after possibly replacing U by a smaller affine neighborhood, Lemma 5.1.1 implies that $H^0(X_U, L_U(D))$ is a free A-modules of finite rank.

The scheme associated with D is finite and flat over U, and $L_U(D)|_D$ is a line bundle on D, so the push forward of $L_U(D)|_D$ to U is a locally free coherent sheaf. Therefore, after possibly replacing U by a smaller affine open neighborhood of s, we have that $H^0(D, L_U(D)|_D)$ is a free A-module of finite rank.

(5.2) Definition of the Determinant of Cohomology: Part 2. Let X be flat and projective over an integral scheme S, with fibers which are all one-dimensional schemes, and let K be the function field of S. Let L be a line bundle on X. We will now extend the definition of (5.1) to the general case. By Lemma 5.1.3, there is a cover $\mathcal{U}$ of S by affine open subschemes, where, for each member $U = \operatorname{Spec} A$ of $\mathcal{U}$, there is an effective horizontal divisor D of X_U such that $H^1(X_U, L_U(D)) = 0$, and such that both $H^0(X_U, L_U(D))$ and $H^0(D, L_U(D)|_D)$ are free A-modules, where L_U is the restriction of L to X_U.

Let L_K be the restriction of L to the generic fiber X_K, and let $\mathcal{C}$ be the constant sheaf of S whose stalks are all equal to the one-dimensional K-vector space $\mathbf{D}(L_K)$.

For each $U \in \mathcal{U}$, Definition 5.1 defines the determinant of cohomology line bundle $\mathbf{D}(L|_{X_U})$ on U as a subsheaf of the constant sheaf $\mathcal{C}|_U$. Lemma 3.5, part 1, together with the description of the stalks given in Lemma 5.1.2 shows that there is a line bundle $\mathbf{D}(L)$ on S, given as a subsheaf of $\mathcal{C}$, such that, for each $U \in \mathcal{U}$,

$$\mathbf{D}(L)|_U = \mathbf{D}(L|_{X_U}).$$

We call $\mathbf{D}(L)$ the *determinant of cohomology line bundle of* L. Clearly, the description of the stalks of $\mathbf{D}(L)$ given in Lemma 5.1.2 applies to this new definition. In particular, the stalks are independent of the choice of the cover $\mathcal{U}$. So, by part 2 of Lemma 3.5, the line bundle $\mathbf{D}(L)$ is independent of the choice of the cover $\mathcal{U}$. Therefore, when Definition 5.1 applies, the present definition agrees with the earlier definition.

(5.3) Proposition. *Let X be flat and projective over an integral scheme S, with fibers which are all one-dimensional schemes, and let K be the function field of S. Let L be a line bundle on X.*

1. The restriction of the line bundle $\mathbf{D}(L)$ to the generic point of S is naturally isomorphic to the K-vector space $\mathbf{D}(L|_{X_K})$.

2. Let U be an open subscheme of S. There is a natural isomorphism

$$\mathbf{D}(L)|_U \xrightarrow{\sim} \mathbf{D}(L|_{X_U}).$$

3. Suppose that S is an affine scheme, $S = \operatorname{Spec} R$, and that both $H^0(X, L)$ and $H^1(X, L)$ are free R-module. Let λ_0 be an ordered R-basis of $H^0(X, L)$, and let λ_1 be an ordered R-basis of $H^1(X, L)$. Let $\langle\lambda_0\rangle$ and $\langle\lambda_1\rangle$ be the associated non-zero vectors of $\operatorname{Det} H^0(X_K, L|_{X_K})$ and $\operatorname{Det} H^1(X_K, L|_{X_K})$ respectively.

Then $\mathbf{D}(L)$ is a trivial line bundle on S with a global, nowhere-zero section δ satisfying the property that, under the natural map mentioned in part 1 above

$$\mathbf{D}(L) \otimes_R K \xrightarrow{\sim} \mathbf{D}(L|_{X_K}) = \frac{\operatorname{Det} H^0(X_K, L|_{X_K})}{\operatorname{Det} H^1(X_K, L|_{X_K})},$$

the image of $\delta \otimes 1$ in $\mathbf{D}(L) \otimes_R K$ is sent to $\frac{\langle\lambda_0\rangle}{\langle\lambda_1\rangle}$.

Proof.

1. As in Lemma 5.1.2, the stalks of the line bundle $\mathbf{D}(L)$ can be considered as submodules of $\mathbf{D}(L|_{X_K})$. Clearly the stalk at the generic point is the whole of $\mathbf{D}(L|_{X_K})$. This identification gives the desired isomorphism.

2. Both line bundles are defined to be $\mathcal{O}_U$-submodules of the same constant sheaf: the constant sheaf whose stalk at every $s \in U$ is $\mathbf{D}(L|_{X_K})$. Lemma 5.1.2 shows that

they both have the same stalks, so part 2 of (3.5) implies that they are in fact the same $\mathcal{O}_U$-submodule.

3. By part 1 of the current proposition, $\mathbf{D}(L) \otimes_R K$ is naturally isomorphic to $\mathbf{D}(L|_{X_K})$. Therefore, the non-zero vector $\frac{\langle\lambda_0\rangle}{\langle\lambda_1\rangle}$ of $\mathbf{D}(L|_{X_K}) = \frac{\operatorname{Det}H^0(X_K, L|_{X_K})}{\operatorname{Det}H^1(X_K, L|_{X_K})}$ determines a meromorphic section of $\mathbf{D}(L)$. We must show that this meromorphic section is defined and non-zero at every point of S.

Let s be a point of S. By Lemma 5.1.3 we can find an affine open neighborhood $U = \operatorname{Spec} A$ of s and an effective horizontal divisor D of X_U such that $H^1(X_U, L_U(D)) = 0$, where $L_U = L|_{X_U}$, and such that both $H^0(X_U, L_U(D))$ and $H^0(D, L_U(D)|_D)$ are free A-modules. The short exact sequence

$$0 \longrightarrow L_U \longrightarrow L_U(D) \longrightarrow L_U(D)|_D \longrightarrow 0$$

yields the following long exact sequence of free A-modules

$$0 \longrightarrow H^0(X_U, L_U) \longrightarrow H^0(X_U, L_U(D)) \longrightarrow H^0(D, L_U(D)|_D) \longrightarrow H^1(X_U, L_U) \longrightarrow 0.$$

Recall that, for $i = 1, 2$, we chose λ_i to be an ordered basis of $H^i(X, L)$. By restricting to X_U, we get ordered bases of $H^i(X_U, L_U)$ which we will also call λ_i. The kernel B of $H^0(D, L_U(D)|_D) \to H^1(X_U, L_U)$ is a finite, flat A-module since this is true of both $H^0(D, L_U(D)|_D)$ and $H^1(X_U, L_U)$. So, after possibly replacing U by a smaller neighborhood of s, we can assume that B is a free A-module of finite rank. Let β be an ordered A-basis of B. We can form an ordered basis of $H^0(D, L_U(D)|_D)$ by starting with β, and then adding to this an inverse image of every member, in order, of λ_1. We will call this ordered basis μ_D. We can form an ordered basis λ_D of $H^0(X_U, L_U(D))$ by starting with the images of the elements of λ_0, and adding to these an inverse image of every member, in order, of β.

Let $\langle\lambda_D\rangle$ be the non-zero vector of $\operatorname{Det}H^0(X_K, L_K(D_K))$ corresponding to λ_D, and let $\langle\mu_D\rangle$ be the non-zero vector of $\operatorname{Det}H^0(D_K, L_K(D_K)|_{D_K})$, corresponding μ_D.

The restriction of the above long exact sequence to X_K yields, by (2.4) of Chapter 1, a canonical isomorphism

$$\frac{\operatorname{Det}H^0(X_K, L_K(D_K))}{\operatorname{Det}H^0(D_K, L_K(D_K)|_{D_K})} = \frac{\mathbf{D}(L_K(D_K))}{\mathbf{D}(L_K(D_K)|_{D_K})} \xrightarrow{\sim} \mathbf{D}(L_K) = \frac{\operatorname{Det}H^0(X_K, L_K)}{\operatorname{Det}H^1(X_K, L_K)}.$$

By the choice of bases, we have that, under the above isomorphism,

$$\frac{\langle\lambda_D\rangle}{\langle\mu_D\rangle} \mapsto \frac{\langle\lambda_0\rangle}{\langle\lambda_1\rangle}.$$

By Proposition 5.1.2, the stalk of $\mathbf{D}(L)$ at $s \in S$ is the $\mathcal{O}_{S,s}$-submodule of $\mathbf{D}(L_K) = \mathbf{D}(L) \otimes_R K$ generated by the image of $\frac{\langle\lambda_D\rangle}{\langle\mu_D\rangle}$ under the above isomorphism, i.e., the stalk is generated by $\frac{\langle\lambda_0\rangle}{\langle\lambda_1\rangle}$. An equivalent way to say this is that the meromorphic section of $\mathbf{D}(L)$ associated with $\frac{\langle\lambda_0\rangle}{\langle\lambda_1\rangle}$ is defined and non-vanishing at s.

(5.4) Definition. We have defined the determinant of cohomology in the case where our scheme has relative dimension 1 over the base, but we will also need to define it in the case where we have relative dimension 0. We now consider this case.

Let D be flat and finite over an integral scheme S, and let K be the function field of S. Let L be a line bundle on D or any coherent sheaf on D flat over S. The expression L_K will denote the restrictions of L to the generic fiber D_K.

First we consider the case when $S = \operatorname{Spec} R$ is an affine scheme and when $H^0(D, L)$ is a free R-module. Choose an R-basis $l_1, \ldots, l_n$ of $H^0(D, L)$, and let $\hat{l}_1, \ldots, \hat{l}_n$ be their restrictions to the generic fiber D_K, so they form a K-basis for $\mathrm{H}^0(D_K, L_K)$. Since D_K is a zero-dimensional scheme over K, $H^i(D_K, L_K) = 0$ for all $i > 0$, so

$$\mathbf{D}(L_K) = \operatorname{Det} H^0(D_K, L_K)$$

is a one-dimensional K-vector space with basis $\hat{l}_1 \wedge \cdots \wedge \hat{l}_n$.

Let $\mathcal{C}$ be the constant sheaf on S whose stalks are the one-dimensional K-vector space $\mathbf{D}(L_K)$. We define $\mathbf{D}(L)$ to be the $\mathcal{O}_S$-submodule of the constant sheaf $\mathcal{C}$ generated by the global section of $\mathcal{C}$ associated to the non-zero vector $\hat{l}_1 \wedge \cdots \wedge \hat{l}_n$ of $\mathbf{D}(L_K)$. Clearly, this is a line bundle on S.

At any $s \in S$, the stalk of $\mathbf{D}(L)$ is thus an $\mathcal{O}_{S,s}$-submodule of $\mathbf{D}(L_K)$. We can describe this stalk as follows: Let $U = \operatorname{Spec} A$ be an affine subscheme of S containing s such that $H^0(D_U, L_U)$ is a free A module of finite rank, where L_U denotes the restriction of L to D_U. Let $l'_1, \ldots, l'_n$ be an A-basis of $H^0(D_U, L_U)$; hence, the restrictions $\hat{l}'_1, \ldots, \hat{l}'_n$ to the generic fiber D_K form a K-basis for $\mathrm{H}^0(D_K, L_K)$. Note that, in $\mathbf{D}(L_K)$, $\hat{l}_1 \wedge \cdots \wedge \hat{l}_n$ and $\hat{l}'_1 \wedge \cdots \wedge \hat{l}'_n$ differ by multiplication by a unit of A, hence by a unit of $\mathcal{O}_{S,s}$. Thus the stalk of $\mathbf{D}(L)$ at s is the $\mathcal{O}_{S,s}$-submodule of $\mathbf{D}(L_K)$ generated by any such vector $\hat{l}'_1 \wedge \cdots \wedge \hat{l}'_n$.

This description shows that the stalks are independent of the particular choice of basis, and hence, by Lemma 3.5 part 2, the same is true of $\mathbf{D}(L)$.

In general, since D is finite and flat over S, we can find a cover $\mathcal{U}$ of S by affine open subschemes such that, for each $U = \operatorname{Spec} A$ in $\mathcal{U}$, the module $H^0(D_U, L|_{D_U})$ is a free A-module. Again, let $\mathcal{C}$ be the constant sheaf of S whose stalks are all equal to the one-dimensional K-vector space $\mathbf{D}(L_K)$.

For each $U \in \mathcal{U}$, the above definition gives a line bundle $\mathbf{D}(L|_{D_U})$ on U as a subsheaf of the constant sheaf $\mathcal{C}\,|_U$. The above description of the stalks, in light of Lemma 3.5 part 1, shows that there is a line bundle $\mathbf{D}(L)$ on all of S, given as a subsheaf of $\mathcal{C}$, such that, for each $U \in \mathcal{U}$,

$$\mathbf{D}(L)|_U = \mathbf{D}(L|_{D_U}).$$

The above description of the stalks shows that the stalks are independent of the choice of the cover $\mathcal{U}$, so, by part 2 of Lemma 3.5, the line bundle $\mathbf{D}(L)$ itself is independent of the choice of this cover.

(5.5) Proposition. *Let D be flat and finite over an integral scheme S, and let K be the function field of S. Let L be a line bundle on D (or any coherent sheaf on D flat over S).*

1. The restriction of the line bundle $\mathbf{D}(L)$ to the generic point of S is naturally isomorphic to the K-vector space $\mathbf{D}(L|_{D_K})$.

2. Let U be an open subscheme of S. There is a natural isomorphism

$$\mathbf{D}(L)|_U \xrightarrow{\sim} \mathbf{D}(L|_{D_U}).$$

3. Suppose that S is an affine scheme, $S = \operatorname{Spec} R$, and that $H^0(D, L)$ is a free R-module. Let $l_1, \ldots, l_n$ be an R-basis of $H^0(D, L)$. Then $\mathbf{D}(L)$ is a trivial line bundle on S, with a global nowhere-zero section δ, such that, under the natural map of part 1 of the current proposition

$$\mathbf{D}(L) \otimes_R K \xrightarrow{\sim} \mathbf{D}(L|_{D_K}) = \operatorname{Det} H^0(D_K, L|_{D_K}),$$

the image of $\delta \otimes 1$ in $\mathbf{D}(L) \otimes_R K$ is $\hat{l}_1 \wedge \ldots \wedge \hat{l}_n$ where $\hat{l}_1, \ldots, \hat{l}_n$ are the restrictions of $l_1, \ldots, l_n$ to D_K.

Proof. These all follow directly from Definition 5.4.

(5.6) Proposition. *Let X be flat and projective over an integral scheme S, with fibers which are all one-dimensional schemes, and let K be the function field of S. Let L be a line bundle on X, and let D be an effective horizontal divisor of X. Let L_K be the restriction of L to the generic fiber X_K, and let D_K be the restriction of D to X_K.*

Consider the isomorphism between one-dimensional K-vector spaces

$$\mathbf{D}(L_K(D_K)) \xrightarrow{\sim} \mathbf{D}(L_K) \otimes \mathbf{D}(L_K(D_K)|_{D_K})$$

obtained, via Definition 2.4 of Chapter 1, from the exact sequence

$$0 \longrightarrow L_K \longrightarrow L_K(D_K) \longrightarrow L_K(D_K)|_{D_K} \longrightarrow 0.$$

This isomorphism induces a canonical isomorphism

$$\mathbf{D}(L(D)) \xrightarrow{\sim} \mathbf{D}(L) \otimes \mathbf{D}(L(D)|_D)$$

between line bundles of S.

Proof. By definition, $\mathbf{D}(L(D))$, $\mathbf{D}(L)$, and $\mathbf{D}(L(D)|_D)$ are subsheaves of the constant sheaves associated to $\mathbf{D}(L_K(D_K))$, $\mathbf{D}(L_K)$, and $\mathbf{D}(L_K(D_K)|_{D_K})$ respectively. Therefore, the sheaf $\mathbf{D}(L) \otimes \mathbf{D}(L(D)|_D)$ is a subsheaf of the constant sheaf associated with $\mathbf{D}(L_K) \otimes \mathbf{D}(L_K(D_K)|_{D_K})$. The canonical isomorphism, mentioned in the statement of the proposition, can be regarded as an isomorphism between the associated constant sheaves. By Lemma 3.5 part 3, to show that this isomorphism restricts to an isomorphism between their respective subsheaves, $\mathbf{D}(L(D))$ and $\mathbf{D}(L) \otimes \mathbf{D}(L(D)|_D)$, we only need to show that it restricts to an isomorphism at the level of stalks.

Let $s \in S$ be a point. By Lemma 5.1.3, we can find an affine open neighborhood $U = \operatorname{Spec} A$ of s and an effective horizontal divisor D' of X_U such that $H^1(X_U, L_U(D')) = 0$, and such that $H^0(X_U, L_U(D'))$ is a free A-module of finite rank, where $L_U = L|_{X_U}$.

From now on, D will denote the restriction of D to X_U. We observe that $H^1(X_U, L_U(D + D')) = 0$. To see this, consider the short exact sequence

$$0 \longrightarrow L_U(D') \longrightarrow L_U(D + D') \longrightarrow L_U(D + D')|_D \longrightarrow 0$$

which gives rise to the exact sequence

$$H^1\big(X_U, L_U(D')\big) \longrightarrow H^1\big(X_U, L_U(D+D')\big) \longrightarrow H^1\big(D, L_U(D+D')|_D\big),$$

but the two outer terms are the zero A-module: $H^1\big(X_U, L_U(D')\big)$ is zero by assumption and $H^1\big(D, L_U(D+D')|_D\big)$ is zero by [H], Chapter III, Corollary 11.2. Using Lemma 5.1.1, we can, without loss of generality, assume that U is small enough so that $H^0\big(X_U, L_U(D+D')\big)$ is a free A-module.

The line bundle $L_U(D)|_D$ is flat over U; this implies that the push forward of $L_U(D)|_D$ to U is locally free. Therefore, after possibly replacing U by a smaller affine open neighborhood of s, we can assume that $H^0\big(D, L_U(D)|_D\big)$ is a free A-module. Similarly, we can assume that both $H^0\big(D+D', L_U(D+D')|_{D+D'}\big)$ and $H^0\big(D', L_U(D+D')|_{D'}\big)$ are also free A-modules.

As in the proof of Lemma 5.1.2 we consider the diagram

$$\begin{array}{cccccccccc}
 & & (\zeta_1) & & (\zeta_2) & & (\zeta_3) & & \\
 & & 0 & & 0 & & 0 & & \\
 & & \downarrow & & \downarrow & & \downarrow & & \\
(\eta_1) & 0 \longrightarrow & L_U & \longrightarrow & L_U(D) & \longrightarrow & L_U(D)|_D & \longrightarrow & 0 \\
 & & \downarrow & & \downarrow & & \downarrow{\scriptstyle I} & & \\
(\eta_2) & 0 \longrightarrow & L_U & \longrightarrow & L_U(D+D') & \longrightarrow & L_U(D+D')|_{D+D'} & \longrightarrow & 0 \\
 & & \downarrow & & \downarrow & & \downarrow & & \\
(\eta_3) & & 0 & \longrightarrow & L_U(D+D')|_{D'} & \longrightarrow & L_U(D+D')|_{D'} & \longrightarrow & 0 \\
 & & & & \downarrow & & \downarrow & & \\
 & & & & 0 & & 0 & &
\end{array}$$

where all the maps are the obvious ones, except I, which is defined as the unique map which makes the diagram commute. Clearly, all the rows and columns are exact.

Consider also the restriction of this diagram to the generic fiber X_K. (We will also use (ζ_1), (ζ_2), (ζ_3), (η_1), (η_2), (η_3) to label the corresponding rows and column of this restriction). Then, by Theorem 2.5.1 of Chapter 1, the following diagram commutes up to sign:

$$\begin{array}{ccc}
\mathbf{D}\big(L_K(D_K+D'_K)\big) & \xrightarrow{\mathbf{D}(\eta_2)} & \mathbf{D}(L_K)\otimes\mathbf{D}\big(L_K(D_K+D'_K)|_{D_K+D'_K}\big) \\
\Big\downarrow{\scriptstyle \mathbf{D}(\zeta_2)} & & \Big\downarrow{\scriptstyle 1\otimes\mathbf{D}(\zeta_3)} \\
\mathbf{D}\big(L_K(D_K)\big)\otimes\mathbf{D}\big(L_K(D_K+D'_K)|_{D'_K}\big) & \xrightarrow{\mathbf{D}(\eta_1)\otimes 1} & \mathbf{D}(L_K)\otimes\mathbf{D}\big(L_K(D_K)|_{D_K}\big)\otimes\mathbf{D}\big(L_K(D_K+D'_K)|_{D'_K}\big)
\end{array}$$

which implies, in turn, that the following commutes up to sign:

$$(*)\qquad\begin{array}{ccc}
\dfrac{\mathbf{D}\big(L_K(D_K+D'_K)\big)}{\mathbf{D}\big(L_K(D_K+D'_K)|_{D'_K}\big)} & \xleftarrow{\frac{1\otimes 1}{\mathbf{D}(\zeta_3)}} & \dfrac{\mathbf{D}\big(L_K(D_K+D'_K)\big)\otimes\mathbf{D}\big(L_K(D_K)|_{D_K}\big)}{\mathbf{D}\big(L_K(D_K+D'_K)|_{D_K+D'_K}\big)} \\
\Big\downarrow{\scriptstyle \frac{\mathbf{D}(\zeta_2)}{1}} & & \Big\downarrow{\scriptstyle \frac{\mathbf{D}(\eta_2)\otimes 1}{1}} \\
\mathbf{D}\big(L_K(D_K)\big) & \xrightarrow{\mathbf{D}(\eta_1)} & \mathbf{D}(L_K)\otimes\mathbf{D}\big(L_K(D_K)|_{D_K}\big)
\end{array}$$

where the maps are as labelled, but then followed, whenever possible, by natural "cancellation" isomorphisms.

Now we set up some notation. If M is a free A-module and $M_K = M \otimes_A K$ is the corresponding K-vector space, then, for any ordered A-basis β of M, the expression $\langle\beta\rangle$ will denote the non-zero vector of $\mathrm{Det}(M_K)$ corresponding to the ordered K-basis of M_K obtained from β. So, if we let $\lambda_{D+D'}$ be an ordered A-basis of $H^0\big(X_U, L_U(D+D')\big)$, then $\langle\lambda_{D+D'}\rangle$ is a non-zero vector of

$$\mathbf{D}\big(L_K(D_K + D'_K)\big) = \mathrm{Det}\, H^0\big(X_K, L_K(D_K + D'_K)\big).$$

Now, since the sequence of free A-modules

$$0 \to H^0\big(D, L_U(D)|_D\big) \to H^0\big(D+D', L_U(D+D')|_{D+D'}\big) \to H^0\big(D', L_U(D+D')|_{D'}\big) \to 0$$

derived from the sequence (ζ_3) above is exact, we can choose ordered A-bases, μ_D a basis of $H^0\big(D, L_U(D)|_D\big)$, $\mu_{D+D'}$ a basis of $H^0\big(D+D', L_U(D+D')|_{D+D'}\big)$, and ν a basis of $\mathrm{Det}H^0\big(D', L_U(D+D')|_{D'}\big)$, such that under the map

$$\mathbf{D}\big(L_K(D_K + D'_K)|_{D_K+D'_K}\big) \xrightarrow{\mathbf{D}(\zeta_3)} \mathbf{D}\big(L_K(D_K)|_{D_K}\big) \otimes \mathbf{D}\big(L_K(D_K+D'_K)|_{D'_K}\big)$$

the vector $\langle\mu_{D+D'}\rangle$ is sent to $\langle\mu_D\rangle \otimes \langle\nu\rangle$.

Let α be the image of $\frac{\langle\lambda_{D+D'}\rangle}{\langle\mu_{D+D'}\rangle}$ under the map

$$\frac{\mathbf{D}\big(L_K(D_K+D'_K)\big)}{\mathbf{D}\big(L_K(D_K+D'_K)|_{D_K+D'_K}\big)} \xrightarrow{\frac{\mathbf{D}(\eta_2)}{1}} \mathbf{D}(L_K)$$

and let β be the image of $\frac{\langle\lambda_{D+D'}\rangle}{\langle\nu\rangle}$ under the map

$$\frac{\mathbf{D}\big(L_K(D_K+D'_K)\big)}{\mathbf{D}\big(L_K(D_K+D'_K)|_{D'_K}\big)} \xrightarrow{\frac{\mathbf{D}(\zeta_2)}{1}} \mathbf{D}\big(L_K(D_K)\big).$$

Now we study the behavior of the maps in the diagram $(*)$ given above:
Under the map

$$\frac{\mathbf{D}\big(L_K(D_K+D'_K)\big)\otimes\mathbf{D}\big(L_K(D_K)|_{D_K}\big)}{\mathbf{D}\big(L_K(D_K+D'_K)|_{D_K+D'_K}\big)} \xrightarrow{\frac{1\otimes 1}{\mathbf{D}(\zeta_3)}} \frac{\mathbf{D}\big(L_K(D_K+D'_K)\big)}{\mathbf{D}\big(L_K(D_K+D'_K)|_{D'_K}\big)}$$

we have

$$\frac{\langle\lambda_{D+D'}\rangle \otimes \langle\mu_D\rangle}{\langle\mu_{D+D'}\rangle} \mapsto \frac{\langle\lambda_{D+D'}\rangle}{\langle\nu\rangle};$$

under the map

$$\frac{\mathbf{D}\big(L_K(D_K+D'_K)\big)\otimes\mathbf{D}\big(L_K(D_K)|_{D_K}\big)}{\mathbf{D}\big(L_K(D_K+D'_K)|_{D_K+D'_K}\big)} \xrightarrow{\frac{\mathbf{D}(\eta_2)\otimes 1}{1}} \mathbf{D}(L_K) \otimes \mathbf{D}\big(L_K(D_K)|_{D_K}\big)$$

we have

$$\frac{\langle\lambda_{D+D'}\rangle \otimes \langle\mu_D\rangle}{\langle\mu_{D+D'}\rangle} \mapsto \alpha \otimes \langle\mu_D\rangle.$$

Therefore, since the above diagram $(*)$ commutes up to sign, under the map

$$\mathbf{D}\big(L_K(D_K)\big) \xrightarrow{\mathbf{D}(\eta_1)} \mathbf{D}\big(L_K\big) \otimes \mathbf{D}\big(L_K(D_K)|_{D_K}\big)$$

we have

$$\beta \mapsto \pm\alpha \otimes \langle \mu_D \rangle.$$

By Lemma 5.1.2, the stalk of $\mathbf{D}\big(L(D)\big)$at $s \in S$ is the $\mathcal{O}_{S,s}$-submodule of $\mathbf{D}\big(L_K(D_K)\big)$ generated by β and the stalk of $\mathbf{D}(L)$ at $s \in S$ is the $\mathcal{O}_{S,s}$-submodule of $\mathbf{D}(L_K)$ generated by α. As mentioned in (5.4), the stalk of $\mathbf{D}\big(L(D)|_D\big)$ at $s \in S$ is the $\mathcal{O}_{S,s}$-submodule of $\mathbf{D}\big(L_K(D_K)|_{D_K}\big)$ generated by $\langle \mu_D \rangle$. Therefore, the stalk of $\mathbf{D}(L) \otimes \mathbf{D}\big(L(D)|_D\big)$ at $s \in S$ is the $\mathcal{O}_{S,s}$-submodule of $\mathbf{D}(L_K) \otimes \mathbf{D}\big(L_K(D_K)|_{D_K}\big)$ generated by $\alpha \otimes \langle \mu_D \rangle$.

Therefore, $\mathbf{D}(\eta_1)$ restricts to an isomorphism between the stalk of $\mathbf{D}\big(L(D)\big)$ at $s \in S$ and the stalk of $\mathbf{D}(L) \otimes \mathbf{D}\big(L(D)|_D\big)$ at $s \in S$.

(5.7) Proposition. *Let X be flat and projective over an integral scheme S, with fibers which are all one-dimensional schemes, and let K be the function field of S. Let L and M be a line bundles on X, and let L_K and M_K be their restriction to the generic fiber X_K.*

Suppose we are given line bundles L_1, L_2, M_1, and M_2 on S, and integers (mod 2) ρ and ρ', such that

$$L = \frac{L_1}{L_2}, \qquad M = \frac{M_1}{M_2}, \qquad \text{and} \qquad \rho + \rho' = \deg L_K \cdot \deg M_K \pmod 2.$$

Then let $\widehat{L}_1$, $\widehat{L}_2$, $\widehat{M}_1$, and $\widehat{M}_2$ be the respective restrictions of L_1, L_2, M_1, and M_2 to the generic fiber X_K. The data

$$\mathcal{D} = \left(\widehat{L}_1, \widehat{L}_2, \widehat{M}_1, \widehat{M}_2, \rho, \rho'\right)$$

provides what we called, in (4.1) of Chapter 1, "modelling data" for L_K and M_K. Recall that the data $\mathcal{D}$ gives a model $\langle L_K, M_K \rangle_{\mathcal{D}}$ for the K-vector space $\langle L_K, M_K \rangle$, which we interpret in this context as an isomorphism

$$\frac{\mathbf{D}\left(\widehat{L}_1 \otimes \widehat{M}_1\right) \otimes \mathbf{D}\left(\widehat{L}_2 \otimes \widehat{M}_2\right)}{\mathbf{D}\left(\widehat{L}_1 \otimes \widehat{M}_2\right) \otimes \mathbf{D}\left(\widehat{L}_2 \otimes \widehat{M}_1\right)} \xrightarrow{\sim} \langle L_K, M_K \rangle$$

depending on ρ and ρ' (mod 2).

This isomorphism induces an isomorphism

$$\frac{\mathbf{D}(L_1 \otimes M_1) \otimes \mathbf{D}(L_2 \otimes M_2)}{\mathbf{D}(L_1 \otimes M_2) \otimes \mathbf{D}(L_2 \otimes M_1)} \xrightarrow{\sim} \langle L, M \rangle$$

between line bundles of S.

Proof. By definition, $\langle L, M \rangle$, $\mathbf{D}(L_1)$, $\mathbf{D}(L_2)$, $\mathbf{D}(M_1)$, and $\mathbf{D}(M_2)$ are subsheaves of the constant sheaves associated to $\langle L_K, M_K \rangle$, $\mathbf{D}(\widehat{L}_1)$, $\mathbf{D}(\widehat{L}_2)$, $\mathbf{D}(\widehat{M}_1)$, and $\mathbf{D}(\widehat{M}_2)$ respectively. It follows that

$$\frac{\mathbf{D}(L_1 \otimes M_1) \otimes \mathbf{D}(L_2 \otimes M_2)}{\mathbf{D}(L_1 \otimes M_2) \otimes \mathbf{D}(L_2 \otimes M_1)}$$

is a subsheaf of the constant sheaf associated to

$$\frac{\mathbf{D}(\widehat{L}_1 \otimes \widehat{M}_1) \otimes \mathbf{D}(\widehat{L}_2 \otimes \widehat{M}_2)}{\mathbf{D}(\widehat{L}_1 \otimes \widehat{M}_2) \otimes \mathbf{D}(\widehat{L}_2 \otimes \widehat{M}_1)}.$$

The isomorphism, mentioned in the statement of the proposition, can be regarded as an isomorphism between the associated constant sheaves. What we want to do is show that this isomorphism restricts to an isomorphism between their respective subsheaves. To do so, it is enough, by Lemma 3.5 part 3, to show that it restricts to an isomorphism at the level of stalks.

Let $s \in S$ be a point. By Proposition 2.4, there is an affine open subscheme $U = \operatorname{Spec} A$ of S containing s, and there are meromorphic sections, l of $L|_{X_U}$ and m of $M|_{X_U}$, whose associated Cartier divisors can be written as horizontal divisors, $(l) = D = D_1 - D_2$ and $(m) = E = E_1 - E_2$, such that D and E have disjoint supports. Let $\widehat{D}_1$, $\widehat{D}_2$, $\widehat{E}_1$, and $\widehat{E}_2$ be the restriction of the effective horizontal divisors D_1, D_2, E_1, and E_2, respectively, to the generic fiber X_K. Let $\hat{l}$ and $\hat{m}$ be the restrictions of l and m respectively to X_K. From now on, we will use L, L_1, L_2, M, M_1, M_2 to denote their respective restrictions to X_U.

In Chapter 1, Definition 4.6, we defined an isomorphism $\Phi_{\hat{l},\hat{m}}$ which allows us to factor the isomorphism mentioned in the statement of this proposition as follows

$$\begin{array}{ccc}
\dfrac{\mathbf{D}(\widehat{L}_1\otimes\widehat{M}_1)\otimes\mathbf{D}(\widehat{L}_2\otimes\widehat{M}_2)}{\mathbf{D}(\widehat{L}_1\otimes\widehat{M}_2)\otimes\mathbf{D}(\widehat{L}_2\otimes\widehat{M}_1)} & \longrightarrow & \langle L_K, M_K\rangle \\
\Big\downarrow \Phi_{\hat{l},\hat{m}} & & \Big\| \\
K & \xrightarrow{\cdot\langle \hat{l},\hat{m}\rangle} & \langle L_K, M_K\rangle
\end{array}$$

Furthermore, in (4.7) of Chapter 1, we defined two isomorphisms, $\Psi_{(\hat{l},\hat{D}_1,\hat{D}_2)}$ and $\Theta_{(\hat{D}_1,\hat{D}_2;\,\hat{m})}$, and showed that $\Phi_{\hat{l},\hat{m}}$ can be factored as follows:

$$\begin{array}{ccc}
\dfrac{\mathbf{D}(\widehat{L}_1\otimes\widehat{M}_1)\otimes\mathbf{D}(\widehat{L}_2\otimes\widehat{M}_2)}{\mathbf{D}(\widehat{L}_1\otimes\widehat{M}_2)\otimes\mathbf{D}(\widehat{L}_2\otimes\widehat{M}_1)} & \xrightarrow{\Phi_{\hat{l},\hat{m}}} & K \\
\Big\downarrow \Psi_{(\hat{l},\hat{D}_1,\hat{D}_2)} & & \Big\| \\
\dfrac{\mathbf{D}(\widehat{L}_1\otimes\widehat{M}_1|_{\hat{D}_1})\otimes\mathbf{D}(\widehat{L}_2\otimes\widehat{M}_2|_{\hat{D}_2})}{\mathbf{D}(\widehat{L}_1\otimes\widehat{M}_2|_{\hat{D}_1})\otimes\mathbf{D}(\widehat{L}_2\otimes\widehat{M}_1|_{\hat{D}_2})} & \xrightarrow{\Theta_{(\hat{D}_1,\hat{D}_2;\,\hat{m})}} & K
\end{array}$$

To get an idea of the behavior of the composition, we study the behavior of each map in the following series

$$\frac{\mathbf{D}(\widehat{L}_1\otimes\widehat{M}_1)\otimes\mathbf{D}(\widehat{L}_2\otimes\widehat{M}_2)}{\mathbf{D}(\widehat{L}_1\otimes\widehat{M}_2)\otimes\mathbf{D}(\widehat{L}_2\otimes\widehat{M}_1)} \xrightarrow{\Psi_{(\hat{l},\hat{D}_1,\hat{D}_2)}} \frac{\mathbf{D}(\widehat{L}_1\otimes\widehat{M}_1|_{\hat{D}_1})\otimes\mathbf{D}(\widehat{L}_2\otimes\widehat{M}_2|_{\hat{D}_2})}{\mathbf{D}(\widehat{L}_1\otimes\widehat{M}_2|_{\hat{D}_1})\otimes\mathbf{D}(\widehat{L}_2\otimes\widehat{M}_1|_{\hat{D}_2})} \xrightarrow{\Theta_{(\hat{D}_1,\hat{D}_2;\hat{m})}} K \xrightarrow{\langle\hat{l},\hat{m}\rangle} \langle L_K, M_K\rangle$$

We will first study $\Psi_{(\hat{l},\hat{D}_1,\hat{D}_2)}$. The meromorphic section l has divisor $D_1 - D_2$, and so defines an isomorphism between L and $\mathcal{O}_{X_U}(D_1 - D_2)$. Thus we have the following isomorphisms

$$\frac{L_1}{L_2} = L \xrightarrow{\sim} \mathcal{O}_{X_U}(D_1 - D_2) = \frac{\mathcal{O}_{X_U}(D_1)}{\mathcal{O}_{X_U}(D_2)} \qquad \text{and so} \qquad L_0 \stackrel{\text{def}}{=} L_1(-D_1) \xrightarrow{\sim} L_2(-D_2).$$

This gives, for each $i = 1, 2$ and $j = 1, 2$, the exact sequence

$$0 \longrightarrow L_0 \otimes M_j \longrightarrow L_i \otimes M_j \longrightarrow L_i \otimes M_j|_{D_i} \longrightarrow 0.$$

By Proposition 5.6, there is an isomorphism

$$\psi_{i,j}\colon \mathbf{D}(L_i \otimes M_j) \xrightarrow{\sim} \mathbf{D}(L_0 \otimes M_j) \otimes \mathbf{D}(L_i \otimes M_j|_{D_i})$$

whose restriction to the generic point of S is the isomorphism

$$\hat{\psi}_{i,j}\colon \mathbf{D}(\widehat{L}_i \otimes \widehat{M}_j) \xrightarrow{\sim} \mathbf{D}(\widehat{L}_0 \otimes \widehat{M}_j) \otimes \mathbf{D}(\widehat{L}_i \otimes \widehat{M}_j|_{\hat{D}_i})$$

induced by the corresponding exact sequence

$$0 \longrightarrow \widehat{L}_0 \otimes \widehat{M}_j \longrightarrow \widehat{L}_i \otimes \widehat{M}_j \longrightarrow \widehat{L}_i \otimes \widehat{M}_j|_{\hat{D}_i} \longrightarrow 0$$

where $\widehat{L}_0$ is the restriction of L_0 to the generic fiber X_K.

The map $\Psi_{(\hat{\iota},\hat{D}_1,\hat{D}_2)}$ was defined in Chapter 1, (4.7), to be ± 1 times $\dfrac{\hat{\psi}_{1,1} \otimes \hat{\psi}_{2,2}}{\hat{\psi}_{1,2} \otimes \hat{\psi}_{2,1}}$ followed by the cancellation map. We can construct an analogous map: consider ± 1 (as above) times $\dfrac{\psi_{1,1} \otimes \psi_{2,2}}{\psi_{1,2} \otimes \psi_{2,1}}$ followed by the cancellation map. This yields an isomorphism

$$\frac{\mathbf{D}(L_1 \otimes M_1) \otimes \mathbf{D}(L_2 \otimes M_2)}{\mathbf{D}(L_1 \otimes M_2) \otimes \mathbf{D}(L_2 \otimes M_1)} \xrightarrow{\sim} \frac{\mathbf{D}(L_1 \otimes M_1|_{D_1}) \otimes \mathbf{D}(L_2 \otimes M_2|_{D_2})}{\mathbf{D}(L_1 \otimes M_2|_{D_1}) \otimes \mathbf{D}(L_2 \otimes M_1|_{D_2})}$$

whose restriction to the generic point of S is $\Psi_{(\hat{\iota},\hat{D}_1,\hat{D}_2)}$.

Now we will study $\Theta_{(\hat{D}_1,\hat{D}_2;\hat{m})}$. Observe that, since $M = \dfrac{M_1}{M_2}$, we can identify M_1 with $M_2 \otimes M$. Outside the support of (m), the meromorphic section m determines an isomorphism between M_1 and M_2. It is characterized as follows: given a meromorphic section m_2 of M_2, it sends the meromorphic section $m_2 \otimes m$ of $M_1 = M_2 \otimes M$ to m_2. Since D_1 and D_2 are disjoint from (m), we can use the above isomorphism between M_1 and M_2 to construct the following isomorphisms for $i = 1, 2$:

$$\beta_i\colon L_i \otimes M_1|_{D_i} \xrightarrow{\sim} L_i \otimes M_2|_{D_i} \qquad \text{where} \qquad l_i \otimes (m_2 \otimes m)\,|_{D_i} \mapsto l_i \otimes m_2|_{D_i}$$

for all meromorphic sections, l_i of L_i and m_2 of M_2, defined and nowhere zero on D_i.

For $i = 1, 2$, let

$$\hat{\beta}_i\colon \widehat{L}_i \otimes \widehat{M}_1|_{\hat{D}_i} \xrightarrow{\sim} \widehat{L}_i \otimes \widehat{M}_2|_{\hat{D}_i}$$

be the restriction of β_i to the generic fiber X_K, and let

$$\mathbf{D}(\hat{\beta}_i)\colon \mathbf{D}\left(\widehat{L}_i \otimes \widehat{M}_1|_{\hat{D}_i}\right) \xrightarrow{\sim} \mathbf{D}\left(\widehat{L}_i \otimes \widehat{M}_2|_{\hat{D}_i}\right)$$

be the corresponding isomorphism.

In Chapter 1, (4.7), we showed that $\Theta_{(\hat{D}_1,\hat{D}_2;\hat{m})}$ is

$$\frac{\mathbf{D}(\hat{\beta}_1)\otimes\mathbf{1}}{\mathbf{1}\otimes\mathbf{D}(\hat{\beta}_2)}\colon \frac{\mathbf{D}\left(\widehat{L}_1 \otimes \widehat{M}_1|_{\hat{D}_1}\right) \otimes \mathbf{D}\left(\widehat{L}_2 \otimes \widehat{M}_2|_{\hat{D}_2}\right)}{\mathbf{D}\left(\widehat{L}_1 \otimes \widehat{M}_2|_{\hat{D}_1}\right) \otimes \mathbf{D}\left(\widehat{L}_2 \otimes \widehat{M}_1|_{\hat{D}_2}\right)} \to \frac{\mathbf{D}\left(\widehat{L}_1 \otimes \widehat{M}_2|_{\hat{D}_1}\right) \otimes \mathbf{D}\left(\widehat{L}_2 \otimes \widehat{M}_2|_{\hat{D}_2}\right)}{\mathbf{D}\left(\widehat{L}_1 \otimes \widehat{M}_2|_{\hat{D}_1}\right) \otimes \mathbf{D}\left(\widehat{L}_2 \otimes \widehat{M}_2|_{\hat{D}_2}\right)}$$

followed by the natural "cancellation" isomorphism.

We can construct an analogous map as follows: for $i = 1, 2$, let

$$\mathbf{D}(\beta_i) : \mathbf{D}(L_i \otimes M_1|_{D_i}) \xrightarrow{\sim} \mathbf{D}(L_i \otimes M_2|_{D_i})$$

be the isomorphism derived from β_i. Consider

$$\frac{\mathbf{D}(\beta_1) \otimes \mathbf{1}}{\mathbf{1} \otimes \mathbf{D}(\beta_2)} : \frac{\mathbf{D}(L_1 \otimes M_1|_{D_1}) \otimes \mathbf{D}(L_2 \otimes M_2|_{D_2})}{\mathbf{D}(L_1 \otimes M_2|_{D_1}) \otimes \mathbf{D}(L_2 \otimes M_1|_{D_2})} \longrightarrow \frac{\mathbf{D}(L_1 \otimes M_2|_{D_1}) \otimes \mathbf{D}(L_2 \otimes M_2|_{D_2})}{\mathbf{D}(L_1 \otimes M_2|_{D_1}) \otimes \mathbf{D}(L_2 \otimes M_2|_{D_2})}$$

followed by the natural "cancellation" isomorphism

$$\frac{\mathbf{D}(L_1 \otimes M_2|_{D_1}) \otimes \mathbf{D}(L_2 \otimes M_2|_{D_2})}{\mathbf{D}(L_1 \otimes M_2|_{D_1}) \otimes \mathbf{D}(L_2 \otimes M_2|_{D_2})} \longrightarrow \mathcal{O}_U.$$

Clearly this gives a map whose restriction to the generic point of S is $\Theta_{(\hat{D}_1, \hat{D}_2; \hat{m})}$.

Finally we study the isomorphism

$$K \xrightarrow{\cdot\langle \hat{l}, \hat{m}\rangle} \langle L_K, M_K \rangle.$$

defined by sending 1 to $\langle \hat{l}, \hat{m}\rangle$. Now $\langle l, m\rangle$ is a global, nowhere vanishing section of $\langle L, M\rangle$ whose image in $\langle L_K, M_K\rangle$ is $\langle \hat{l}, \hat{m}\rangle$. Hence,

$$\mathcal{O}_U \xrightarrow{\cdot\langle l, m\rangle} \langle L, M\rangle,$$

defined by taking the global section $\mathbf{1}$ to $\langle l, m\rangle$, is an isomorphism whose restriction to the generic point of S gives the above isomorphism from K to $\langle L_K, M_K\rangle$.

The three isomorphisms we have just constructed between various line bundles on U compose to give an isomorphism

$$\frac{\mathbf{D}(L_1 \otimes M_1) \otimes \mathbf{D}(L_2 \otimes M_2)}{\mathbf{D}(L_1 \otimes M_2) \otimes \mathbf{D}(L_2 \otimes M_1)} \longrightarrow \langle L, M\rangle$$

whose restriction to the generic point of S is

$$\frac{\mathbf{D}(\widehat{L_1} \otimes \widehat{M_1}) \otimes \mathbf{D}(\widehat{L_2} \otimes \widehat{M_2})}{\mathbf{D}(\widehat{L_1} \otimes \widehat{M_2}) \otimes \mathbf{D}(\widehat{L_2} \otimes \widehat{M_1})} \longrightarrow \langle L_K, M_K\rangle.$$

The existence of such an isomorphism implies that the stalk of

$$\frac{\mathbf{D}(L_1 \otimes M_1) \otimes \mathbf{D}(L_2 \otimes M_2)}{\mathbf{D}(L_1 \otimes M_2) \otimes \mathbf{D}(L_2 \otimes M_1)}$$

at s is mapped isomorphically onto the stalk of $\langle L, M\rangle$ at s under the isomorphism between the stalks of the two corresponding constant sheaves.

6. Flat Base Change

Now we will study how the intersection pairing behaves under flat base change.

(6.1) Notation. Let X be a scheme over S, and let $\sigma: S' \to S$ be a flat morphism.

• The scheme X' is defined to be the scheme obtained from X by change of base from S to S', i.e.,

$$X' = X \times_S S'.$$

In addition, σ_X will signify the natural morphism from X' to X. So we have the fiber diagram

$$\begin{array}{ccc} X' & \xrightarrow{\sigma_X} & X \\ \downarrow & & \downarrow \\ S' & \xrightarrow{\sigma} & S \end{array}$$

If X is projective and flat over S with fibers all of dimension n, then the same is true of X' over S'.

• Let L be a line bundle on X, and let l be a meromorphic section of L. Since $\sigma_X: X' \to X$ is flat, the associated points of X' map to associated points of X. This allows us to define the pull-back $\sigma_X^*(l)$ of l as a meromorphic section of $\sigma_X^* L$.

Similarly if f is an invertible meromorphic function of X, then we can define $\sigma_X^*(f)$, an invertible meromorphic function of X'. Note that if l and f are as above, then $\sigma_X^*(f\,l) = \sigma_X^*(f)\,\sigma_X^*(l)$.

• Let D be a Cartier divisor of X. Since X' is flat over X, local non-zero divisors pull-back to a local non-zero divisors of X'. This allows us to define the pull back D' of D as a Cartier divisor of X'.

Given a line bundle L on X and a meromorphic section l of L, if D is the Cartier divisor associated with l, then D' is the Cartier divisor associated with $\sigma_X^*(l)$.

• If D is an effective Cartier divisor of X then the notation D' appears at first sight to be ambiguous: it might refer to the pull-back of the Cartier divisor D, or, considering D as a scheme, it might refer to the scheme obtained by changing base. However, the scheme $D' = D \times_S S'$ is naturally a closed subscheme of $X' = X \times_S S'$, and it is, in fact, the subscheme associated with the effective Cartier divisor D'.

• Consider the case when D is an effective horizontal divisor of X, i.e., where, as a scheme, D is finite and flat over S. Then the pull-back $D' = \sigma_X^* D$ is an effective Cartier divisor of X' whose associated scheme D' is finite and flat over S'. Therefore, the Cartier divisor D' is an effective horizontal divisor of X'.

Since a horizontal divisor D of X is, by definition, the difference of two effective horizontal divisors, its pull-back D' is a horizontal divisor of X'.

• Now suppose S and S' are integral schemes, and K' and K their respective function fields. Then the generic point of S' maps to the generic point of S, so K' is a field extension of K. Furthermore, the generic fibers X_K and $X'_{K'}$ of X and X' are related by the formula

$$X'_{K'} = X_K \times_K \operatorname{Spec} K'.$$

(6.2) Proposition. *Let σ be a flat morphism between integral schemes S' and S. Let X be a scheme, flat and projective over S, all of whose fibers are one-dimensional* *schemes. Therefore, the same is true of $X' = X \times_S S'$ over S'. Let K be the function field of S, and let K' be the function field of S'.*

Let L and M be line bundles on X. By Proposition 8.4 of Chapter 1, there is a canonical isomorphism,

$$\langle L|_{X_K}, M|_{X_K}\rangle \otimes K' \xrightarrow{\sim} \langle \sigma_X^* L|_{X'_{K'}}, \sigma_X^* M|_{X'_{K'}}\rangle.$$

This isomorphism induces a canonical isomorphism

$$\sigma^*\langle L, M\rangle \xrightarrow{\sim} \langle \sigma_X^* L, \sigma_X^* M\rangle$$

between line bundles of S'.

Proof. By definition, $\langle L, M\rangle$ is a subsheaf of the constant sheaf which has stalks equal to $\langle L|_{X_K}, M|_{X_K}\rangle$. This allows us to regard the pull-back $\sigma^*\langle L, M\rangle$ as a subsheaf of the constant sheaf which has stalks equal to $\langle L|_{X_K}, M|_{X_K}\rangle \otimes_K K'$. Note also that the sheaf $\langle \sigma_X^* L, \sigma_X^* M\rangle$ is a subsheaf of the constant sheaf associated to $\langle \sigma_X^* L|X'_{K'}, \sigma_X^* M|X'_{K'}\rangle$. The canonical isomorphism mentioned in the statement of the theorem can be regarded as an isomorphism between these two constant sheaves. In light of Lemma 3.5 part 3, to show that this isomorphism restricts to an isomorphism between their respective subsheaves $\sigma^*\langle L, M\rangle$ and $\langle \sigma_X^* L, \sigma_X^* M\rangle$ we only need to show that it restricts to an isomorphism at the level of stalks.

Let s' be a point of S' and s its image in S. By Proposition 2.4, there is an open subscheme U of S containing s and meromorphic sections, l of $L|_{X_U}$ and m of $M|_{X_U}$, such that their associated Cartier divisors can be written as horizontal divisors, D and E respectively, where D and E have disjoint supports. Let U' be the inverse image of U in S'. As mentioned in (6.1), the meromorphic sections l and m pull back to meromorphic sections $\sigma_{X_U}^*(l)$ and $\sigma_{X_U}^*(m)$ of $\sigma_X^* L|_{X'_{U'}}$ and $\sigma_X^* M|_{X'_{U'}}$ respectively. The Cartier divisor associated with $\sigma_{X_U}^*(l)$ is the horizontal divisor D', the Cartier divisor associated with $\sigma_{X_U}^*(m)$ is the horizontal divisor E', and D' and E' are horizontal divisors of $X'_{U'}$ with disjoint supports.

By Lemma 4.1.1, the stalk of $\langle L, M\rangle$ at s is the $\mathcal{O}_{S,s}$-submodule of $\langle L|_{X_K}, M|_{X_K}\rangle$ generated by $\langle l|_{X_K}, m|_{X_K}\rangle$, so the stalk of $\sigma^*\langle L, M\rangle$ at s' is the $\mathcal{O}_{S',s'}$-submodule of $\langle L|_{X_K}, M|_{X_K}\rangle \otimes K'$ generated by $\langle l|_{X_K}, m|_{X_K}\rangle \otimes 1$. Similarly, by Lemma 4.1.1, the stalk of $\langle \sigma_X^* L, \sigma_X^* M\rangle$ at s' is the $\mathcal{O}_{S',s'}$-submodule of $\langle \sigma_X^* L|_{X'_{K'}}, \sigma_X^* M|_{X'_{K'}}\rangle$ generated by $\langle \sigma_{X_U}^*(l)|_{X'_{K'}}, \sigma_{X_U}^*(m)|_{X'_{K'}}\rangle$. By Proposition 8.4 of Chapter 1, the isomorphism between $\langle L|_{X_K}, M|_{X_K}\rangle \otimes K'$ and $\langle \sigma_X^* L, \sigma_X^* M\rangle$ sends $\langle l|_{X_K}, m|_{X_K}\rangle \otimes 1$ to $\langle \sigma_{X_U}^*(l)|_{X'_{K'}}, \sigma_{X_U}^*(m)|_{X'_{K'}}\rangle$, so the stalks these elements generate are isomorphic under this isomorphism.

Chapter 3
The Riemann-Roch Isomorphism

This chapter is concerned with the relative dualizing sheaf, the associated duality isomorphisms, and the associated Riemann-Roch isomorphism. In keeping with the stated aims of this paper, I have tried to make this account as concrete and self-contained as possible. I do invoke S. Kleiman's paper "Relative Duality for Quasi-Coherent Sheaves" [K] for the existence of the relative dualizing sheaf, but from this basic starting point I provide all the constructions and proofs for the needed duality theory.

1. The Relative Dualizing Sheaf

After stating, without proof, the main existence theorem, we construct the duality maps and give conditions for them to be isomorphisms.

(1.1) Theorem. *Let $\pi: X \to S$ be a projective, flat morphism such that, above any point $s \in S$, the fiber X_s is Cohen-Macaulay with components all of dimension r.*
1. There is a coherent sheaf $\mathcal{K}_\pi$ on X, flat over S, and a morphism

$$t_\pi: R^r\pi_*(\mathcal{K}_\pi) \to \mathcal{O}_S$$

characterized by the following property: for any coherent sheaf $\mathcal{F}$ on X and any morphism $\beta: R^r\pi_*(\mathcal{F}) \to \mathcal{O}_S$ there is a unique morphism $\alpha: \mathcal{F} \to \mathcal{K}_\pi$ making the following commute:

$$\begin{array}{ccc} R^r\pi_*(\mathcal{F}) & \xrightarrow{\beta} & \mathcal{O}_S \\ \big\downarrow{\scriptstyle R^r\pi_*(\alpha)} & & \big\| \\ R^r\pi_*(\mathcal{K}_\pi) & \xrightarrow{t_\pi} & \mathcal{O}_S. \end{array}$$

Note that this implies that $\mathcal{K}_\pi$ and t_π are unique up to canonical isomorphism. The sheaf $\mathcal{K}_\pi$, usually written $\mathcal{K}_{X/S}$ when π is clear from context, is called the **relative dualizing sheaf of X over S** *or sometimes* **the canonical sheaf of X over S**. *The map t_π is sometimes called the* **trace map**. *When S is the spectrum of a field k, and both k and π are clear from context, then we usually write $\mathcal{K}_X$ for $\mathcal{K}_\pi$, and call this sheaf the* **dualizing sheaf** *or the* **canonical sheaf** *of X.*
2. Let $\sigma: S' \to S$ be a morphism and let $X' = X \times_S S'$. Let $\sigma': X' \to X$ and $\pi': X' \to S'$ be the projection maps. These maps form a cartesian square:

$$\begin{array}{ccc} X' & \xrightarrow{\sigma'} & X \\ \big\downarrow{\scriptstyle \pi'} & & \big\downarrow{\scriptstyle \pi} \\ S' & \xrightarrow{\sigma} & S. \end{array}$$

In this situation there is a relative dualizing sheaf $\mathcal{K}_{X'/S'}$ on X' and a trace map $t_{\pi'}: R^r\pi'_*(\mathcal{K}_{X'/S'}) \to \mathcal{O}_{S'}$ satisfying the property described in statement 1 above. Furthermore, (i) the sheaf $\mathcal{K}_{X'/S'}$ is canonically isomorphic to the pull-back of $\mathcal{K}_{X/S}$ via σ', (ii) the natural map

$$\sigma^* R^r\pi_*(\mathcal{K}_{X/S}) \to R^r\pi'_*(\sigma'^*\mathcal{K}_{X/S}) \xrightarrow{\sim} R^r\pi'_*(\mathcal{K}_{X'/S'})$$

is an isomorphism, and (iii) the map $t_{\pi'}$ is the composition

$$R^r\pi'_*(\mathcal{K}_{X'/S'}) \xrightarrow{\sim} \sigma^* R^r\pi_*(\mathcal{K}_{X/S}) \xrightarrow{\sigma^*(t_\pi)} \mathcal{O}_{S'}.$$

Proof. For part 1 see [K], Theorem 21, Definition 10, Proposition 9 (i) and (ii), and Proposition 2 (i). For part 2 (i) see [K], Proposition 9 (iii); for part 2 (ii) see [K], Lemma 3 (iii); and for part 2 (iii) see [K], Theorem 5 and Proposition 9 (i).

(1.2) Proposition. *Let $\pi: X \to S$ be a projective, flat morphism, and assume every fiber X_s of π is Cohen-Macaulay with components all of dimension r. Let $\mathcal{F}$ be a coherent sheaf on X. Then the morphism of sheaves on S*

$$\pi_*\mathcal{H}om_X(\mathcal{F}, \mathcal{K}_{X/S}) \to \mathcal{H}om_S(R^r\pi_*(\mathcal{F}), \mathcal{O}_S),$$

induced by the trace map $t_\pi: R^r\pi_*(\mathcal{K}_{X/S}) \to \mathcal{O}_S$, is an isomorphism. We call this isomorphism the **0th degree duality map**. *Furthermore, this map is functorial with respect to $\mathcal{F}$ and is compatible with base change.*

Proof. By Theorem 1.1 part 2, the trace map is well behaved with respect to restrictions to open subschemes, so we can restrict to the case where S is affine. Then, since both sheaves are coherent, it is enough to show that the associated map on global sections is an isomorphism, i.e., that the map

$$\operatorname{Hom}_X(\mathcal{F}, \mathcal{K}_{X/S}) \to \operatorname{Hom}_S(R^r\pi_*(\mathcal{F}), \mathcal{O}_S),$$

induced by the trace map $t_\pi: R^r\pi_*(\mathcal{K}_{X/S}) \to \mathcal{O}_S$, is an isomorphism. This follows from Theorem 1.1, part 1.

That this map is functorial with respect to $\mathcal{F}$ follows straight from the definition. Compatibility with respect to base change follows from Theorem 1.1 part 2 (and the fact that the functor $R^r\pi_*$ is well-behaved with respect to base change, see [K] Lemma 3).

(1.3) The Relative Ext Functor. Let $\pi: X \to S$ be a projective morphism. We define the bi-functor $\mathcal{E}xt^i_\pi$ to be the ith derived functor of $\pi_*\mathcal{H}om_X$. It take pairs of $\mathcal{O}_X$-modules to $\mathcal{O}_S$-modules. This functor is in some sense intermediate to $\mathcal{E}xt^i_X$, the ith derived functor of $\mathcal{H}om_X$, and Ext^i_X, the ith derived functor of Hom_X. As one expects, the functor $\mathcal{E}xt^i_\pi$ shares some of the properties of the other two functors; for example, if L is a locally free coherent sheaf, and if $\mathcal{F}$ and $\mathcal{G}$ are $\mathcal{O}_X$-modules, then $\mathcal{E}xt^i_\pi(\mathcal{F} \otimes L, \mathcal{G})$ is naturally isomorphic to $\mathcal{E}xt^i_\pi(\mathcal{F}, \mathcal{G} \otimes L^\vee)$. One important property of this functor is that, for any $\mathcal{O}_X$-module $\mathcal{G}$, $\mathcal{E}xt^i_\pi(\mathcal{O}_X, \mathcal{G})$ is naturally isomorphic to $R^i\pi_*\mathcal{G}$. Another important property is that if U is an open subscheme of X, then $\mathcal{E}xt^i_\pi(\mathcal{F}, \mathcal{G})|_U$ is naturally isomorphic to $\mathcal{E}xt^i_{\tilde{\pi}}(\mathcal{F}|_U, \mathcal{G}|_U)$ where $\tilde{\pi}$ is the restriction of π to the inverse image of U in X. Like the other two ext functors, $\mathcal{E}xt^i_\pi(\mathcal{F}, \mathcal{G})$ is well-behaved if $\mathcal{F}$ is coherent and $\mathcal{G}$ is quasi-coherent.

For example, in this situation (1) $\mathcal{E}xt^i_\pi(\mathcal{F},\mathcal{G})$ is well-behaved with respect to flat base changes; (2) if $S = \operatorname{Spec} A$ is affine, then $\mathcal{E}xt^i_\pi(\mathcal{F},\mathcal{G})$ is the quasi-coherent sheaf associated to the A-module $\operatorname{Ext}^i{}_X(\mathcal{F},\mathcal{G})$; (3) thus, for general S, $\mathcal{E}xt^i_\pi(\mathcal{F},\mathcal{G})$ is quasi-coherent.

(1.4) Definition. Let $S = \operatorname{Spec} A$ be an affine scheme, and let $\pi: X \to S$ be a projective, flat morphism whose fibers are Cohen-Macaulay with components all of dimension 1. Let $\mathcal{F}$ be a coherent sheaf on X. There is a canonical A-module morphism, a particular case of the Yoneda pairing,

$$\operatorname{Ext}^1_X(\mathcal{F},\mathcal{K}_{X/S}) \to \operatorname{Hom}_A\big(H^0(X,\mathcal{F}), H^1(X,\mathcal{K}_{X/S})\big)$$

which we define as follows:

Let $0 \to \mathcal{K}_{X/S} \to \mathcal{I}^0 \to \mathcal{I}^1 \to \dots$ be an injective resolution of $\mathcal{K}_{X/S}$. Let a be an element of $\operatorname{Ext}^1_X(\mathcal{F},\mathcal{K}_{X/S})$ and let $\alpha: \mathcal{F} \to \mathcal{I}^1$ be a representative of a in $\operatorname{Hom}_X(\mathcal{F},\mathcal{I}^1)$. In particular, the composition of α with $\mathcal{I}^1 \to \mathcal{I}^2$ is the zero map. By taking global sections, we get the map $\Gamma(\alpha): H^0(X,\mathcal{F}) \to \Gamma(\mathcal{I}^1)$ whose composition with $\Gamma(\mathcal{I}^1) \to \Gamma(\mathcal{I}^2)$ is the zero map. Hence, $\Gamma(\alpha)$ factors through the kernel of $\Gamma(\mathcal{I}^1) \to \Gamma(\mathcal{I}^2)$ and so yields an associated map $a': H^0(X,\mathcal{F}) \to H^1(X,\mathcal{K}_{X/S})$. It is straightforward to check that this morphism is independent of the choice of α representing a, and that the rule $a \mapsto a'$ defines an A-module homomorphism from $\operatorname{Ext}^1_X(\mathcal{F},\mathcal{K}_{X/S})$ to $\operatorname{Hom}_X\big(H^0(X,\mathcal{F}), H^1(X,\mathcal{K}_{X/S})\big)$. We also leave it to the reader to verify the following:

1. This morphism is independent of the choice of resolution $\{\mathcal{I}^i\}$.
2. This morphism is functorial in $\mathcal{F}$.
3. This morphism is well-behaved with respect to restriction to open affine subschemes, or, more generally, with respect to any flat base change.
4. If $0 \to \mathcal{F}_1 \to \mathcal{F}_2 \to \mathcal{F}_3 \to 0$ is an exact sequence of coherent sheaves, then the following anti-commutes:

$$\begin{array}{ccc} \operatorname{Hom}_X(\mathcal{F}_1,\mathcal{K}_{X/S}) & \longrightarrow & \operatorname{Ext}^1_X(\mathcal{F}_3,\mathcal{K}_{X/S}) \\ \downarrow & & \downarrow \\ \operatorname{Hom}_A\big(H^1(X,\mathcal{F}_1), H^1(X,\mathcal{K}_{X/S})\big) & \longrightarrow & \operatorname{Hom}_A\big(H^0(X,\mathcal{F}_3), H^1(X,\mathcal{K}_{X/S})\big) \end{array}$$

where the top horizontal map is the boundary map, the bottom horizontal map is associated with a boundary map, the left vertical map is the map induced by the functor H^1, and the right vertical map is the morphism just defined.

(1.5) Definition. The A-module morphism in the above definition induces a morphism at the level of sheaves on S:

$$\mathcal{E}xt^1_\pi(\mathcal{F},\mathcal{K}_{X/S}) \to \mathcal{H}om_S\big(R^0\pi_*(\mathcal{F}), R^1\pi_*(\mathcal{K}_{X/S})\big).$$

When S is not affine, we define the above morphism by defining the isomorphism locally on the affine open subschemes. Since the morphism is canonically defined on such subschemes, and the maps are well-behaved with respect to restrictions to smaller affine subschemes, these maps patch together to form a global isomorphism.

The trace map $t_\pi: R^1\pi_*(\mathcal{K}_{X/S}) \to \mathcal{O}_S$ allows us to form, from the above map, the map

$$\mathcal{E}xt^1_\pi(\mathcal{F}, \mathcal{K}_{X/S}) \to \mathcal{H}om_S(R^0\pi_*(\mathcal{F}), \mathcal{O}_S).$$

We call this map the *1st degree duality map.*

Note that by definition, and by the properties mentioned in (1.4), we have the following:

1. this morphism is functorial in $\mathcal{F}$.
2. this morphism is well-behaved with respect to open restrictions, or, more generally, with respect to flat base changes.
3. If $0 \to \mathcal{F}_1 \to \mathcal{F}_2 \to \mathcal{F}_3 \to 0$ is an exact sequence of coherent sheaves, then the following anti-commutes:

$$\begin{array}{ccc} \pi_*\mathcal{H}om_X(\mathcal{F}_1, \mathcal{K}_{X/S}) & \longrightarrow & \mathcal{E}xt^1_\pi(\mathcal{F}_3, \mathcal{K}_{X/S}) \\ \downarrow & & \downarrow \\ \mathcal{H}om_S(R^1\pi_*(\mathcal{F}_1), \mathcal{O}_S) & \longrightarrow & \mathcal{H}om_S(R^0\pi_*(\mathcal{F}_3), \mathcal{O}_S) \end{array}$$

where the top horizontal map is the boundary map, the bottom horizontal map is associated with a boundary map, the left vertical map is the 0th order duality map, and the right vertical map is the 1st order duality map.

(1.6) Proposition. *Let $\pi: X \to S$ be a projective, flat morphism, and assume every fiber X_s of π is Cohen-Macaulay with components all of dimension 1. Let $\mathcal{F}$ be a coherent sheaf on X, flat over S. If $R^1\pi_*(\mathcal{F})$ is a locally free coherent sheaf, then the 1st degree duality map*

$$\mathcal{E}xt^1_\pi(\mathcal{F}, \mathcal{K}_{X/S}) \to \mathcal{H}om_S(R^0\pi_*(\mathcal{F}), \mathcal{O}_S)$$

is an isomorphism.

Proof. The map is well-behaved with respect to open restrictions, so we can assume that S is affine. Let $\mathcal{O}_X(1)$ be a sheaf on X which is very ample with respect to $\pi: X \to S$. Choose a positive integer n such that (i) $R^1\pi_*(\mathcal{K}_{X/S}(n)) = 0$, (ii) $R^0\pi_*\mathcal{O}_X(-n) = 0$, (iii) $R^1\pi_*\mathcal{O}_X(-n)$ is locally free, and (iv) the sheaf $\mathcal{F}(n)$ is generated by a finite number of global sections. Such an n exists by Lemma 1.6.3 (which we will prove below) and by standard results (See [H], III, Theorem 8.8(c) and II, Theorem 5.17). The last requirement implies that there is a surjection from $(\mathcal{O}_X)^m$ to $\mathcal{F}(n)$ and hence a surjection from $(\mathcal{O}_X(-n))^m$ to $\mathcal{F}$. Let $\mathcal{G}$ be the kernel of this surjection; it must be a coherent sheaf on X flat over S because this is true of the other terms of the following short exact sequence:

$$0 \longrightarrow \mathcal{G} \longrightarrow \mathcal{O}_X(-n)^m \longrightarrow \mathcal{F} \longrightarrow 0. \qquad (*)$$

Let X_s be a fiber of π. By Theorem 12.11 in Ch. III of [H], $H^0(X_s, \mathcal{O}_X(-n)|_{X_s})$ is zero. Since $\mathcal{F}$ is flat over S, the restriction of the above exact sequence (*) to the fiber X_s is also exact. This implies that, for any fiber X_s, $H^0(X_s, \mathcal{G}|_{X_s}) = 0$. Note that $H^2(X_s, \mathcal{G}|_{X_s}) = 0$. Therefore, by Theorem 12.11 of [H], Chapter III, we have that $R^1\pi_*\mathcal{G}$ is locally free and (using Nakayama's lemma) that $R^2\pi_*\mathcal{G} = 0$. Also, by the choice of n, $R^0\pi_*(\mathcal{O}_X(-n)^m) = R^0\pi_*(\mathcal{O}_X(-n))^m = 0$. Hence, we have the following exact sequence derived from (*) above:

$$0 \longrightarrow R^0\pi_*(\mathcal{F}) \longrightarrow R^1\pi_*(\mathcal{G}) \longrightarrow R^1\pi_*(\mathcal{O}_X(-n)^m) \longrightarrow R^1\pi_*(\mathcal{F}) \longrightarrow 0.$$

We are assuming that $R^1\pi_*(\mathcal{F})$ is locally free, so every term in the above sequence is known to be a locally free coherent sheaf, except perhaps $R^0\pi_*(\mathcal{F})$. However, since the above sequence is exact, $R^0\pi_*(\mathcal{F})$ must also be a locally free coherent sheaf. Consequently,

$$\mathcal{H}om_S(R^1\pi_*\mathcal{O}_X(-n)^m, \mathcal{O}_S) \to \mathcal{H}om_S(R^1\pi_*(\mathcal{G}), \mathcal{O}_S) \to \mathcal{H}om_S(R^0\pi_*(\mathcal{F}), \mathcal{O}_S) \to 0$$

is exact.

The choice of n was such that

$$\begin{aligned} \mathcal{E}xt^1_\pi(\mathcal{O}_X(-n)^m, \mathcal{K}_{X/S}) &\cong \mathcal{E}xt^1_\pi(\mathcal{O}_X, \mathcal{K}_{X/S}(n)^m) \cong R^1\pi_*(\mathcal{K}_{X/S}(n)^m) \\ &\cong R^1\pi_*(\mathcal{K}_{X/S}(n))^m = 0; \end{aligned}$$

therefore, the following sequence is exact:

$$\pi_*\mathcal{H}om_X(\mathcal{O}_X(-n)^m, \mathcal{K}_{X/S}) \longrightarrow \pi_*\mathcal{H}om_X(\mathcal{G}, \mathcal{K}_{X/S}) \longrightarrow \mathcal{E}xt^1_\pi(\mathcal{F}, \mathcal{K}_{X/S}) \longrightarrow 0$$

We can combine the two above exact sequences in the following commutative diagram:

$$\begin{array}{ccccccc} \pi_*\mathcal{H}om_X(\mathcal{O}_X(-n)^m, \mathcal{K}_{X/S}) & \to & \pi_*\mathcal{H}om_X(\mathcal{G}, \mathcal{K}_{X/S}) & \to & \mathcal{E}xt^1_\pi(\mathcal{F}, \mathcal{K}_{X/S}) & \to & 0 \\ \downarrow D^0 & & \downarrow D^0 & & \downarrow -D^1 & & \\ \mathcal{H}om_S(R^1\pi_*\mathcal{O}_X(-n)^m, \mathcal{O}_S) & \to & \mathcal{H}om_S(R^1\pi_*(\mathcal{G}), \mathcal{O}_S) & \to & \mathcal{H}om_S(R^0\pi_*(\mathcal{F}), \mathcal{O}_S) & \to & 0 \end{array}$$

The vertical maps labelled D^0 are the 0th degree duality isomorphisms, and the vertical map labelled $-D^1$ is the negative of the 1st degree duality map. The properties of the duality maps imply that this diagram commutes.

By Proposition 1.2, the two leftmost vertical maps are isomorphisms. Therefore, the rightmost vertical map must also be an isomorphism. This proves the proposition, but we used a lemma, Lemma 1.6.3, which we will now prove via the following chain of lemmas.

(1.6.1) Lemma. *Let $\mathcal{O}$ be a noetherian local ring, and let f be a non-zero divisor in its maximal ideal. Then, for every non-zero element $g \in \mathcal{O}$, there is a positive integer d such that g is not in the ideal generated by f^d.*

Proof. For each positive integer n, let $\mathcal{F}_n$ be the ideal generated by f^n. Let $\mathcal{F}_\infty$ be the intersection of all the $\mathcal{F}_n$. As you would expect, $f\mathcal{F}_\infty = \mathcal{F}_\infty$; to see this, observe that if b is in $\mathcal{F}_\infty$ it is in $\mathcal{F}_1$ so we can write $b = f\,c$ for some c. For any integer n, b is in $\mathcal{F}_{n+1}$, so we can write $b = f^{n+1}\,e$ for some e. Since f is not a zero divisor, $b = f\,c = f^{n+1}\,e$ implies that $c = f^n\,e$. Therefore, c is in $\mathcal{F}_n$ for any n, i.e. c is in $\mathcal{F}_\infty$, and b is in $f\mathcal{F}_\infty$.

By Nakayama's lemma, $f\mathcal{F}_\infty = \mathcal{F}_\infty$ implies that $\mathcal{F}_\infty = 0$, which yields the result.

(1.6.2) Lemma. *Suppose that X is projective over a field K, that X has components which are all one-dimensional, and that X is Cohen-Macaulay, i.e., that X has no embedded points. Let $\mathcal{O}_X(1)$ be a very ample line-bundle on X. Then there is a positive integer N such that $H^0(X, \mathcal{O}_X(-n)) = 0$ for all $n \geq N$.*

Proof. By Lemma 2.3 of Chapter 2, $\mathcal{O}_X(N_0) = \mathcal{O}_X(D)$ for some integer N_0 and some effective divisor D. Note that $H^0(X, \mathcal{O}_X(-D))$ is a subspace of $H^0(X, \mathcal{O}_X)$. For any element g of $H^0(X, \mathcal{O}_X)$, the following gives a criterion for whether g is in $H^0(X, \mathcal{O}_X(-D))$:

Let $x_1, \ldots, x_m$ be the points of the support of D and, for each x_i, let $f_i \in \mathcal{O}_{X,x_i}$ be a local generator of D at x_i. Then g is in $H^0(X, \mathcal{O}_X(-D))$ if and only if, for each x_i, the image of g in $\mathcal{O}_{X,x_i}$ is in the ideal generated by f_i.

Let g be an element of $H^0(X, \mathcal{O}_X)$, and let x_i be a point of D. By Lemma 1.6.1, either the image of g in $\mathcal{O}_{X,x_i}$ is zero or there is a positive integer d_i such that the image of g is not in the ideal generated by $f_i^{d_i}$, i.e., g is not an element of $H^0(X, \mathcal{O}_X(-d_i\, D))$. So either the image of g is zero in each $\mathcal{O}_{X,x_i}$, i.e., g is zero on a neighborhood of D, or there is a positive integer d such that g is not an element of $H^0(X, \mathcal{O}_X(-d\, D))$.

The sheaf $\mathcal{O}_X(D) = \mathcal{O}_X(N_0)$ is ample, so the restriction of $\mathcal{O}_X(D)$ to each component of X is ample. Therefore, D must intersect each component. So if g is zero on a neighborhood of D, then g is zero on a dense open subscheme of X. We are assuming that X has no embedded points, so any function which is zero on an open dense subscheme of X is zero everywhere. Therefore, if g is zero in a neighborhood of D, then g is identically zero.

We conclude that if g is a non-zero element of $H^0(X, \mathcal{O}_X)$, there is a positive integer d such that g is not an element of the subspace $H^0(X, \mathcal{O}_X(-d\, D))$. Since $H^0(X, \mathcal{O}_X)$ is a finite dimensional K-vector space, we can choose d large enough so that

$$H^0(X, \mathcal{O}_X(-d\, N_0)) = H^0(X, \mathcal{O}_X(-d\, D)) = 0.$$

By Lemma 2.3 of Chapter 2, there is an integer N_1 such that, if $n \geq N_1$, then $\mathcal{O}_X(n) = \mathcal{O}_X(D_n)$ for some effective divisor D_n. Let $N = d\, N_0 + N_1$.

If $n \geq N$, then let $n_1 = n - d\, N_0$. Since $n_1 \geq N_1$, we have $\mathcal{O}_X(n_1) = \mathcal{O}_X(D_{n_1})$. So

$$\mathcal{O}_X(-n) = \mathcal{O}_X(-d\, D - D_{n_1}).$$

This sheaf maps injectively, in an obvious way, into the sheaf $\mathcal{O}_X(-d\, D)$. Therefore, $H^0(X, \mathcal{O}_X(-n))$ maps injectively into $H^0(X, \mathcal{O}_X(-d\, D)) = 0$, and so must be zero.

(1.6.3) Lemma. *Let $\pi\colon X \to S$ be a projective, flat morphism such that each fiber of π is Cohen-Macaulay and has components which are all one-dimensional. Let $\mathcal{O}_X(1)$ be a line bundle on X which is very ample with respect to π. There is a positive integer N such that, for all positive integers n, $R^0\pi_*(\mathcal{O}_X(-nN)) = 0$ and $R^1\pi_*(\mathcal{O}_X(-nN))$ is locally free.*

Proof. For any integer n and any point $s \in S$, let $h^0(n, s)$ be the dimension of $H^0(X_s, \mathcal{O}_X(n)\,|_{X_s})$.

Let σ be a closed point of S. By Lemma 2.3 of Chapter 2, there is an integer N_σ and an open subscheme U_σ of S containing σ, such that $\mathcal{O}_X(N_\sigma)\,|X_{U_\sigma}$ is isomorphic to $\mathcal{O}_{X_{U_\sigma}}(D_\sigma)$ for some effective horizontal divisor D_σ on X_{U_σ}. By choosing N_σ

large enough, we can also assume, by Lemma 1.6.2, that $h^0(-N_\sigma, \sigma) = 0$. The function $h^0(-N_\sigma, s)$ is upper semicontinuous as s varies over S (see [H], III, Theorem 12.8) and so must be zero on a neighborhood of σ, i.e., after possibly replacing U_σ by a smaller neighborhood of σ, we can assume that $h^0(-N_\sigma, s) = 0$ for all $s \in U_\sigma$.

For any integer $n \geq 2$ and any point s of U_σ, the restriction of the exact sequence

$$0 \to \mathcal{O}_X(-nD) \to \mathcal{O}_X(-D) \to \mathcal{O}_X(-D)|_{(n-1)D} \to 0$$

to the fiber X_s is exact. This is due to the fact that D, and any positive multiple of D, is a horizontal effective divisor, so the rightmost non-zero term of the above exact sequence is flat over S. We conclude therefore that

$$h^0(-nN_\sigma, s) \leq h^0(-N_\sigma, s) = 0.$$

Since $h^0(-nN_\sigma, s) = 0$ for all positive n and all points s of U_σ, Theorem 12.11 of [H], Chapter III, together with Nakayama's Lemma implies that the restriction $R^0\pi_*(\mathcal{O}_X(-nN_\sigma)) \mid U_\sigma$ is zero. Consequently, Theorem 12.11 of [H], Chapter III, also implies that $R^1\pi_*(\mathcal{O}_X(-nN_\sigma)) \mid U_\sigma$ is locally free.

Since S is Noetherian, there is a finite set $\{\sigma_i\}$ of closed points of S such that the corresponding set of open subschemes $\{U_{\sigma_i}\}$ covers S. The condition of the lemma is satisfied for $N = \prod_i N_{\sigma_i}$.

Note. With a little more work we can show that there is an N_0 such that, if $n \geq N_0$, then $R^0\pi_*(\mathcal{O}_X(-n)) = 0$ and $R^1\pi_*(\mathcal{O}_X(-n))$ is locally free.

2. The Adjunction Formula

Let X be projective and flat over S with fibers which are Cohen-Macaulay and have only one-dimensional components. In this section we will show that, for any effective horizontal divisor D of X, the relative dualizing sheaves of X and D over the base S are related by the formula

$$\mathcal{K}_{D/S} = \mathcal{K}_{X/S}(D)\,|_D.$$

(2.1) Lemma. *Let $\pi: X \to S$ be a flat, projective morphism with fibers which are Cohen-Macaulay and have components all of dimension 1. Let D be an effective horizontal divisor on X, and let $\mathcal{K} = \mathcal{K}_{X/S}$ be the relative dualizing sheaf of X over S. Then the following sequence is exact*

$$0 \longrightarrow \mathcal{K} \longrightarrow \mathcal{K}(D) \longrightarrow \mathcal{K}(D)|_D \longrightarrow 0.$$

Proof. The sheaf $\mathcal{K}$ is well-behaved with respect to restrictions to open subschemes of S, so we can, without loss of generality, assume that S is an affine scheme.

The tensor product of $\mathcal{K}$ with the terms of the short exact sequence

$$0 \longrightarrow \mathcal{O}_X \longrightarrow \mathcal{O}_X(D) \longrightarrow \mathcal{O}_X(D)|_D \longrightarrow 0$$

gives the exact sequence

$$\mathcal{K} \longrightarrow \mathcal{K}(D) \longrightarrow \mathcal{K}(D)|_D \longrightarrow 0.$$

So we only need to show that the map $\mathcal{K} \longrightarrow \mathcal{K}(D)$ is injective.

We will begin by showing that, for any line bundle L, the following map is injective:

$$\mathrm{Hom}_X\big(L, \mathcal{K}\big) \longrightarrow \mathrm{Hom}_X\big(L, \mathcal{K}(D)\big) = \mathrm{Hom}_X\big(L(-D), \mathcal{K}\big).$$

Consider the diagram

$$\begin{array}{ccccc} & & \mathrm{Hom}_X\big(L, \mathcal{K}\big) & \longrightarrow & \mathrm{Hom}_X\big(L(-D), \mathcal{K}\big) \\ & & \big\downarrow & & \big\downarrow \\ 0 & \longrightarrow & \mathrm{Hom}_S\big(R^1\pi_* L, \mathcal{O}_S\big) & \longrightarrow & \mathrm{Hom}_S\big(R^1\pi_* L(-D), \mathcal{O}_S\big) \end{array}$$

where the vertical maps are obtained from the 0th degree duality isomorphisms. This diagram commutes by Proposition 1.2. The bottom row is exact since the exact sequence

$$0 \to L(-D) \to L \to L|_D \to 0$$

gives rise to the long exact sequence

$$\ldots \to R^1\pi_*(L(-D)) \to R^1\pi_*(L) \to R^1\pi_*(L|_D) = 0,$$

and the functor $\mathrm{Hom}_S(\bullet, \mathcal{O}_S)$ is contravariant and left-exact. The equality in the rightmost term, $R^1\pi_*(L|_D) = 0$, follows from [H], Chapter III, Corollary 11.2. We conclude that the map of the top row must also be injective.

Consider the exact sequence

$$0 \to \mathcal{F} \to \mathcal{K} \to \mathcal{K}(D)$$

where $\mathcal{F}$ is the kernel of the rightmost map. This yields, for any coherent sheaf L, the exact sequence

$$0 \to \mathrm{Hom}_X(L, \mathcal{F}) \to \mathrm{Hom}_X(L, \mathcal{K}) \to \mathrm{Hom}_X\big(L, \mathcal{K}(D)\big).$$

When L is a line bundle, we saw above that the rightmost map is injective, and so

$$\mathrm{Hom}_X(L, \mathcal{F}) = 0.$$

Let $\mathcal{O}(1)$ be a sheaf on X which is very ample with respect to π. We are assuming that S is affine, so, by Serre's theorem (see [H], II, Theorem 5.17), there is an integer n such that $\mathcal{F}(n)$ is generated by global sections. But note that

$$H^0\big(X, \mathcal{F}(n)\big) = \mathrm{Hom}_X\big(\mathcal{O}_X, \mathcal{F}(n)\big) \xrightarrow{\sim} \mathrm{Hom}_X\big(\mathcal{O}(-n), \mathcal{F}\big) = 0.$$

Therefore, $\mathcal{F} = 0$.

(2.2) Definition. Let $\pi \colon X \to S$ be a flat, projective morphism with fibers having components all of dimension 1. Assume the fibers of π are Cohen-Macaulay, i.e., that they do not contain any embedded points. Let $\mathcal{K} = \mathcal{K}_{X/S}$ be the relative dualizing sheaf of X over S, and let D be an effective horizontal divisor.

For any coherent sheaf $\mathcal{F}$ on X we define the map

$$\Theta(\mathcal{F}) \colon \pi_* \mathcal{H}om_X\big(\mathcal{F}, \mathcal{K}(D)|_D\big) \longrightarrow \mathcal{E}xt^1_\pi\big(\mathcal{F}|_D, \mathcal{K}\big)$$

as the composition of the natural isomorphism

$$\pi_* \mathcal{H}om_X\big(\mathcal{F}, \mathcal{K}(D)|_D\big) \xrightarrow{\sim} \pi_* \mathcal{H}om_X\big(\mathcal{F}|_D, \mathcal{K}(D)|_D\big)$$

with the boundary map

$$\pi_* \mathcal{H}om_X\big(\mathcal{F}|_D, \mathcal{K}(D)|_D\big) \longrightarrow \mathcal{E}xt^1_\pi\big(\mathcal{F}|_D, \mathcal{K}\big)$$

derived from the exact sequence of Lemma 2.1:

$$0 \to \mathcal{K} \to \mathcal{K}(D) \to \mathcal{K}(D)|_D \to 0.$$

Clearly, $\Theta(\mathcal{F})$ is functorial in $\mathcal{F}$, and is well-behaved with respect to restrictions to open subschemes of S.

(2.3) Let $\pi\colon X \to S$ be as in the above definition. Let $\mathcal{K} = \mathcal{K}_{X/S}$ be the relative dualizing sheaf of X over S, and let D be an effective horizontal divisor.

Let L be a locally free coherent sheaf on X. Consider the following diagram:

$$\begin{array}{ccccccccc}
\pi_*\mathcal{H}om_X(L,\mathcal{K}) & \to & \pi_*\mathcal{H}om_X(L,\mathcal{K}(D)) & \to & \pi_*\mathcal{H}om_X\big(L,\mathcal{K}(D)|_D\big) & \to & \mathcal{E}xt^1_\pi(L,\mathcal{K}) & \to & \mathcal{E}xt^1_\pi(L,\mathcal{K}(D)) \\
\| & & \downarrow & & \downarrow{\scriptstyle\Theta(L)} & & \| & & \downarrow \\
\pi_*\mathcal{H}om_X(L,\mathcal{K}) & \to & \pi_*\mathcal{H}om_X(L(-D),\mathcal{K}) & \to & \mathcal{E}xt^1_\pi\big(L|_D,\mathcal{K}\big) & \to & \mathcal{E}xt^1_\pi(L,\mathcal{K}) & \to & \mathcal{E}xt^1_\pi(L(-D),\mathcal{K}) \\
\downarrow & & \downarrow & & \downarrow & & \downarrow & & \downarrow \\
\big(R^1\pi_*(L)\big)^\vee & \to & \big(R^1\pi_*L(-D)\big)^\vee & \to & \big(R^0\pi_*(L|_D)\big)^\vee & \to & \big(R^0\pi_*(L)\big)^\vee & \to & \big(R^0\pi_*L(-D)\big)^\vee
\end{array}$$

where

- the map $\Theta(L)$ is the map defined in (2.2);
- the other vertical maps between the first and second row are the natural isomorphisms;
- for any coherent sheaf $\mathcal{G}$ on S, the sheaf $\mathcal{G}^\vee$ is defined to be $\mathcal{H}om_S(\mathcal{G}, \mathcal{O}_S)$;
- the vertical maps between the middle and bottom rows are the respective duality maps;
- the top row is part of the long exact sequence coming from the short exact sequence
$$0 \to \mathcal{K} \to \mathcal{K}(D) \to \mathcal{K}(D)|_D \to 0;$$
- the middle row and bottom rows are derived from the short exact sequence
$$0 \to L(-D) \to L \to L|_D \to 0.$$

We claim that
(i) the above diagram commutes up to sign; more specifically, all the squares in the above diagram are commutative except for two squares, those directly left of center, which are anti-commutative.
(ii) the map Θ is an isomorphism.

Note that (ii) follows from (i) by the five lemma. The bottom row of squares in the above diagram commute up to sign by the properties of the duality maps (see

Proposition 1.2 and Definition 1.5). Thus all we need to show is that the top row of squares in the above diagram commute up to sign.

This is a local problem, so we can assume that S is affine: $S = \operatorname{Spec} A$. Since all sheaves involved are coherent, we only need to check commutivity in the corresponding situation for the associated A-modules of global sections:

$$
\begin{array}{ccccccccc}
\operatorname{Hom}_X(L,\mathcal{K}) & \to & \operatorname{Hom}_X(L,\mathcal{K}(D)) & \to & \operatorname{Hom}_X\big(L,\mathcal{K}(D)|_D\big) & \to & \operatorname{Ext}^1_X(L,\mathcal{K}) & \to & \operatorname{Ext}^1_X(L,\mathcal{K}(D)) \\
\| & & \downarrow & & \downarrow{\scriptstyle \Gamma(\Theta(L))} & & \| & & \downarrow \\
\operatorname{Hom}_X(L,\mathcal{K}) & \to & \operatorname{Hom}_X(L(-D),\mathcal{K}) & \to & \operatorname{Ext}^1_X\big(L|_D,\mathcal{K}\big) & \to & \operatorname{Ext}^1_X(L,\mathcal{K}) & \to & \operatorname{Ext}^1_X(L(-D),\mathcal{K})
\end{array}
$$

The leftmost square clearly commutes. The rightmost square commutes by a standard argument (I will leave this to the reader).

Now consider the square

$$
\begin{array}{ccc}
\operatorname{Hom}_X\big(L,\mathcal{K}(D)\big) & \longrightarrow & \operatorname{Hom}_X\big(L,\mathcal{K}(D)|_D\big) \\
\downarrow & & \downarrow{\scriptstyle \Gamma(\Theta(L))} \\
\operatorname{Hom}_X\big(L(-D),\mathcal{K}\big) & \longrightarrow & \operatorname{Ext}^1_X\big(L|_D,\mathcal{K}\big).
\end{array}
$$

To study this square, first we choose an exact sequence of injective resolutions:

$$
\begin{array}{ccccccccccc}
 & & 0 & & 0 & & 0 & & 0 & & \\
 & & \downarrow & & \downarrow & & \downarrow & & \downarrow & & \\
0 & \longrightarrow & \mathcal{K} & \longrightarrow & \mathcal{I}^0 & \stackrel{\alpha}{\longrightarrow} & \mathcal{I}^1 & \longrightarrow & \mathcal{I}^2 & \longrightarrow & \cdots \\
 & & \downarrow & & \downarrow & & \downarrow & & \downarrow & & \\
0 & \longrightarrow & \mathcal{K}(D) & \longrightarrow & \mathcal{I}^0 \oplus \mathcal{J}^0 & \longrightarrow & \mathcal{I}^1 \oplus \mathcal{J}^1 & \longrightarrow & \mathcal{I}^2 \oplus \mathcal{J}^2 & \longrightarrow & \cdots \\
 & & \downarrow & & \downarrow & & \downarrow & & \downarrow & & \\
0 & \longrightarrow & \mathcal{K}(D)|_D & \longrightarrow & \mathcal{J}^0 & \stackrel{\beta}{\longrightarrow} & \mathcal{J}^1 & \longrightarrow & \mathcal{J}^2 & \longrightarrow & \cdots \\
 & & \downarrow & & \downarrow & & \downarrow & & \downarrow & & \\
 & & 0 & & 0 & & 0 & & 0 & &
\end{array}
$$

It follows that there is a morphism $\gamma\colon \mathcal{J}^0 \to \mathcal{I}^1$ such that the image under the map $\mathcal{I}^0 \oplus \mathcal{J}^0 \to \mathcal{I}^1 \oplus \mathcal{J}^1$ of a pair (u,v) from a given stalk of $\mathcal{I}^0 \oplus \mathcal{J}^0$ is the pair $\big(\alpha(u)+\gamma(v),\beta(v)\big)$ in the corresponding stalk of $\mathcal{I}^1 \oplus \mathcal{J}^1$.

Let $\phi\colon L \to \mathcal{K}(D)$ be an element of $\operatorname{Hom}_X\big(L,\mathcal{K}(D)\big)$ and let $\phi'\colon L(-D) \to \mathcal{K}$ be the corresponding element of $\operatorname{Hom}_X\big(L(-D),\mathcal{K}\big)$. This gives the following commutative diagram

$$\begin{array}{ccccccccccc}
 & 0 & & 0 & & 0 & & 0 & & 0 & \\
 & \downarrow & & \downarrow & & \downarrow & & \downarrow & & \downarrow & \\
 & L(-D) & \xrightarrow{\phi'} & \mathcal{K} & \longrightarrow & \mathcal{I}^0 & \xrightarrow{\alpha} & \mathcal{I}^1 & \longrightarrow & \mathcal{I}^2 & \longrightarrow \cdots \\
 & \downarrow & & \downarrow & & \downarrow & & \downarrow & & \downarrow & \\
 & L & \xrightarrow{\phi} & \mathcal{K}(D) & \longrightarrow & \mathcal{I}^0 \oplus \mathcal{J}^0 & \longrightarrow & \mathcal{I}^1 \oplus \mathcal{J}^1 & \longrightarrow & \mathcal{I}^2 \oplus \mathcal{J}^2 & \longrightarrow \cdots \\
 & \downarrow & & \downarrow & & \downarrow & & \downarrow & & \downarrow & \\
 & L|_D & \xrightarrow{\phi|_D} & \mathcal{K}(D)|_D & \longrightarrow & \mathcal{J}^0 & \xrightarrow{\beta} & \mathcal{J}^1 & \longrightarrow & \mathcal{J}^2 & \longrightarrow \cdots \\
 & \downarrow & & \downarrow & & \downarrow & & \downarrow & & \downarrow & \\
 & 0 & & 0 & & 0 & & 0 & & 0 &
\end{array}$$

The image of ϕ under the composition

$$\mathrm{Hom}_X\big(L, \mathcal{K}(D)\big) \longrightarrow \mathrm{Hom}_X\big(L, \mathcal{K}(D)|_D\big) \xrightarrow{\Gamma(\Theta(L))} \mathrm{Ext}^1_X\big(L|_D, \mathcal{K}\big)$$

is, by construction, the image of $\phi|_D$ under the boundary map

$$\mathrm{Hom}_X\big(L|_D, \mathcal{K}(D)|_D\big) \to \mathrm{Ext}^1_X\big(L|_D, \mathcal{K}\big).$$

It is easy to see that this image is represented by the composition

$$L|_D \xrightarrow{\phi|_D} \mathcal{K}(D)|_D \longrightarrow \mathcal{J}^0 \xrightarrow{\gamma} \mathcal{I}^1.$$

In order to calculate the image of ϕ' in $\mathrm{Ext}^1_X\big(L|_D, \mathcal{K}\big)$ under the map

$$\mathrm{Hom}_X\big(L(-D), \mathcal{K}\big) \to \mathrm{Ext}^1_X\big(L|_D, \mathcal{K}\big)$$

we must first lift the composition

$$L(-D) \xrightarrow{\phi'} \mathcal{K} \longrightarrow \mathcal{I}^0$$

to a map from L to $\mathcal{I}^0$. We do this with the composition

$$L \xrightarrow{\phi} \mathcal{K}(D) \longrightarrow \mathcal{I}^0 \oplus \mathcal{J}^0 \longrightarrow \mathcal{I}^0$$

where the last map is the natural projection map. It follows that $L(-D)$ is in the kernel of the composition

$$L \xrightarrow{\phi} \mathcal{K}(D) \longrightarrow \mathcal{I}^0 \oplus \mathcal{J}^0 \longrightarrow \mathcal{I}^0 \longrightarrow \mathcal{I}^1,$$

i.e., this map can be defined on $L|_D$, and the resulting map represents the image of ϕ' in $\mathrm{Ext}^1_X\big(L|_D, \mathcal{K}\big)$.

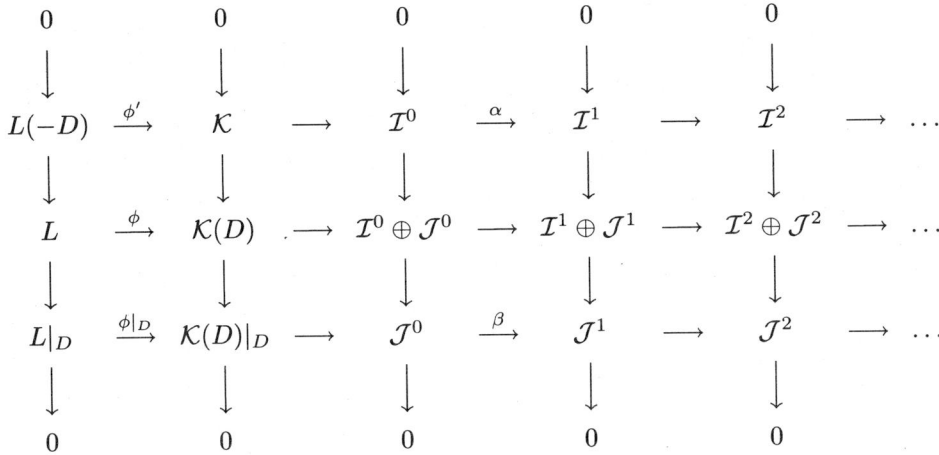

Fix a point x of X. Let l be an element of the stalk of $L|_D$ at x and let $\hat{l}$ be an element of the stalk of L at x which maps to l. Let (u, v) be the image of $\hat{l}$ in the stalk of $\mathcal{I}^0 \oplus \mathcal{J}^0$ at x. By definition, the above map sends l to $\alpha(u)$. Note that the composition of $\mathcal{K}(D) \to \mathcal{I}^0 \oplus \mathcal{J}^0$ with the map $\mathcal{I}^0 \oplus \mathcal{J}^0 \to \mathcal{I}^1 \oplus \mathcal{J}^1$ is the zero map, so the image of $\hat{l}$ under the composition

$$L \longrightarrow \mathcal{K}(D) \longrightarrow \mathcal{I}^0 \oplus \mathcal{J}^0 \longrightarrow \mathcal{I}^1 \oplus \mathcal{J}^1$$

is zero, but in terms of u and v it is $\big(\alpha(u)+\gamma(v), \beta(v)\big)$. So we conclude that l is sent to $-\gamma(v) = \alpha(u)$. In other words, the image of ϕ' in $\operatorname{Ext}^1_X(L|_D, \mathcal{K})$ is represented by the composition

$$L|_D \xrightarrow{\phi|_D} \mathcal{K}(D)|_D \longrightarrow \mathcal{J}^0 \xrightarrow{-\gamma} \mathcal{I}^1.$$

Note that this is (-1) times the map whose class represents the other image of ϕ in $\operatorname{Ext}^1_X(L|_D, \mathcal{K})$, i.e., the square in question is anti-commutative.

Now we will show that the following square commutes:

$$\begin{array}{ccc} \operatorname{Hom}_X\big(L, \mathcal{K}(D)|_D\big) & \longrightarrow & \operatorname{Ext}^1_X(L, \mathcal{K}) \\ \Big\downarrow{\scriptstyle \Gamma(\Theta(L))} & & \Big\| \\ \operatorname{Ext}^1_X(L|_D, \mathcal{K}) & \longrightarrow & \operatorname{Ext}^1_X(L, \mathcal{K}). \end{array}$$

Let $\phi: L \to \mathcal{K}(D)|_D$ be an element of $\operatorname{Hom}_X\big(L, \mathcal{K}(D)|_D\big)$. We factor ϕ as follows:

$$L \longrightarrow L|_D \xrightarrow{\phi|_D} \mathcal{K}(D)|_D.$$

The image of ϕ in $\operatorname{Ext}^1_X(L|_D, \mathcal{K})$ is by definition the image of $\phi|_D$ under the boundary map

$$\operatorname{Hom}_X\big(L|_D, \mathcal{K}(D)|_D\big) \to \operatorname{Ext}^1_X(L|_D, \mathcal{K}).$$

Using the same injective resolutions as before, this image is represented by the composition

$$L|_D \xrightarrow{\phi|_D} \mathcal{K}(D)|_D \longrightarrow \mathcal{J}^0 \xrightarrow{\gamma} \mathcal{I}^1;$$

on the other hand, the image of ϕ under the boundary map

$$\operatorname{Hom}_X\big(L, \mathcal{K}(D)|_D\big) \to \operatorname{Ext}^1_X(L, \mathcal{K})$$

is represented by the composition

$$L \longrightarrow L|_D \xrightarrow{\phi|_D} \mathcal{K}(D)|_D \longrightarrow \mathcal{J}^0 \xrightarrow{\gamma} \mathcal{I}^1.$$

It follows that the above square commutes.

(2.4) Lemma. *Let $\pi: X \to S$ be a flat, projective morphism. Assume that the fibers of π are Cohen-Macaulay and have components which are all one-dimensional. Let D be an effective horizontal divisor on X, and let $\mathcal{F}$ be a coherent sheaf on X. If $\mathcal{F}$ is flat over S, then the map*

$$\Theta(\mathcal{F}):\ \pi_*\mathcal{H}om_X\big(\mathcal{F}, \mathcal{K}_{X/S}(D)|_D\big) \longrightarrow \mathcal{E}xt^1_\pi\big(\mathcal{F}|_D, \mathcal{K}_{X/S}\big)$$

of Definition 2.2 is an isomorphism. Also, even if $\mathcal{F}$ is not flat over S, the composition of $\Theta(\mathcal{F})$ with the 1st degree duality map is an isomorphism.

Proof. We can assume that S is affine. By (2.3) the result is true for $\mathcal{F}$ locally free. Let $\mathcal{O}(1)$ be a sheaf on X which is very ample with respect to π. There is an integer n such that $\mathcal{F}(n)$ is generated by a finite number of global sections. This implies that there is a surjection from $\mathcal{O}(-n)^m$ to $\mathcal{F}$ for some integer m. By also applying the same reasoning to the kernel of this map, we get an exact sequence

$$M \longrightarrow L \longrightarrow \mathcal{F} \longrightarrow 0$$

where L and M are locally free coherent sheaves. Let $\mathcal{K} = \mathcal{K}_{X/S}$. Consider the commutative diagram

$$\begin{array}{ccccccc}
0 \to & \pi_*\mathcal{H}om_X\big(\mathcal{F},\mathcal{K}(D)|_D\big) & \to & \pi_*\mathcal{H}om_X\big(L,\mathcal{K}(D)|_D\big) & \to & \pi_*\mathcal{H}om_X\big(M,\mathcal{K}(D)|_D\big) \\
& \big\downarrow{\scriptstyle \Theta(\mathcal{F})} & & \big\downarrow{\scriptstyle \Theta(L)} & & \big\downarrow{\scriptstyle \Theta(M)} \\
& \mathcal{E}xt^1_\pi\big(\mathcal{F}|_D,\mathcal{K}\big) & \to & \mathcal{E}xt^1_\pi\big(L|_D,\mathcal{K}\big) & \to & \mathcal{E}xt^1_\pi\big(M|_D,\mathcal{K}\big) \\
& \big\downarrow & & \big\downarrow & & \big\downarrow \\
0 \to & \mathcal{H}om_S\big(R^0\pi_*(\mathcal{F}|_D),\mathcal{O}_S\big) & \to & \mathcal{H}om_S\big(R^0\pi_*(L|_D),\mathcal{O}_S\big) & \to & \mathcal{H}om_S\big(R^0\pi_*(M|_D),\mathcal{O}_S\big)
\end{array}$$

where the lower vertical maps are the 1st degree duality maps. The top row is clearly exact; the bottom row is exact since the functor $R^0\pi_*(\bullet|_D)$ is right exact (note: D is finite over S, so the higher derived functors of π_* vanish for coherent sheaves with support on D). Proposition 1.6 is applicable to $L|_D$ since $R^1\pi_*(L|_D) = 0$ is trivially locally free, likewise for $M|_D$. So all the vertical maps of the rightmost two columns have been shown to be isomorphisms, therefore the composition of the two vertical maps of the left column must also be an isomorphism.

If $\mathcal{F}$ is flat over S, then, by Proposition 1.6, the bottom map of the left column is an isomorphism. Therefore, $\Theta(\mathcal{F})$ must also be an isomorphism.

(2.5) Proposition. *Let $\pi\colon X \to S$ be a flat, projective morphism with fibers which are Cohen-Macaulay and have components all of dimension 1. Let D be an effective horizontal divisor of X, and let π' be the restriction of π to D. Then there is a canonical isomorphism*

$$\mathcal{K}_{D/S} \xrightarrow{\sim} \mathcal{K}_{X/S}(D)\,|_D$$

such that, for each coherent sheaf $\mathcal{F}$ on D, considered also as a coherent sheaf on X, the following commutes

$$\begin{array}{ccc}
\pi'_*\mathcal{H}om_D\big(\mathcal{F},\mathcal{K}_{D/S}\big) & \xrightarrow{\sim} & \pi_*\mathcal{H}om_X\big(\mathcal{F},\mathcal{K}_{X/S}(D)|_D\big) \\
\big\downarrow{\scriptstyle D_0} & & \big\downarrow \\
\mathcal{H}om_S\big(R^0\pi'_*(\mathcal{F}),\mathcal{O}_S\big) & = & \mathcal{H}om_S\big(R^0\pi_*(\mathcal{F}),\mathcal{O}_S\big)
\end{array}$$

where the top map is the isomorphism induced by the above canonical isomorphism between $\mathcal{K}_{D/S}$ and $\mathcal{K}_{X/S}(D)\,|_D$, the left map is the 0th degree duality isomorphism on D, and the right map is the composition of the map $\Theta(\mathcal{F})$ (of Definition 2.2) followed by the 1st degree duality map on X.

Proof. For each coherent sheaf $\mathcal{F}$ on X, define $\Phi(\mathcal{F})$ to be the map

$$\Phi(\mathcal{F})\colon \operatorname{Hom}_X\big(\mathcal{F},\mathcal{K}_{X/S}(D)|_D\big) \longrightarrow \operatorname{Hom}_S\big(R^0\pi_*(\mathcal{F}|_D),\mathcal{O}_S\big)$$

formed by taking global sections of the composition of $\Theta(\mathcal{F})$ with the 1st degree duality map. The map $\Phi(\mathcal{F})$ is an isomorphism by Lemma 2.4, and is functorial with respect to $\mathcal{F}$.

In particular, $\Phi\big(\mathcal{K}_{X/S}(D)|_D\big)$ is an isomorphism

$$\operatorname{Hom}_X\Big(\mathcal{K}_{X/S}(D)|_D,\ \mathcal{K}_{X/S}(D)|_D\Big) \longrightarrow \operatorname{Hom}_S\Big(R^0\pi_*\big(\mathcal{K}_{X/S}(D)|_D\big),\ \mathcal{O}_S\Big).$$

Let $t: R^0\pi_*\big(\mathcal{K}_{X/S}(D)|_D\big) \to \mathcal{O}_S$ be the image of the identity map under this isomorphism.

Let $\mathcal{F}$ be a coherent sheaf on D. By the functorial nature of Φ, any morphism $\alpha: \mathcal{F} \to \mathcal{K}_{X/S}(D)|_D$ relates $\Phi(\mathcal{F})$ and $\Phi\big(\mathcal{K}_{X/S}(D)|_D\big)$ by the following commutative diagram:

$$\begin{array}{ccc}
\operatorname{Hom}_X\Big(\mathcal{K}_{X/S}(D)|_D,\ \mathcal{K}_{X/S}(D)|_D\Big) & \xrightarrow{\Phi\left(\mathcal{K}_{X/S}(D)|_D\right)} & \operatorname{Hom}_S\Big(R^0\pi_*(\mathcal{K}_{X/S}(D)|_D),\ \mathcal{O}_S\Big) \\
\downarrow & & \downarrow \\
\operatorname{Hom}_X\Big(\mathcal{F},\ \mathcal{K}_{X/S}(D)|_D\Big) & \xrightarrow{\Phi(\mathcal{F})} & \operatorname{Hom}_S\Big(R^0\pi_*(\mathcal{F}|_D),\ \mathcal{O}_S\Big)
\end{array}$$

where the vertical maps are induced by α. By considering the images of the identity map on $\mathcal{K}_{X/S}(D)|_D$ under the maps of the above diagram, we conclude that the following commutes

$$\begin{array}{ccc}
R^0\pi_*(\mathcal{F}) & \xrightarrow{\Phi(\mathcal{F})(\alpha)} & \mathcal{O}_S \\
\big\downarrow{\scriptstyle R^0\pi_*(\alpha)} & & \big\| \\
R^0\pi_*\big(\mathcal{K}_{X/S}(D)|_D\big) & \xrightarrow{t} & \mathcal{O}_S.
\end{array}$$

Note that this gives a characterization of Φ in terms of t. The fact that $\Phi(\mathcal{F})$ is an isomorphism means that $\mathcal{K}_{X/S}(D)|_D$ and t satisfy the same universal property characterizing $\mathcal{K}_{D/S}$ and $t_{\pi'}$ in Theorem 1.1. Therefore, there is a unique isomorphism δ between $\mathcal{K}_{X/S}(D)|_D$ and $\mathcal{K}_{D/S}$ making the following commute:

$$\begin{array}{ccc}
R^0\pi_*\big(\mathcal{K}_{X/S}(D)|_D\big) & \xrightarrow{t} & \mathcal{O}_S \\
\big\downarrow{\scriptstyle R^0\pi_*(\delta)} & & \big\| \\
R^0\pi'_*(\mathcal{K}_{D/S}) & \xrightarrow{t_{\pi'}} & \mathcal{O}_S.
\end{array}$$

This allows us to identify $\mathcal{K}_{X/S}(D)|_D$ with $\mathcal{K}_{D/S}$ and t with $t_{\pi'}$. Under this identification, and for any coherent sheaf $\mathcal{F}$ on D, we have two isomorphisms from $\pi'_*\mathcal{H}om_D(\mathcal{F}, \mathcal{K}_{D/S})$ to $\mathcal{H}om_S(R^0\pi'_*(\mathcal{F}), \mathcal{O}_S)$: (i) the composition of $\Theta(\mathcal{K}_{D/S})$ with the 1st degree duality map on X, and (ii) the 0th degree duality isomorphism on D. Both these isomorphisms, as well as the relative dualizing sheaves, are well-behaved with respect to restrictions to open subschemes. So, since showing these maps are equal is a local problem, we can assume that S is affine: $S = \operatorname{Spec} A$. Then, to show that these maps are equal, we only need to show that they induce the same map between the associated A-modules, $\operatorname{Hom}_D(\mathcal{F}, \mathcal{K}_{D/S})$ and $\operatorname{Hom}_S(R^0\pi_*(\mathcal{F}), \mathcal{O}_S)$.

The first isomorphism induces, by definition, the map $\Phi(\mathcal{F})$. As we showed above using different notation, $\Phi(\mathcal{F})$ takes any element $\alpha: \mathcal{F} \to \mathcal{K}_{D/S}$ of $\operatorname{Hom}_D(\mathcal{F}, \mathcal{K}_{D/S})$ to the composition

$$R^0\pi'_*(\mathcal{F}) \xrightarrow{R^0\pi'_*(\alpha)} R^0\pi'_*(\mathcal{K}_{D/S}) \xrightarrow{t_{\pi'}} \mathcal{O}_S.$$

This is exactly the map induced by the 0th degree duality isomorphism. Therefore, the two maps are equal.

3. The Duality Isomorphism on Determinants

In this section we will use the duality maps developed in the first section to define duality isomorphisms for determinants of cohomology. In (3.1) – (3.3) we consider the case of one or zero dimensional schemes defined over a field. In (3.4) we consider the case of relative dimension zero. Finally, in (3.5) we consider the case of relative dimension one. In this last case, since determinants of cohomology have only been defined for line bundles, we will assume that the relative dualizing sheaf is a line bundle; as we will point out in the introduction of the next section, this is true in many interesting situations.

(3.1) Definition. Let X be projective over a field K. Assume that X is Cohen-Macaulay and that every component of X is of dimension 1. Let L be a locally free coherent sheaf on X. We define the *duality isomorphism*

$$\mathbf{D}(\mathcal{K}_X \otimes L^\vee) \xrightarrow{\sim} \mathbf{D}(L)$$

as follows:

Consider the duality isomorphisms of (1.2) and (1.5), which in this situation can be written in the form

$$H^0(X, \mathcal{K}_X \otimes L^\vee) = \mathrm{Hom}_X(L, \mathcal{K}_X) \xrightarrow{\sim} H^1(X, L)^\vee$$

and

$$H^1(X, \mathcal{K}_X \otimes L^\vee) = \mathrm{Ext}^1_X(L, \mathcal{K}_X) \xrightarrow{\sim} H^0(X, L)^\vee.$$

Taking determinants, we get isomorphisms

$$\mathrm{Det}\, H^0(X, \mathcal{K}_X \otimes L^\vee) \xrightarrow{\sim} \mathrm{Det}\big(H^1(X, L)^\vee\big) = \mathrm{Det}\big(H^1(X, L)\big)^{-1}$$

and

$$\mathrm{Det}\, H^1(X, \mathcal{K}_X \otimes L^\vee) \xrightarrow{\sim} \mathrm{Det}\big(H^0(X, L)^\vee\big) = \mathrm{Det}\big(H^0(X, L)\big)^{-1}.$$

We use these maps to construct the desired isomorphism:

$$\mathbf{D}(\mathcal{K}_X \otimes L^\vee) \stackrel{\text{def}}{=} \frac{\mathrm{Det}\, H^0(X, \mathcal{K}_X \otimes L^\vee)}{\mathrm{Det}\, H^1(X, \mathcal{K}_X \otimes L^\vee)} \longrightarrow \frac{\mathrm{Det}\big(H^1(X, L)\big)^{-1}}{\mathrm{Det}\big(H^0(X, L)\big)^{-1}} = \frac{\mathrm{Det}\, H^0(X, L)}{\mathrm{Det}\, H^1(X, L)} \stackrel{\text{def}}{=} \mathbf{D}(L).$$

(3.2) Definition. Let D be finite over a field K. For example, D could be an effective Cartier divisor on a one-dimensional projective scheme. Let L be a locally free coherent sheaf on D. We define the *duality isomorphism*

$$\mathbf{D}(\mathcal{K}_D \otimes L^\vee) \xrightarrow{\sim} \mathbf{D}(L)^{-1}$$

as follows:

Consider the duality isomorphism of Proposition 1.2, which in this situation can be written in the form

$$H^0(D, \mathcal{K}_D \otimes L^\vee) = \mathrm{Hom}_D(L, \mathcal{K}_D) \xrightarrow{\sim} H^0(D, L)^\vee.$$

We get the desired isomorphism by taking determinants:

$$\mathbf{D}(\mathcal{K}_D \otimes L^\vee) \overset{\text{def}}{=} \operatorname{Det} H^0(D, \mathcal{K}_D \otimes L^\vee) \to \operatorname{Det}(H^0(D,L)^\vee) = \operatorname{Det}(H^0(D,L))^{-1} \overset{\text{def}}{=} \mathbf{D}(L)^{-1}.$$

(3.3) Proposition. *Let X be projective over a field K. Assume that X is Cohen-Macaulay, and that every component of X is of dimension 1. Let L be a locally free coherent sheaf on X, and let D be an effective Cartier divisor on X. By Lemma 2.1 the following is exact:*

$$0 \longrightarrow \mathcal{K}_X \longrightarrow \mathcal{K}_X(D) \longrightarrow \mathcal{K}_X(D)|_D \longrightarrow 0.$$

By tensoring with $L^\vee$, and using the identification $\mathcal{K}_D = \mathcal{K}_X(D)|_D$ of (2.5), we get the exact sequence

$$0 \longrightarrow \mathcal{K}_X \otimes L^\vee \longrightarrow \mathcal{K}_X \otimes L^\vee(D) \longrightarrow \mathcal{K}_D \otimes L^\vee|_D \longrightarrow 0.$$

We also have the exact sequence

$$0 \longrightarrow L(-D) \longrightarrow L \longrightarrow L|_D \longrightarrow 0.$$

As in (2.4) of Chapter 1, these exact sequences give isomorphisms

$$\mathbf{D}\big(\mathcal{K}_X \otimes L^\vee(D)\big) \xrightarrow{\sim} \mathbf{D}(\mathcal{K}_X \otimes L^\vee) \otimes \mathbf{D}(\mathcal{K}_D \otimes L^\vee|_D)$$

and

$$\mathbf{D}(L(-D)) \xrightarrow{\sim} \mathbf{D}(L) \otimes \mathbf{D}(L|_D)^{-1}.$$

In this situation, the following commutes:

$$\begin{array}{ccc} \mathbf{D}\big(\mathcal{K}_X \otimes L^\vee(D)\big) & \longrightarrow & \mathbf{D}(\mathcal{K}_X \otimes L^\vee) \otimes \mathbf{D}(\mathcal{K}_D \otimes L^\vee|_D) \\ \downarrow & & \downarrow \\ \mathbf{D}(L(-D)) & \longrightarrow & \mathbf{D}(L) \otimes \mathbf{D}(L|_D)^{-1} \end{array}$$

where the horizontal maps are the above isomorphisms, the left vertical map is the duality isomorphism, and the right vertical map is the tensor product of the two duality isomorphisms.

Proof. The diagram of (2.3) yields, in this situation, the following commutative diagram:

$$\begin{array}{ccccccccc} H^0(X, \mathcal{K}_X \otimes L^\vee) & \to & H^0(X, \mathcal{K}_X \otimes L^\vee(D)) & \to & H^0(D, \mathcal{K}_D \otimes L^\vee|_D) & \to & H^1(X, \mathcal{K}_X \otimes L^\vee) & \to & H^1(X, \mathcal{K}_X \otimes L^\vee(D)) \\ \downarrow & & \downarrow & & \downarrow & & \downarrow & & \downarrow \\ H^1(X,L)^\vee & \to & H^1(X, L(-D))^\vee & \to & H^0(D, L|_D)^\vee & \to & H^0(X,L)^\vee & \to & H^0(X, L(-D))^\vee \end{array}$$

where, using Proposition 2.5, we are identifying $\mathcal{K}_D$ with $\mathcal{K}_X(D)|_D$. Also by Proposition 2.5, we conclude that the central vertical map is the 0th degree duality map on D. All the other vertical maps are, by definition, the duality isomorphisms. The rows are the long exact sequences obtained from the short exact sequences given in the statement of the proposition.

We choose compatible bases, in the sense of (2.2) of Chapter 1, for the spaces of the top row. The duality isomorphisms induce compatible bases for the spaces of the bottom row. With such bases, the theorem follows from the definitions of the various isomorphisms.

(3.4) Proposition. *Let D be finite and flat over an integral scheme S, and let K be the function field of S. Let L be a line bundle on D, and let L_K be the restriction of L to the generic fiber D_K. Then there is a unique isomorphism*

$$\mathbf{D}\left(\mathcal{K}_{D/S} \otimes L^{-1}\right) \xrightarrow{\sim} \mathbf{D}(L)^{-1}$$

whose restriction to the generic point of S gives the duality isomorphism of Definition 3.2:

$$\mathbf{D}\left(\mathcal{K}_{D_K} \otimes L_K^{-1}\right) \xrightarrow{\sim} \mathbf{D}(L_K)^{-1}.$$

Proof. This is a local question on S, so, as in Definition 5.4 of Chapter 2, we can assume that S is affine, $S = \operatorname{Spec} A$, and that both $H^0(D, \mathcal{K}_{D/S} \otimes L^{-1})$ and $H^0(D, L)$ are free A-modules.

Let λ be an ordered A-basis of $H^0(D, L)$, and let $\lambda^\vee$ be the ordered basis of $H^0(D, L)^\vee$ which is dual to λ. We have the duality isomorphism of Proposition 1.2, which, in this situation, yields an isomorphism

$$H^0\left(D, \mathcal{K}_{D/S} \otimes L^{-1}\right) \xrightarrow{\sim} H^0(D, L)^\vee.$$

Let μ be the ordered A-basis of $H^0(D, \mathcal{K}_{D/S} \otimes L^{-1})$ which is sent to $\lambda^\vee$ under the above duality map.

Let $\langle \lambda \,|_{D_K} \rangle$ be the element of $\operatorname{Det} H^0(D_K, L_K) = \mathbf{D}(L_K)$ associated with the basis $\lambda|_{D_K}$ of $H^0(D_K, L_K)$. Likewise, let $\langle \mu|_{D_K} \rangle$ be the element of $\mathbf{D}(\mathcal{K}_{D_K} \otimes L_K^{-1})$ associated with $\mu|_{D_K}$.

By part 3 of Proposition 5.5 in Chapter 2, $\mathbf{D}(L)$ is a trivial line bundle with a nowhere vanishing global section, which we will call $\langle \lambda \rangle$, which has the property that the restriction of $\langle \lambda \rangle$ to the generic fiber D_K is $\langle \lambda \,|_{D_K} \rangle$. Likewise, $\mathbf{D}(\mathcal{K}_{D/S} \otimes L^{-1})$ is a trivial line bundle with nowhere vanishing global section $\langle \mu \rangle$ which has the property that the restriction of $\langle \mu \rangle$ to the generic fiber D_K is $\langle \mu \,|_{D_K} \rangle$.

By definition, the duality isomorphism

$$\mathbf{D}\left(\mathcal{K}_{D_K} \otimes L_K^{-1}\right) \xrightarrow{\sim} \mathbf{D}(L_K)^{-1}$$

sends the generator $\langle \mu \,|_{D_K} \rangle$ to the generator $\langle \lambda \,|_{D_K} \rangle^{-1}$, and is characterized by this fact. Therefore, the isomorphism

$$\mathbf{D}\left(\mathcal{K}_{D/S} \otimes L^{-1}\right) \xrightarrow{\sim} \mathbf{D}(L)^{-1}$$

defined by sending $\langle \mu \rangle$ to $\langle \lambda \rangle^{-1}$ has the property that its restriction to the generic point of S is the duality isomorphism. Since this isomorphism is between line bundles, it is the only such isomorphism.

(3.5) Proposition. *Let $\pi \colon X \to S$ be a flat, projective morphism. Assume that the base S is integral with function field K, and that every fiber of π is Cohen-Macaulay with components of dimension 1. In addition, assume that the relative dualizing sheaf $\mathcal{K}_{X/S}$ is a line bundle on X. Let L be a line bundle on X, and let L_K be the restriction of L to the generic fiber X_K.*

Then there is a unique isomorphism

$$\mathbf{D}\left(\mathcal{K}_{X/S} \otimes L^{-1}\right) \xrightarrow{\sim} \mathbf{D}(L)$$

whose restriction to the generic point of S gives the duality isomorphism of Definition 3.1:

$$\mathbf{D}(\mathcal{K}_{X_K} \otimes L_K^{-1}) \xrightarrow{\sim} \mathbf{D}(L_K).$$

Proof. This question is local on S, so by Lemma 5.1.3 of Chapter 2, we can assume that S is affine, $S = \operatorname{Spec} A$, and that there is an effective horizontal divisor D on X with the property that $H^1(X, L(D)) = 0$, and that $H^0(X, L(D))$ is a free A-module of finite rank.

Theorem 12.11 of [H], Chapter III, (for $i = 2$) implies that $H^1(X_s, L(D)|_{X_s}) = 0$ for every point $s \in S$; so, by 0th degree duality on X_s,

$$H^0\Big(X_s, (\mathcal{K}_{X/S} \otimes L(D)^{-1})\big|_{X_s}\Big) = H^0\Big(X_s, \mathcal{K}_{X_s} \otimes \big(L(D)|_{X_s}\big)^{-1}\Big) = H^1\Big(X_s, L(D)|_{X_s}\Big)^{\vee} = 0$$

for each $s \in S$. So, by Corollary 12.9 of [H], Chapter III, together with Nakayama's lemma, we have that

$$H^0(X, \mathcal{K}_{X/S} \otimes L(D)^{-1}) = 0.$$

Let λ be an ordered A-basis of $H^0(X, L(D))$. The 1st degree duality map of (1.5) gives the isomorphism

$$H^1(X, \mathcal{K}_{X/S} \otimes L(D)^{-1}) \xrightarrow{\sim} H^0(X, L(D))^{\vee}.$$

Let $\lambda^{\vee}$ be the ordered basis of $H^0(D, L(D))^{\vee}$ which is dual to λ, and let μ be the ordered A-basis of $H^1(X, \mathcal{K}_{X/S} \otimes L(D)^{-1})$ which is sent to $\lambda^{\vee}$ under the above duality isomorphism.

Let D_K be the restriction of D to X_K. Let $\langle \lambda\, |_{X_K} \rangle$ be the non-zero element of $Det\ H^0(X_K, L_K(D_K)) = \mathbf{D}(L_K(D_K))$ associated with the K-basis $\lambda|_{X_K}$ of $H^0(X_K, L_K(D_K))$. Likewise, let $\langle \mu|_{X_K} \rangle \in Det\ H^1(X_K, \mathcal{K}_{X_K} \otimes L_K(D_K)^{-1})$ be the non-zero element associate with the K-basis $\mu|_{X_K}$ of $H^1(X_K, \mathcal{K}_{X_K} \otimes L_K(D_K)^{-1})$. Thus we can regard $\langle \mu|_{X_K} \rangle^{-1}$ as a non-zero element of

$$\begin{aligned} \mathbf{D}(\mathcal{K}_{X_K} \otimes L_K(D_K)^{-1}) &= \frac{\operatorname{Det} H^0(X_K, \mathcal{K}_{X_K} \otimes L_K(D_K)^{-1})}{\operatorname{Det} H^1(X_K, \mathcal{K}_{X_K} \otimes L_K(D_K)^{-1})} \\ &= \operatorname{Det} H^1(X_K, \mathcal{K}_{X_K} \otimes L_K(D_K)^{-1})^{-1}. \end{aligned}$$

By part 3 of Proposition 5.3 in Chapter 2, $\mathbf{D}(L(D))$ is a trivial line bundle with a nowhere vanishing global section, which we will call $\langle \lambda \rangle$, which has the property that the restriction of $\langle \lambda \rangle$ to the generic fiber X_K is $\langle \lambda\, |_{X_K} \rangle$. Likewise, $\mathbf{D}(\mathcal{K}_{X/S} \otimes L(D)^{-1})$ is a trivial line bundle with a nowhere vanishing global section, which we call $\langle \mu \rangle^{-1}$, which has the property that the restriction of $\langle \mu \rangle^{-1}$ to X_K is $\langle \mu\, |_{X_K} \rangle^{-1}$.

By definition, the duality isomorphism

$$\mathbf{D}(\mathcal{K}_{X_K} \otimes L_K(D_K)^{-1}) \xrightarrow{\sim} \mathbf{D}(L_K(D_K))$$

sends the generator $\langle \mu\, |_{X_K} \rangle^{-1}$ to the generator $\langle \lambda\, |_{X_K} \rangle$, and is characterized by this fact. Therefore, the isomorphism

$$\mathbf{D}(\mathcal{K}_{X/S} \otimes L(D)^{-1}) \xrightarrow{\sim} \mathbf{D}(L(D))$$

defined by sending $\langle\mu\rangle^{-1}$ to $\langle\lambda\rangle$ has the property that its restriction to the generic point of S is the above duality isomorphism.

Now consider the following diagram which is commutative by Proposition 3.3:

$$\begin{array}{ccc} \mathbf{D}(\mathcal{K}_{X_K}\otimes L_K^{-1}) & \longrightarrow & \mathbf{D}(\mathcal{K}_{X_K}\otimes L_K(D_K)^{-1}) \otimes \mathbf{D}(\mathcal{K}_{D_K}\otimes L_K(D_K)^{-1}|_{D_K}) \\ \downarrow & & \downarrow \\ \mathbf{D}(L_K) & \longrightarrow & \mathbf{D}(L_K(D_K))\otimes \mathbf{D}(L_K(D_K)|_{D_K})^{-1} \end{array}$$

where the left vertical map is the duality isomorphism, the right vertical map is the tensor product of the two duality isomorphisms, and the horizontal maps are the isomorphisms which result from the exact sequences

$$0 \to \mathcal{K}_{X_K}\otimes L_K(D_K)^{-1} \to \mathcal{K}_{X_K}\otimes L_K^{-1} \to \mathcal{K}_{D_K}\otimes L_K(D_K)^{-1}|_{D_K} \to 0$$

and

$$0 \to L_K \to L_K(D_K) \to L_K(D_K)|_{D_K} \to 0.$$

Our goal is to extend the duality isomorphism, the leftmost vertical isomorphism in the above diagram, to an isomorphism between the corresponding sheaves on S. By Proposition 5.6 of Chapter 2, the isomorphisms

$$\mathbf{D}(\mathcal{K}_{X_K}\otimes L_K^{-1}) \longrightarrow \mathbf{D}(\mathcal{K}_{X_K}\otimes L_K(D_K)^{-1}) \otimes \mathbf{D}(\mathcal{K}_{D_K}\otimes L_K(D_K)^{-1}|_{D_K})$$

and

$$\mathbf{D}(L_K) \longrightarrow \mathbf{D}(L_K(D_K))\otimes \mathbf{D}(L_K(D_K)|_{D_K})^{-1}$$

extend to isomorphisms

$$\mathbf{D}(\mathcal{K}_{X/S}\otimes L^{-1}) \longrightarrow \mathbf{D}(\mathcal{K}_{X/S}\otimes L(D)^{-1}) \otimes \mathbf{D}(\mathcal{K}_{D/S}\otimes L(D)^{-1}|_{D})$$

and

$$\mathbf{D}(L) \longrightarrow \mathbf{D}(L(D))\otimes \mathbf{D}(L(D)|_{D})^{-1}.$$

By Proposition 3.4, the isomorphism

$$\mathbf{D}(\mathcal{K}_{D_K}\otimes L_K(D_K)^{-1}|_{D_K}) \xrightarrow{\sim} \mathbf{D}(L_K(D_K)|_{D_K})^{-1}$$

extends to an isomorphism

$$\mathbf{D}(\mathcal{K}_{D/S}\otimes L(D)^{-1}|_{D}) \xrightarrow{\sim} \mathbf{D}(L(D)|_{D})^{-1}.$$

So we have shown that all the other isomorphisms in the above diagram extend to isomorphisms between the corresponding line bundles on S. Using these other isomorphisms we can construct the desired isomorphism: an isomorphism

$$\mathbf{D}(\mathcal{K}_{X/S}\otimes L^{-1}) \longrightarrow \mathbf{D}(L)$$

whose restriction to the generic fiber of S is

$$\mathbf{D}(\mathcal{K}_{X_K}\otimes L_K^{-1}) \longrightarrow \mathbf{D}(L_K).$$

Clearly, such an isomorphism is unique.

4. The Riemann-Roch Isomorphism

Let $\pi: X \to S$ be a flat, projective morphism. Assume that the base S is integral with function field K, and that every fiber of π is Cohen-Macaulay with components of dimension 1. In addition, assume that the relative dualizing sheaf $\mathcal{K}_{X/S}$ is a line bundle on X, i.e., an invertible sheaf.

This last assumption, that the relative dualizing sheaf $\mathcal{K}_{X/S}$ be an invertible sheaf, is true in many important situations. Using Nakayama's Lemma and the compatability of the relative dualizing sheaf with respect to base change, it is easy to show that when X is reduced the relative dualizing sheaf $\mathcal{K}_{X/S}$ is an invertible sheaf if and only if, for each fiber X_s of π, the dualizing sheaf of X_s is an invertible sheaf. Actually, since $\mathcal{K}_{X/S}$ is flat over S, this is true even when X is not reduced. Now the dualizing sheaf of a projective curve defined over a field is known to be an invertible sheaf if the curve is non-singular, or if its singularities are "not too bad". For example, if the singularities are all ordinary double points, if the curve can be embedded as a subvariety of a smooth projective surface, or, more generally, if the curve is Gorenstein, its dualizing sheaf is known to be an invertible sheaf. In particular, for one important case, namely semi-stable families of curves, the relative dualizing sheaf is indeed an invertible sheaf.

(4.1) Definition. Let L be a line bundles on X. We now define a pair of isomorphisms, each called the *Riemann-Roch Isomorphism*,

$$\mathbf{D}(L)^{\otimes 2} \otimes \mathbf{D}(\mathcal{O}_X)^{\otimes -2} \xrightarrow{\sim} \langle L, L \otimes \mathcal{K}_{X/S}^{-1} \rangle$$

as follows:

Let ρ and ρ' be a solution, mod 2, to the equation

$$\rho + \rho' = \deg(L|_{X_K}) \cdot \big(\deg \mathcal{K}_{X_K} + 1\big).$$

Consider the isomorphism

$$\frac{\mathbf{D}(\mathcal{K}_{X/S}) \otimes \mathbf{D}(\mathcal{O}_X)}{\mathbf{D}(L) \otimes \mathbf{D}(\mathcal{K}_{X/S} \otimes L^{-1})} \xrightarrow{\sim} \langle L, \mathcal{K}_{X/S} \otimes L^{-1} \rangle$$

of Proposition 5.7 of Chapter 2 (where we take $L_1 = L$, $L_2 = \mathcal{O}_X$, $M_1 = \mathcal{K}_{X/S} \otimes L^{-1}$, $M_2 = \mathcal{O}_X$, and ρ and ρ' as above). The bilinearity of the intersection pairing, Proposition 4.6 of Chapter 2, allows us to form from this isomorphism the following isomorphism

$$\frac{\mathbf{D}(L) \otimes \mathbf{D}(\mathcal{K}_{X/S} \otimes L^{-1})}{\mathbf{D}(\mathcal{K}_{X/S}) \otimes \mathbf{D}(\mathcal{O}_X)} \xrightarrow{\sim} \langle L, L \otimes \mathcal{K}_{X/S}^{-1} \rangle.$$

Proposition 3.5 gives us the duality isomorphisms:

$$\mathbf{D}\big(\mathcal{K}_{X/S} \otimes L^{-1}\big) \xrightarrow{\sim} \mathbf{D}(L) \qquad \text{and} \qquad \mathbf{D}\big(\mathcal{K}_{X/S}\big) \xrightarrow{\sim} \mathbf{D}(\mathcal{O}_X).$$

We use the inverses of these to form the isomorphism

$$\frac{\mathbf{D}(L) \otimes \mathbf{D}(L)}{\mathbf{D}(\mathcal{O}_X) \otimes \mathbf{D}(\mathcal{O}_X)} \xrightarrow{\sim} \frac{\mathbf{D}(L) \otimes \mathbf{D}(\mathcal{K}_{X/S} \otimes L^{-1})}{\mathbf{D}(\mathcal{K}_{X/S}) \otimes \mathbf{D}(\mathcal{O}_X)}$$

which, when we compose it with the above isomorphism, gives us the desired isomorphism.

If we choose ρ and ρ' to be the other solution to the above equation, we obtain a similar isomorphism. By Proposition 5.7 of Chapter 2 [together with Definition 4.6 of Chapter 1], this isomorphism differs from the other by multiplication by -1.

(4.2) Note that each Riemann-Roch isomorphism is well-behaved with respect to restriction to the generic point of S. I will leave it to the reader to check that, if S is the spectrum of a field, then these isomorphisms are well-behaved with respect to extensions of the base field.

Chapter 4

Intersection Functions on Complex Curves

1. Motivation: The Non-Archimedean Situation

In the previous chapter we discussed the Riemann-Roch isomorphism for families of (possibly) singular curves. If the base scheme B of such a family is arithmetic, i.e., if $B = \operatorname{Spec} R$ where R is the ring of integers of a number field K, then the usual procedure in algebraic number theory is to complete B using the archimedean places of K. This is done by formally adding to B a finite number of complex points, one for each embedding of K into the complex numbers $\mathbf{C}$. Call this new object $\overline{B}$, and call the new points added to complete B the *archimedean points* of $\overline{B}$.

Our ultimate goal, which we pursue in the next chapter, is to extend the Riemann-Roch isomorphism of the previous chapter to all of $\overline{B}$. In order to do so we will need to (1) define the notion of a family of curves over $\overline{B}$, (2) define the notion of a line bundle on $\overline{B}$ and on a family of curves over $\overline{B}$, (3) define an intersection pairing between line bundles of a family of curves over $\overline{B}$ which yields a line bundle of $\overline{B}$, (4) define the determinant of cohomology of a line bundle on a family of curves over $\overline{B}$, and (5) show that these definitions result in a Riemann-Roch isomorphism similar to that of the previous chapter.

All this will be done in Chapter 5. The goal of the current chapter is to motivate and develop the theory which will be used in the final chapter. In particular we will investigate the archimedean analogue of a local family of curves near a non-archimedean point of B (i.e., a curve defined over a discrete valuation ring). This analogue will center on the concept of an intersection function on the generic fiber of a local family.

(1.1) Notation. We will be dealing with two analogous situations. The first is the non-archimedean situation which will be the main object of study in this section. Throughout this section let K be a field complete with respect to a discrete valuation ν. Let R be the discrete valuation ring of K, and let k be its residue field. Fix a real constant $\gamma > 1$, for example, if the residue field k is finite of order q, we usually take γ to be q. The valuation ν gives an absolute value on K sending any non-zero $a \in K$ to $|a|_\nu = \gamma^{-\nu(a)}$.

The second situation is the archimedean situation. In this case the field of complex numbers $\mathbf{C}$ plays the role of K, and the standard absolute value $|z| = (z\overline{z})^{\frac{1}{2}}$ on $\mathbf{C}$ plays the role of $|\ |_\nu$. There is in this situation, of course, no scheme analogous to $\operatorname{Spec} R$, but you could formally define a set with two elements: a generic element representing $\operatorname{Spec} \mathbf{C}$, and another element representing the structure of the absolute value on $\mathbf{C}$.

(1.2) Line Bundles on the Base. Let $\mathcal{M}$ be a line bundle on $\operatorname{Spec} R$. We can identify $\mathcal{M}$ with $M = \Gamma(\mathcal{M})$, a free R-module of rank 1. Let $M_K = M \otimes_R K$ be the restriction to $\operatorname{Spec} K$. So M_K is as a one-dimensional K-vector space and M can be regarded as a R-submodule of M_K. We can give M_K a norm $|\,|_{\mathcal{M}}$ as follows. Choose a generating element m_0 of $M = \Gamma(\mathcal{M})$. Then, for any element m of M_K, we can write m as $b\, m_0$ for some $b \in K$. Define $|m|_{\mathcal{M}}$ to be $|b|_\nu$. This is clearly independent of the choice of m_0.

Note that the functor which takes $\mathcal{M}$ to $(M_K, |\,|_{\mathcal{M}})$ defines an equivalence of categories between (1) the category of line bundles on $\operatorname{Spec} R$ whose arrows are isomorphisms between such line bundles and (2) the category of one-dimensional normed K-vector spaces whose arrows are norm-preserving vector space isomorphisms.

This suggests the analogous object in the archimedean situation is a one-dimensional normed $\mathbf{C}$-vector space. An isomorphism between two such objects is a norm-preserving vector space isomorphism.

(1.3) Local Families of Curves in the Non-Archimedean Case. As mentioned above, we would like to form an archimedean analogue to the idea of a local family of curves near a non-archimedean point. For the remainder of this section we will concentrate on the non-archimedean situation, studying aspects of it which might generalize to the archimedean situation. Since this material is motivational in nature, we have decided to leave many of the details of the proofs to the reader.

Let $\mathcal{X}$ be projective and flat over $\operatorname{Spec} R$ of relative dimension 1, i.e., every fiber has components all of dimension 1. There are two fibers: the generic fiber X, and the special fiber X_k. We assume that both fibers, X and X_k, are connected and non-singular except for possibly a finite number of ordinary double points. We assume that $\mathcal{X}$ has "optimal reduction", by which we mean the following: we assume that each component of the normalization $\widetilde{\mathcal{X}}$ of $\mathcal{X}$ is smooth over $\operatorname{Spec} R$, and that there are $2n$ sections $R_1, \ldots, R_n, S_1, \ldots, S_n$ on $\widetilde{\mathcal{X}}$, each intersecting none of the others, such that $\mathcal{X}$ is formed from $\widetilde{\mathcal{X}}$ by identifying each R_i with S_i. Thus the singular locus of $\mathcal{X}$ is the image of the sections $R_1, \ldots, R_n, S_1, \ldots, S_n$ in $\mathcal{X}$.

The valuation ν on K induces a topology on the K-rational points of the generic fiber X which we will call the *ν-adic topology*. It is characterized by being the weakest topology satisfying the following.

1. The K-rational points of any Zariski open subset of X forms an open set in the ν-adic topology.
2. Let W be an affine Zariski open subset of X, and let A be its coordinate ring. All the elements of A, considered as functions from the K-rational points of W to the field K, are continuous functions with respect to the ν-adic topology, where K is given the topology induced by the absolute value $|\,|_\nu$.

In the present situation we can think of an invertible meromorphic function on $\mathcal{X}$ (or on the generic fiber X) as a section of the sheaf $\mathcal{O}_{\mathcal{X}}$ (or $\mathcal{O}_X$) defined and nowhere zero on a Zariski dense open subscheme U (where U is maximal with this property). The set of invertible meromorphic functions form a group under multiplication. Note that the group of invertible meromorphic functions on $\mathcal{X}$ is canonically isomorphic to the group of invertible meromorphic functions on X. More generally, given a line bundle L on $\mathcal{X}$ (or on X), an invertible meromorphic section of L is a section of L defined and nowhere zero on a (Zariski) dense open subscheme U

(where U is maximal with this property). The set of invertible meromorphic sections of a line bundle is a principle homogeneous space for the group of invertible meromorphic functions. Note that, if L is a line bundle on $\mathcal{X}$, the set of invertible meromorphic sections of L can naturally be identified with the set of invertible meromorphic sections of $L|_X$. Also note that any invertible meromorphic section l_0 gives rise to a Cartier divisor (l_0), and any Cartier divisor gives rise to a Weil divisor.

Let Z be the set of all closed points of the generic fiber X whose Zariski closure in $\mathcal{X}$ intersects the Zariski closure in $\mathcal{X}$ of the singular locus of X. Let L be a line bundle on X. A *Z-regular meromorphic section* of L is defined to be an invertible meromorphic section of L which is defined and nowhere vanishing on a (Zariski) dense open subscheme of X which contains Z. It can be shown that, if L arises as the restriction of a line bundle on $\mathcal{X}$ to the generic fiber X, then the set of Z-regular meromorphic sections of L is not empty. A *Z-regular meromorphic function* is a Z-regular meromorphic section of $\mathcal{O}_X$.

(1.4) The Intersection Function. Let D and E be two Cartier divisors on X whose supports are disjoint and do not intersect Z. This implies that their closures, $\overline{D}$ and $\overline{E}$ in $\mathcal{X}$ lie in the regular locus of $\mathcal{X}$. So the intersection multiplicity $g(D, E)$ of $\overline{D}$ and $\overline{E}$ in $\mathcal{X}$ is well-defined.

We call $g(D, E)$ the *intersection function* of X induced by $\mathcal{X}$. The following proposition follows directly from the above definition.

(1.4.1) Proposition.

1. *The intersection function g is symmetric:*

$$g(D,E) = g(E,D).$$

2. *The intersection function g is bi-additive, i.e.,*

$$g(D_1+D_2,E) = g(D_1,E)+g(D_2,E) \qquad \text{and} \qquad g(E,D_1+D_2) = g(E,D_1)+g(E,D_2)$$

for all divisors D_1, D_2 and E of X whose supports are disjoint from Z, and where the supports of D_1 and D_2 are disjoint from the support of E.

If we restrict g to prime divisors which are K-rational points, we get a pairing on disjoint K-rational points of X lying outside of Z. The following proposition shows that this pairing is continuous in the ν-adic topology.

(1.4.2) Proposition. *Let Y be the set of K-rational points of $X - Z$. The intersection function g, restricted to the points of Y, is defined and continuous in the ν-adic topology on all of $Y \times Y - \Delta$ where Δ is the diagonal.*

The asymptotic behavior of g near the diagonal can be described as follows. Let U be a subset of Y which is open in the ν-adic topology, and let z be a local coordinate on U, i.e., an invertible meromorphic function z which is defined on all of U, such that, for any p in U, the function $z - z(p)$ has a simple zero at p and no other zeros in U. Note that Y can be covered by such open subsets. For any such U and z there is a continuous function h defined on all of $U \times U$ such that

$$g(p,q) \;=\; \nu\big(z(p) - z(q)\big) + h(p,q)$$

for p and q distinct points in U.

Proof. Let p_0 be a K-rational point of $X - Z$. Since p_0 is not in Z, the closure $\overline{p_0}$ of p_0 in $\mathcal{X}$ lies in the regular locus of X. Hence, $\overline{p_0}$ is an effective Cartier divisor in $\mathcal{X}$. So there is an affine open neighborhood W of $\overline{p_0}$ and a regular function f on W whose associated Cartier divisor in W is $\overline{p_0}$. Let U_0 be the subset of Y, open in the ν-adic topology, consisting of all the K-rational points p of $W|_X$ such that $|f(p)|_\nu < 1$. In particular, (1) p_0 is in U_0, and (2) any other point p of Y is in U_0 if and only if $g(p, p_0) > 0$.

Now let p be a point of U_0. Its closure $\overline{p}$ in $\mathcal{X}$ is an effective Cartier divisor contained in W. In fact, there is a Zariski open subscheme W_p of W containing both $\overline{p}_0$ and $\overline{p}$ such that the Cartier divisor associated to $f - f(p)$ in W_p is exactly $\overline{p}$. In particular, the set of K-rational points of $W_p|_X$ contains U_0.

If q is any other point of U_0, its closure in $\mathcal{X}$ must be contained in W_p. Hence, the intersection multiplicity between the closures of q and p in $\mathcal{X}$ can be determined by restricting $f - f(p)$ to the closure of q. In other words,

$$g(p,q) = \overline{p} \cdot \overline{q} = \nu\big(f(p) - f(q)\big).$$

Now, if q is a point outside of U_0, its closure in $\mathcal{X}$ does not intersect the closure of p_0 in $\mathcal{X}$, and so does not intersect the closure of p in $\mathcal{X}$. So, in this case,

$$g(p,q) = 0.$$

Since U_0 is both open and closed in Y, the above shows that g is continuous for all p in U_0 and for all q in Y not equal to p. The set U_0 is a neighborhood of p_0, and p_0 is arbitrary, so g is continuous on all of $Y \times Y - \Delta$.

Now let U be a subset of U_0, open in the ν-adic topology, and let z be a local parameter of U in the sense mentioned in the statement of the proposition. In particular, the differentials $\mathrm{d}f$ and $\mathrm{d}z$ are both defined on all of U, and neither have any zeros in U. This, together with the fact that for any point p of U both $z - z(p)$ and $f - f(p)$ both have a simple zero at p and no other zeros, implies that the function

$$\frac{f(p) - f(q)}{z(p) - z(q)}$$

can be extended to a continuous nowhere zero function on all of $U \times U$. So there is a continuous function h, defined on all of $U \times U$, such that, for p and q disjoint points on U,

$$\begin{aligned} h(p,q) &= \nu\left(\frac{f(p) - f(q)}{z(p) - z(q)}\right) \\ &= \nu\big(f(p) - f(q)\big) - \nu\big(z(p) - z(q)\big) \\ &= g(p,q) - \nu\big(z(p) - z(q)\big). \end{aligned}$$

So the proposition is true, at least if U is contained in the neighborhood U_0 of p_0. Since p_0 is arbitrary, the proposition is true for general U.

(1.5) Divisors. Let D be a Weil divisor of the generic fiber X; hence, we can express D as a finite integral combination of prime divisors on X:

$$D = \sum_i n_i P_i$$

We define $\overline{D}$, the closure of D in $\mathcal{X}$, to be the following Weil divisor on $\mathcal{X}$:

$$\overline{D} \stackrel{\text{def}}{=} \sum_i n_i \overline{P_i}$$

where $\overline{P_i}$ is the closure in $\mathcal{X}$ of the codimension 1 subscheme P_i of X. I leave the proof of the following to the reader.

(1.5.1) Lemma. *Let G_1 be the group of Cartier divisors on $\mathcal{X}$ whose restriction to the the generic fiber X has support disjoint from Z. Let G_2 be the group of Weil divisors on $\mathcal{X}$ of the form*

$$\overline{D} + n \sum_{i=1}^{m} C_i$$

where D is a Weil divisor of $X - Z$, n is an integer, and $C_1, \ldots, C_m$ are the components of the special fiber X_k considered as prime Weil divisors on $\mathcal{X}$. Then the natural homomorphism from the group of Cartier divisors on $\mathcal{X}$ to the group of Weil divisors on $\mathcal{X}$ sends G_1 into G_2. Moreover, the restriction of this homomorphism to G_1 is an isomorphism between G_1 and G_2.

(1.6) Line Bundles and Order Functions. Let L be a line bundle on $\mathcal{X}$, and let $L|_X$ be its restriction to the generic fiber X. The *order function on $L|_X$ associated to L*, written Φ_L, is a real-valued function on the (non-empty) set of Z-regular meromorphic section of $L|_X$ defined as follows. Let l be a Z-regular meromorphic section of $L|_X$. We can regard l also as an invertible meromorphic section of L, and so we have the associated Cartier divisor (l) on $\mathcal{X}$. The above lemma implies that the Weil divisor associated to (l) is of the form

$$\overline{D} + n\,[X_k]$$

where D is a Weil divisor of X, n is an integer, and $[X_k]$ is the Weil divisor associated with the special fiber X_k. We define $\Phi_L(l)$ to be the integer n. Note that if L and M are two line bundles on $\mathcal{X}$, then the order function $\Phi_{L\otimes M}$ on the line bundle $L|_X \otimes M|_X = (L \otimes M)|_X$ satisfies the formula

$$\Phi_{L\otimes M}(l \otimes m) = \Phi_L(l) + \Phi_M(m) \tag{1}$$

where l is any Z-regular meromorphic section of $L|_X$, and m is any Z-regular meromorphic section of $M|_X$.

The *order function Φ on X* is defined to be the order function on $\mathcal{O}_X$ associated to $\mathcal{O}_{\mathcal{X}}$. In other words, we identify $\mathcal{O}_X$ with $\mathcal{O}_{\mathcal{X}}|_X$, and define Φ to be $\Phi_{\mathcal{O}_{\mathcal{X}}}$. Equation (1) above implies that the order function Φ is a homomorphism from multiplicative group of Z-regular meromorphic functions of X to the additive group of real numbers. Also note that, for any non-zero element c of K, we have that $\Phi(c) = \nu(c)$. Equation (1) also implies that, if L is a line bundle on $\mathcal{X}$, l is a Z-regular meromorphic section of the restriction $L|_X$, and f is a Z-regular meromorphic function of X, then

$$\Phi_L(fl) = \Phi(f) + \Phi_L(l). \tag{2}$$

Let D be a Weil divisor of X with support disjoint from Z. The *order function Φ_D associated with D* is defined to be the order function on $\mathcal{O}_X(D)$ associated to the line bundle $\mathcal{O}_{\mathcal{X}}(\overline{D})$. Equation (2) above implies that if f is a Z-regular meromorphic function on X, then

$$\Phi_D(f\,\mathbf{1}_D) \;=\; \Phi(f) \tag{3}$$

where $\mathbf{1}_D$ is the canonical meromorphic section of $\mathcal{O}_X(D)$ whose divisor is D.

Let L be a line bundle on the generic fiber X which has Z-regular meromorphic sections; for example, the restriction of a line bundle on $\mathcal{X}$ to the generic fiber X always has Z-regular meromorphic sections. An *abstract order function on L* is an integer-valued function Φ_L on the set of Z-regular meromorphic sections of L which satisfies the equation

$$\Phi_L(f\,l) = \Phi(f) + \Phi_L(l)$$

for any Z-regular meromorphic section l of L and any Z-regular meromorphic function f.

Let $\mathcal{D}$ be the category whose objects are pairs (L, Φ_L) where L is a line bundle on X which has Z-regular meromorphic sections, and where Φ_L is an abstract order function on L. Given two such objects (L, Φ_L) and (M, Φ_M) of $\mathcal{D}$, a morphism between these two objects is defined to be an isomorphism between L and L such that the pull-back of Φ_M under this isomorphism is equal to Φ_L.

Let $\mathcal{C}$ be the category whose objects are line bundles of $\mathcal{X}$ and whose morphisms are isomorphisms between line bundles.

(1.6.1) Proposition. *The functor F from $\mathcal{C}$ to $\mathcal{D}$ which takes a line bundle L on $\mathcal{X}$ to the pair $(L|_X, \Phi_L)$ is an equivalence of categories.*

Proof. In order to define an inverse functor we have to choose, for each pair (L, Φ_L) in $\mathcal{D}$, a Z-regular meromorphic section l. By replacing l by $c\,l$ for a suitable element c of K we can assume that $\Phi_L(l) = 0$. The rule $l \mapsto \mathbf{1}_D$ defines an isomorphism between (L, Φ_L) and $(\mathcal{O}_X(D), \Phi_D)$ where D is (l), the Cartier divisor on X associated with l. We define a functor G by sending the pair (L, Φ_L) to the line bundle $\mathcal{O}_{\mathcal{X}}(\overline{D})$. We leave it to the reader to fill in the details and check that the functors $F \circ G$ and $G \circ F$ are naturally equivalent to the respective identity functors.

(1.6.2) Proposition. *Let f be a Z-regular meromorphic function on the generic fiber X, and let p be a K-rational point of X not in Z nor in the support of the divisor (f). Then*

$$\nu\big(f(p)\big) \;=\; g\big((f), p\big) \;+\; \Phi(f).$$

Proof. To see this, consider f not only as a meromorphic function of X, but also as a meromorphic function of $\mathcal{X}$. Let E' be the Weil divisor in $\mathcal{X}$ associated with f when regarded as a meromorphic function of $\mathcal{X}$, and let E be the divisor in X associated with f when regarded as a meromorphic function of X. We have the relationship

$$E' \;=\; \overline{E} \;+\; \Phi(f)\,[X_k]$$

where $[X_k]$ is the Weil divisor associated with the special fiber X_k. So the intersection number of E' with $\overline{p}$ is $g(E, p) + \Phi(f)$. On the other hand, this intersection number represents the order vanishing of $f\,|\,\overline{p}$ at the closed point of $\overline{p}$, but this is just $\nu\big(f(p)\big)$.

One corollary of this proposition is that Φ is completely determined by the intersection function g, at least if there exists suitable K-rational points. (Note: even if no such K-rational points exist, a similar argument applied to divisors instead of K-rational points shows that Φ is completely determined by g.)

(1.7) The Dualizing Sheaf. One important property of the intersection function is its behavior with respect to the relative dualizing sheaf. To develop this relationship we will need to establish some notation. Let $\mathcal{K}_{\mathcal{X}/R}$ be the relative dualizing sheaf of $\mathcal{X}$ over *Spec(R)*. The singularities of $\mathcal{X}$ are mild enough to insure that $\mathcal{K}_{\mathcal{X}/R}$ exists as a line bundle on $\mathcal{X}$. Furthermore, the restriction of the relative dualizing sheaf to the smooth locus U of $\mathcal{X}$ over R is canonically isomorphic to the sheaf of relative differentials of U over R. In addition, the restriction of $\mathcal{K}_{\mathcal{X}/R}$ to the generic fiber X is canonically isomorphic to the dualizing sheaf $\mathcal{K}_X$ of X.

Let p be a K-valued point of X not in Z, and let $\overline{p}$ be its Zariski closure in $\mathcal{X}$. We have the standard residue map which assigns to any local differential ω with a simple pole at p its residue $res_p(\omega)$. The residue map can be regarded as a K-vector space isomorphism from the space $\mathcal{K}_X(p)\,|\,p$ to K itself. We can regard the sections of $\mathcal{K}_{\mathcal{X}/R}(\overline{p})\,|\,\overline{p}$ as an R-submodule of $\mathcal{K}_X(p)\,|\,p$, and we can restrict the residue isomorphism to this R-submodule. This results in an R-module isomorphism from $\mathcal{K}_{\mathcal{X}/R}(\overline{p})\,|\,\overline{p}$ to R itself.

Since the dualizing sheaf $\mathcal{K}_X$ can be identified with $\mathcal{K}_{\mathcal{X}/R}\,|_X$, there is a canonically determined order function Ψ on $\mathcal{K}_X$, namely the order function $\Psi \stackrel{\text{def}}{=} \Phi_{\mathcal{K}_{\mathcal{X}/R}}$ associated to $\mathcal{K}_{\mathcal{X}/R}$. Similarly, there is a canonical order function Ψ_p on $\mathcal{K}_X(p)$, namely the order function associated with $\mathcal{K}_{\mathcal{X}/R}(\overline{p})$.

(1.7.1) Proposition. *Let p be a K-valued point of X not in Z, and let ω be a Z-regular meromorphic section of $\mathcal{K}_X(p)$ which is defined and not zero at the point p. If κ is the divisor on X associated to ω, then*

$$\nu(\mathrm{res}_p(\omega)) \;=\; g(\kappa, p) + \Psi_p(\omega).$$

Proof. We can regard ω as both a Z-regular meromorphic section of $\mathcal{K}_X(p)$ and as an invertible meromorphic section of $\mathcal{K}_{\mathcal{X}/R}(\overline{p})$. Let κ be the divisor in X of ω when regarded as a meromorphic section of $\mathcal{K}_X(p)$, and let κ' be the Weil divisor in $\mathcal{X}$ of ω when regarded as a meromorphic section of $\mathcal{K}_{\mathcal{X}/R}(\overline{p})$. So, as Weil divisors,

$$\kappa' \;=\; \overline{\kappa} \,+\, \Psi_p(\omega)\,[X_k]$$

where $[X_k]$ is the Weil divisor in X associated with the special fiber X_k. Thus, the intersection number of $\overline{p}$ with κ' is

$$g(\kappa, p) \;+\; \Psi_p(\omega),$$

but this intersection number is exactly the order vanishing at the closed point of $\overline{p}$ of $\omega\,|\,\overline{p}$ regarded as a meromorphic section of $\mathcal{K}_{\mathcal{X}/R}(\overline{p})\,|\,\overline{p}$. By the residue isomorphism this is just $\nu(res_p(\omega))$.

(1.7.2) Corollary (The Residue Formula). *Let ω be a Z-regular meromorphic section of $\mathcal{K}_X$ with a simple pole at a K-rational point p of $X - Z$. If κ is the divisor on X associated to ω, then*

$$\nu(\mathrm{res}_p(\omega)) \;=\; g(\kappa + p, p) + \Psi(\omega).$$

Proof. We can regard ω as both a Z-regular meromorphic section of $\mathcal{K}_X$ and as a Z-regular meromorphic section of $\mathcal{K}_X(p)$. In the former case ω has divisor κ and in the latter case $\kappa + p$. The result follows from the observation that $\Psi_p(\omega) = \Psi(\omega)$.

(1.8) The Intersection Pairing. Let $\mathcal{L}$ and $\mathcal{M}$ be two line-bundles on $\mathcal{X}$. By Proposition 1.6.1, we can think of $\mathcal{L}$ as $(L, \Phi_{\mathcal{L}})$ and $\mathcal{M}$ as $(M, \Phi_{\mathcal{M}})$ where L and M are the respective restriction of $\mathcal{L}$ and $\mathcal{M}$ to the generic fiber X, and where $\Phi_{\mathcal{L}}$ and $\Phi_{\mathcal{M}}$ are the order functions associated to $\mathcal{L}$ and $\mathcal{M}$ respectively.

Consider the intersection pairing $\langle L, M\rangle$ of L and M (defined in Section 4 of Chapter 1). The pairing $\langle L, M\rangle$ is a one-dimensional K-vector space. As in Proposition 4.3 of Chapter 2 we can identify $\langle L, M\rangle$ with the restriction of $\langle \mathcal{L}, \mathcal{M}\rangle$ to the generic point of $\operatorname{Spec} R$. So the remarks of (1.2) show how we can think of $\mathcal{I} = \langle \mathcal{L}, \mathcal{M}\rangle$ as $\langle L, M\rangle$ together with a norm $|\,|_{\mathcal{I}}$ on $\langle L, M\rangle$.

(1.8.1) Proposition. *In the above situation we have the following formula for* $|\,|_{\mathcal{I}}$. *Let l and m be Z-regular meromorphic sections of L and M whose associated divisors (l) and (m) in X are disjoint. Then the non-zero vector $\langle l, m\rangle$ of $\langle L, M\rangle$ has norm given by the following formula:*

$$-\log_{\gamma}\big|\langle l, m\rangle\big|_{\mathcal{I}} = g\big((l),(m)\big) + \deg L\cdot \Phi_{\mathcal{M}}(m) + \deg M\cdot \Phi_{\mathcal{L}}(l).$$

Proof. We can always find meromorphic sections l_0 and m_0 of $\mathcal{L}$ and $\mathcal{M}$ respectively whose associated Cartier divisors in $\mathcal{X}$ have supports which are disjoint from each other, disjoint from the singular locus of $\mathcal{X}$, and do not contain any components of the special fiber X_k. This implies that if we consider l_0 and m_0 as meromorphic sections of L and M respectively, then $\Phi_{\mathcal{L}}(l_0) = \Phi_{\mathcal{M}}(m_0) = 0$, and, if we consider their associated divisors (l_0) and (m_0) in X, then $g\big((l_0),(m_0)\big) = 0$. By Proposition 4.3 of Chapter 2, the section $\langle l_0, m_0\rangle$ generates the line bundle $\langle \mathcal{L}, \mathcal{M}\rangle$, so, thinking of $\langle l_0, m_0\rangle$ as a vector of $\langle L, M\rangle$,

$$-\log_{\gamma}\big|\langle l, m\rangle\big|_{\mathcal{I}} = 0.$$

This implies that the proposition is trivially true for the pair l_0, m_0.

Now suppose the proposition is true for a particular pair l and m of Z-regular meromorphic sections of L and M respectively with divisors (l) and (m) which are disjoint. Let $l' = fl$ where f is a Z-regular meromorphic function on X whose associated divisor (f) is disjoint from (m). Then, by Proposition 6.7 of Chapter 1, $\langle l', m\rangle = f\big((m)\big)\langle l, m\rangle$ where $f\big((m)\big)$ is as in Definition 6.3 of Chapter 1. Thus

$$\begin{aligned}
-\log_{\gamma}\big|\langle l', m\rangle\big|_{\mathcal{I}} &= \nu\Big(f\big((m)\big)\Big) - \log_{\gamma}\big|\langle l', m\rangle\big|_{\mathcal{I}}\\
&= \nu\Big(f\big((m)\big)\Big) + g\big((l),(m)\big) + \deg L\cdot \Phi_{\mathcal{M}}(m) + \deg M\cdot \Phi_{\mathcal{L}}(l)
\end{aligned}$$

However, by a minor generalization of Proposition 1.6.2,

$$\nu\Big(f\big((m)\big)\Big) = g\big((f),(m)\big) + \deg M\cdot \Phi(f).$$

So, by Proposition 1.4.1 and equation (2) in (1.6),

$$\begin{aligned}
-\log_{\gamma}\big|\langle l', m\rangle\big|_{\mathcal{I}} &= g\big((f),(m)\big) + g\big((l),(m)\big) + \deg L\cdot \Phi_{\mathcal{M}}(m)\\
&\qquad + \deg M\cdot \Phi(f) + \deg M\cdot \Phi_{\mathcal{L}}(l)\\
&= g\big((l'),(m)\big) + \deg L\cdot \Phi_{\mathcal{M}}(m) + \deg M\cdot \Phi_{\mathcal{L}}(l')
\end{aligned}$$

Therefore, the proposition is true of the pair l' and m.

In a similar way, we could show that if the proposition is true of a pair l and m then it is true when we replace m by any suitable m'. We can then conclude that the proposition is true for all suitable pairs l, m.

(1.9) The Induced Metric on a Line Bundle. Let $\mathcal{L}$ be a line-bundle on $\mathcal{X}$. By Proposition 1.6.1 we can think of $\mathcal{L}$ as $(L, \Phi_{\mathcal{L}})$ where L is the restriction of $\mathcal{L}$ to the generic fiber X and where $\Phi_{\mathcal{L}}$ is the order function of L associated to $\mathcal{L}$. Let p be any K-rational point of X outside of Z. By (6.3) of Chapter 1, there is a canonical isomorphism between the intersection pairing $\langle L, \mathcal{O}_X(p)\rangle$ and the one-dimensional K-vector space $L|_p$ (using the basic fact that the norm space $\mathbf{N}_p(L)$ is canonically isomorphic to $L|_p$). Therefore, the norm on $\langle L, \mathcal{O}_X(p)\rangle$, discussed in (1.8) above, induces a norm on $L|_p$; call this norm $|\,|_{\mathcal{L},p}$. We can get simple formula for this norm as follows: let l be a Z-regular merormorpic section of L which is defined and not zero at p. Then Proposition 1.8.1 together with the fact that $\Phi_p(\mathbf{1}_p) = 0$ (see equation (3) of (1.6)) implies that

$$-\log_\gamma \Big| l|_p \Big|_{\mathcal{L},p} = g\big((l), p\big) + \Phi_{\mathcal{L}}(l).$$

The continuity of g implies that this defines a continuous family of norms (i.e., a continuous metric) for L on the set of K-rational points of $X - Z$. We call this the *metric of L induced by $\mathcal{L}$*. Note that the right-hand side of the above equation represents the intersection of the divisor of l, considered as a meromorphic section of $\mathcal{X}$, with the closure of p in $\mathcal{X}$.

2. The Archimedean Case: Basic Definitions

We will now develop a theory for the archimedean case analogous to the non-archimedean situation considered in the previous section. In the previous section we considered a (possibly singular) curve $\mathcal{X}$ defined over a discrete valuation ring. The results of the previous section indicate that most of the structure of $\mathcal{X}$ (or at least the parts we will be interested in) can be recovered by knowing only (a) the generic fiber X, and (b) the intersection function on X induced by $\mathcal{X}$. This suggests that the archimedean analogue of such an $\mathcal{X}$ is a (possibly singular) complex projective curve, representing the generic fiber, and an intersection function defined on this curve.

In what follows let X be a projective complex curve, connected but not necessarily irreducible, and let Z be the set of singular points of X. We assume that all the points of Z are ordinary double points. This set Z is supposed to be analogous to the set Z defined in the previous section for the non-archimedean case (1.3). In the non-archimedean case the set Z consists of a set of points of X which are ν-adically close to the singular set of Z. We will adopt the stance that in the archimedean case Z should *only* contain the singular points of X (there is no other natural alternative).

A *Z-regular meromorphic function* on X is a meromorphic function on X which is defined and non-zero on a Zariski dense open set containing Z. More generally, if L is a line bundle on X, a *Z-regular meromorphic section of L* is a meromorphic section of L which is defined and nowhere zero on a Zariski dense open subset of X containing Z.

(2.1) The Definition of an Intersection Function.

Let g be a real-valued function defined on pairs (D, E) where D and E are divisors of $X - Z$ with disjoint supports. The function g is said to be an *intersection function* if it satisfies the following five properties:

I. The function g is symmetric:

$$g(D,E) = g(E,D)$$

where D and E are divisors of $X - Z$ with disjoint supports.

II. The function g is bi-additive:

$$g(D_1+D_2,E) = g(D_1,E)+g(D_2,E) \qquad \text{and} \qquad g(E,D_1+D_2) = g(E,D_1)+g(E,D_2)$$

where D_1, D_2 and E are divisors with supports disjoint from Z, and where D_1 and D_2 have supports disjoint from that of E. This implies that g is completely determined by its values on disjoint pairs of point divisors of $X - Z$. So we will often think of $g(p,q)$ as a function on $X \times X$, defined when p and q are distinct and not contained in Z.

III. The function $g(p,q)$, defined for all pairs of distinct points of X not contained in Z, is a continuous function. Its asymptotic behavior near the diagonal is as follows: if U is a neighborhood of X disjoint from Z and parameterized by a local coordinate z, then there is a continuous function h defined on all of $U \times U$ (even when $p = q$) such that

$$g(p,q) = -\log\big|z(p) - z(q)\big| + h(p,q)$$

for p and q distinct points of U.

IV. Let f be any Z-regular meromorphic function of X. Then the quantity

$$\Phi(f) \stackrel{\text{def}}{=} -\log\big|f(p)\big| - g\big((f),p\big),$$

defined for all non-singular points p outside the support of (f), is a constant independent of p. We call Φ the *order function* on X associated to g.

V. Let ω be a Z-regular meromorphic section of the dualizing sheaf $\mathcal{K}_X$, and let κ be the divisor of ω. Let U be a neighborhood of X, disjoint from Z and the support of κ, which is parameterized by a local coordinate z. On U we can write $\omega = \psi\, dz$ for some nowhere-zero holomorphic function ψ on U. Then the quantity

$$\Psi(\omega) \stackrel{\text{def}}{=} -\Big(g(\kappa,p) + \log\big|\psi(p)\big|\Big) - \lim_{q\to p}\Big(g(p,q) + \log\big|z(p) - z(q)\big|\Big)$$

is a constant depending only on ω (i.e. it is independent of p, U, and z).

Remarks The Archimedean analog to the discrete valuation ν of the previous section is the function

$$x \;\mapsto\; -\log|x|$$

defined for all non-zero complex numbers. Note that this function takes on all real values, not just integer values. Similarly, in order for the desired continuity properties to hold, an intersection function on the complex curve X must be a real-valued, not an integer-valued, function.

That Properties I and II should hold follows from Proposition 1.4.1. Property III is the analogy of Proposition 1.4.2. Note that in Proposition 1.4.2, the function z is required to be meromorphic on X; however, it is easy to see that in the current situation Property III for meromorphic z is equivalent to Property III for any local analytic coordinate. Property IV is the analog of Proposition 1.6.2. Note that Property IV together with Property II implies that the function sending f to $\Phi(f)$ is a homomorphism from the multiplicative group of Z-regular meromorphic functions to the additive group of real numbers. Property V is basically analogous to the residue formula (Corollary 1.7.2); we will demonstrate this below in Proposition 2.3.1.

(2.2) Admissible Line Bundles. Now suppose that X has been endowed with a fixed intersection function g, and let Φ be as defined in Propery IV of the above definition. The discussion in (1.6), and especially Proposition 1.6.1, of the previous section suggests that we should not consider just line bundles on X, but pairs (L, Φ_L) where L is a line bundle on X and where Φ_L is an order function on L compatable with Φ, i.e., Φ_L is a real-valued function on the set of Z-regular sections of L which satisfies

$$\Phi_L(fl) = \Phi(f) + \Phi_L(l)$$

for all Z-regular meromorphic sections l of L and all Z-regular meromorphic functions f. In particular, Φ_L is completely determined by its value on a single Z-regular meromorphic section of L. We call such pairs (L, Φ_L) *g-admissible line bundles.*

The discussion in (1.9) of the previous section suggests that, for any g-admissible line bundle (L, Φ_L), we define a metric (i.e., a Hermitian structure) ρ on L by the formula

$$-\log \Big| l|_p \Big|_{\rho,p} \overset{\text{def}}{=} g((l), p) + \Phi_L(l)$$

where l is any Z-regular meromorphic section of L. Note that this defines a Hermitian structure for L only on $X - Z$.

(2.3) The Dualizing Sheaf. Let $\mathcal{K}_X$ be the dualizing sheaf on X. Since all the singularities of X are ordinary double points, $\mathcal{K}_X$ exists as an invertible sheaf (i.e., a line bundle) on X. Furthermore, outside of the singular set Z the sheaf $\mathcal{K}_X$ is canonically isomorphic to the sheaf of holomorphic differentials. The discussion in (1.7) of the previous section suggests that $\mathcal{K}_X$ should have a canonical order function. It is easy to see that the function Ψ defined by Property V is an order function on $\mathcal{K}_X$. The following proposition will prove that Ψ is only order function of $\mathcal{K}_X$ which satisfies the residue formula, and hence is the only reasonable choice for a canonical order function for $\mathcal{K}_X$. We call the pair $(\mathcal{K}_X, \Psi)$ the *canonical g-admissible dualizing sheaf.* The following proposition will also show that, in fact, Property V of the definition of the intersection function is equivalent to the residue formula.

(2.3.1) Proposition. *Let g be a function which satisfies all the properties of an intersection function on X except possibly Property V. In particular, the notion of an order function on a line bundle can be defined with g. Suppose the dualizing sheaf $\mathcal{K}_X$ has an order function Ψ satisfying the residue formula:*

$$-\log |\operatorname{res}_p(\omega)| = g(\kappa + p, p) + \Psi(\omega)$$

where ω is any Z-regular meromorphic section of $\mathcal{K}_X$ with a simple pole at a point p, and where κ is the divisor of ω. Then g must satisfy Property V and Ψ must be equal to the quantity defined in Property V.

Conversely, if g is an intersection function on X and so satisfies Property V above, and if the dualizing sheaf $\mathcal{K}_X$ is given the order function Ψ defined in Property V, then the residue formula is satisfied.

Proof. Assume that the dualizing sheaf $\mathcal{K}_X$ has an order function Ψ which satisfies the residue formula. Let ω be a Z-regular meromorphic section of the dualizing sheaf $\mathcal{K}_X$, let κ be the divisor of ω, and let p be a point of X outside Z and the support of κ. Choose a Z-regular meromorphic function z with a simple zero at z, and let U be a small enough neighborhood of p (open in the standard topology) so that, for each q in U, the function $z - z(q)$ is a Z-regular meromorphic function which has a simple zero at q, and is defined and non-zero at p. In particular, z is a local coordinate of U. Also assume that U is small enough so that it doesn't intersect Z or the support of κ.

On U we can write $\omega = \psi\, dz$ for some nowhere-zero holomorphic function ψ on U. Thus, for any point q in U, the differential $\omega/\big(z - z(q)\big)$ has residue $\psi(q)$, and so the residue formula takes the form

$$-\log\big|\psi(q)\big| \;=\; g\big(\kappa - D(q), q\big) \;+\; \Psi\left(\frac{\omega}{z - z(q)}\right)$$

where $D(q)$ is a divisor defined by the property that $z - z(q)$ has divisor $q + D(q)$. So

$$\begin{aligned}
\Psi(\omega) &= \Psi\left(\frac{\omega}{z - z(q)}\right) + \Phi\big(z - z(q)\big) \\
&= -\log\big|\psi(q)\big| \;-\; g\big(\kappa - D(q), q\big) \;-\; \log\big|z(p) - z(q)\big| \;-\; g\big(q + D(q), p\big) \\
&= -\log\big|\psi(q)\big| \;-\; g(\kappa, q) \;-\; \log\big|z(p) - z(q)\big| \;-\; g(q, p) \;+\; g\big(D(q), q - p\big).
\end{aligned}$$

By the continuity of g,

$$\Psi(\omega) \;=\; -\Big(g(\kappa, p) + \log\big|\psi(p)\big|\Big) - \lim_{q \to p}\Big(g(p, q) + \log\big|z(p) - z(q)\big|\Big).$$

This forces the right hand side of the equation to be independent of the choice of p, z, and U. Also the fact that this formula is true for any choice z of a Z-regular meromorphic function with a simple zero at p implies the truth of the formula for any local holomorphic coordinate z parameterizing a neighborhood of p. Therefore, we can dispense with the assumption that z be a Z-regular meromorphic function on X, and we can assume z is any local holomorphic coordinate of a neighborhood of p.

This shows that g satisfies Property V, and Ψ is what it is supposed to be. The converse can be shown by a similar argument. I leave this to the reader.

(2.4) Tensor Products. Let (L, Φ_L) and (M, Φ_M) be two g-admissible line bundles on X. By analogy with (1.6) we define the *tensor product* of these two g-admissible line bundle to be $(L \otimes M, \Phi_{L\otimes M})$ where $\Phi_{L\otimes M}$ is the order function which sends $l \otimes m$ to $\Phi_L(l) + \Phi_M(m)$ for any pair l, m of Z-regular meromorphic sections of L and M respectively.

Let (L, Φ_L) be a g-admissible line bundle of X. We define the *inverse of* (L, Φ_L) to be the g-admissible line bundle $(L^{-1}, \Phi_{L^{-1}})$ where $\Phi_{L^{-1}}$ is the order function which sends l^{-1} to $-\Phi_L(l)$ for any Z-regular meromorphic section l of L.

(2.5) Divisors. Let D be a divisor of X with support disjoint from Z. The *g-admissible line bundle on X associated to D* is defined to be the pair $(\mathcal{O}_X(D), \Phi_D)$ where Φ_D is defined, by analogy with formula (3) of (1.6), to be the order function on $\mathcal{O}_X(D)$ which sends $f\,\mathbf{1}_D$ to $\Phi(f)$. Here $\mathbf{1}_D$ is the canonical meromorphic section on $\mathcal{O}_X(D)$ whose divisor is D and f is any Z-regular meromorphic section of X.

(2.6) The Normed Intersection Pairing. Let (L, Φ_L) and (M, Φ_M) be two g-admissible line bundles on X. By analogy with (1.8) we define a norm on the intersection pairing $\langle L, M\rangle$ of L and M as follows. Let l and m be Z-regular meromorphic sections of L and M whose associated divisors (l) and (m) in X are disjoint. Thus $\langle l, m\rangle$ is a non-zero vector of $\langle L, M\rangle$; we define its norm by the formula

$$-\log\bigl|\langle l, m\rangle\bigr| \stackrel{\text{def}}{=} g\bigl((l), (m)\bigr) + \deg L \cdot \Phi_M(m) + \deg M \cdot \Phi_L(l).$$

The intersection pairing $\langle L, M\rangle$ together with the above norm is called the *normed intersection pairing* of (L, Φ_L) with (M, Φ_M).

Let D and E be two divisor of X whose supports are disjoint from Z and from each other. Note that $-\log\bigl|\langle \mathbf{1}_D, \mathbf{1}_E\rangle\bigr| = g(D, E)$ where the norm on the left is the norm associated to the normed intersection pairing of $(\mathcal{O}_X(D), \Phi_D)$ with $(\mathcal{O}_X(E), \Phi_E)$. This quantity represents the "intersection" between D and E.

(2.6.1) Proposition. *Assume that there is an intersection function g, fixed once and for all, on X.*

1. Let (L, Φ_L) and (M, Φ_M) be g-admissible line bundles on X. Then the canonical isomorphism

$$\Bigl\langle L\,,\,M\Bigr\rangle \to \Bigl\langle M\,,\,L\Bigr\rangle$$

of Proposition 7.3, Chapter 1 is an isometry.

2. Let (L_1, Φ_{L_1}), (L_2, Φ_{L_2}) and (M, Φ_M) be g-admissible line bundles on X. Then the canonical isomorphisms

$$\Bigl\langle L_1 \otimes L_2\,,\,M\Bigr\rangle \to \Bigl\langle L_1\,,\,M\Bigr\rangle \otimes \Bigl\langle L_2\,,\,M\Bigr\rangle$$

and

$$\Bigl\langle M\,,\,L_1 \otimes L_2\Bigr\rangle \to \Bigl\langle M\,,\,L_1\Bigr\rangle \otimes \Bigl\langle M\,,\,L_2\Bigr\rangle$$

of Proposition 7.4, Chapter 1 are isometries.

3. Let p be a point of X outside Z, and let $(\mathcal{K}_X(p), \Psi_p)$ be the tensor product of the canonical g-admissible dualizing sheaf $(\mathcal{K}_X, \Psi)$ with the g-admissible line bundle $(\mathcal{O}_X(p), \Phi_p)$. Then the canonical isomorphism

$$\Bigl\langle \mathcal{K}_X(p)\,,\,\mathcal{O}_X(p)\Bigr\rangle \to N_p(\mathcal{K}_X(p)) = \mathcal{K}_X(p)|_p \to \mathbf{C},$$

where the last map is the residue map, is an isometry. Here $\mathbf{C}$ is given the standard norm.

Proof.

1. This follows from the definition of the norm of the intersection pairing together with Property I of the definition of an intersection function.

2. This follows from the definition of the norm of the intersection pairing, the definition of the tensor product of two g-admissible line bundles, and Property II of the definition of an intersection function.

3. The definition, in (2.4), of the order function of the tensor product, and the definition of Φ_p, in (2.5), together imply that, for all Z-regular meromorphic sections ω of $\mathcal{K}_X$,

$$\Psi(\omega) = \Psi_p(\omega)$$

where on the left ω is regarded as a Z-regular meromorphic section of $\mathcal{K}_X$, and on the right ω is regarded as a Z-regular meromorphic section of $\mathcal{K}_X(p)$.

Let ω be any Z-regular meromorphic section of $\mathcal{K}_X(p)$ defined and not zero at p, and let κ be its divisor. We can also think of ω as a Z-regular meromorphic section of $\mathcal{K}_X$ with with a simple pole at a point p, and with divisor $\kappa - p$. By the residue formula, Proposition 2.3.1,

$$-\log|\operatorname{res}_p(\omega)| = g(\kappa, p) + \Psi(\omega).$$

So, thinking of ω as a Z-regular section of $\mathcal{K}_X(p)$,

$$-\log|\operatorname{res}_p(\omega)| = g(\kappa, p) + \Psi_p(\omega).$$

By the definition of the normed intersection pairing (2.6), the non-zero vector $\langle \omega, \mathbf{1}_p \rangle$ of $\langle \mathcal{K}_X(p), \mathcal{O}_X(p) \rangle$ has norm given by the formula

$$-\log\big|\langle \omega, \mathbf{1}_p \rangle\big| = g\big(\kappa, p\big) + \Psi_p(\omega).$$

Under the canonical isomorphism in question $\langle \omega, \mathbf{1}_p \rangle$ is sent to the complex number $\operatorname{res}_p(\omega)$, and the above equations show they have the same norms in their respective spaces; therefore, the isomorphism in question is an isometry.

3. Intersection Functions on Nonsingular Curves

In this section we review the theory of intersection functions for nonsingular curves, i.e., Arakelov's theory of the canonical Green's function. The treatment presented here is somewhat different than other treatments of Arakelov theory: it shows that the theory of the canonical Green's function follows directly from the theory of the Neron pairing on pairs of disjoint divisors of degree 0. The main technique is the use of intersection functions with non-trivial exceptional sets.

Throughout this section X will be a nonsingular projective curve over the complex numbers $\mathbf{C}$.

(3.1) Definitions. In this section it will be useful to consider various generalizations of the intersection function. A function which satisfies all the properties of an intersection function except possibly property V will be called a *quasi-intersection function.* Recall that in Section 2 we defined the standard exceptional set Z to be the set of singular points of X. In this section it will be useful to allow Z to be other finite sets of points containing the singular points of X. We call such sets *extended exceptional sets.* If Z is an extended exceptional set and g is a function which satisfies all the properties of an intersection function except with respect to this exceptional set Z instead of the standard exceptional set, then we say that g is an *intersection function with exceptional set* Z. We define a *quasi-intersection function with exceptional set* Z in the obvious way. Likewise, we extend the definition of Z-regular meromorphic sections or functions to the case where Z is an extended exceptional set.

(3.2) The Genus 0 Case. First we study intersection functions in the genus 0 case, which behaves differently than the general case. First we will construct examples of, and then classify, intersection functions with extended exceptional sets. Then will will discuss intersection functions proper, i.e., intersection functions with an empty exceptional set.

(3.2.1) An Example in Genus 0. Let X be the projective line $\mathbf{P}^1$ over $\mathbf{C}$ with coordinate z, and let Z be the set $\{\infty\}$. Consider the homomorphism Φ from the the multiplicative group of Z-regular meromorphic functions to the additive group of real numbers defined by sending f to $\Phi(f) \stackrel{\text{def}}{=} -\log\left|f(\infty)\right|$. The map Φ is a good candidate for an order function.

What intersection functions give rise to Φ as an order function? Suppose g is such an intersection function. Let e be the point with coordinate 0, i.e., $z(e) = 0$. For p and q any two distinct points of X not equal to e or ∞, consider the function $f = \big(z - z(p)\big)/z$. This function has divisor $p - e$, and satisfies $\Phi(f) = 0$. So, by property IV of (2.1), we should have

$$g(p-e,q) = -\log\left|f(q)\right| = -\log\left|z(q) - z(p)\right| + \log\left|z(q)\right|;$$

in other words

$$g(p,q) = -\log\left|z(q) - z(p)\right| + \log\left|z(q)\right| + g(e,q).$$

By property I of (2.1), $g(p,q) = g(q,p)$, which, together with the above equation, implies that

$$g(p,q) = -\log\left|z(q) - z(p)\right| + c$$

for some real constant c. By continuity, this formula holds even if either $p = e$ or $q = e$.

This intersection function clearly satisfies properties I to IV of (2.1). To show that g is an intersection function with (extended) exceptional set Z, we only need to show property V.

Let ω be a Z-regular meromorphic section of the canonical sheaf $\mathcal{K}_X$, hence its divisor κ has support disjoint from ∞. Let U be an analytic neighborhood of X, disjoint from ∞ and κ. We can write $\omega = h\,dz/z^2$ for some holomorphic function h on U. So $\kappa = (h) - 2e$. Consider the expression

$$\begin{aligned}
\Psi(\omega, p) &\stackrel{\text{def}}{=} -\Big(g(\kappa, p) + \log\big|h(p)/z(p)^2\big|\Big) - \lim_{q\to p}\Big(g(p,q) + \log\big|z(p) - z(q)\big|\Big)\\
&= -\Big(g\big((h), p\big) - 2g(e,p) + \log\big|h(p)\big| - 2\log\big|z(p)\big|\Big) - c\\
&= -g\big((h), p\big) - \log\big|h(p)\big| - c\\
&= \Phi(h) - c.
\end{aligned}$$

$\Psi(\omega, p)$ is clearly independent of p and U. This implies that Property V of (2.1) is satisfied.

We conclude that the functions of the form

$$g(p,q) = -\log\big|z(q) - z(p)\big| + c$$

are exactly the intersection functions with exceptional set $Z = \{\infty\}$ such that $\Phi(f) = -\log\big|f(\infty)\big|$.

(3.2.2) Genus 0 Case with Extended Exceptional Sets. We just showed that

$$g_0(p,q) = -\log\big|z(q) - z(p)\big|$$

is an intersection function of $X = \mathbf{P}^1$ with (extended) exceptional set $Z_0 = \{\infty\}$. Let g be another intersection function of X with exceptional set Z_1, and consider the difference

$$h(p,q) \stackrel{\text{def}}{=} g(p,q) - g_0(p,q).$$

By Property III of (2.1), the function $h(p,q)$ is continuous and defined for all p and q outside $Z_1 \cup Z_0$, even when $p = q$.

Since all divisors of degree 0 on $X = \mathbf{P}^1$ are principal, Property IV of (2.1) implies that

$$h(D, D') = 0$$

for all divisors D and D' both of degree 0.

Let e be a fixed point outside of Z_1 and Z_0. Then, for any p and q outside Z_1 and Z_0,

$$\begin{aligned}
h(p,q) &= h(p-e, q-e) + h(p,e) + h(q,e) - h(e,e)\\
&= 0 + \Big(h(p,e) - h(e,e)/2\Big) + \Big(h(q,e) - h(e,e)/2\Big)\\
&= \gamma(p) + \gamma(q)
\end{aligned}$$

where γ is a continuous function on X defined outside of Z_1 and Z_0. Therefore,

$$g(p,q) = -\log\big|z(q) - z(p)\big| + \gamma(p) + \gamma(q).$$

Conversely, it is easily shown that any function of the above form, where in general γ is a continuous function defined outside some finite set Z_1, is an intersection function with exceptional set $Z_1 \cup Z_0$. Properties I to IV of (2.3) are easily verified. What is special about the genus 0 case is that remaining property, Property V, is automatically satisfied. This follows from the fact that any divisor of the canonical sheaf $\mathcal{K}_X$ has degree -2.

Thus we have classified all intersection functions (with finite exceptional sets Z) when X is nonsingular of genus 0. Notice that (when g is C^∞) the Chern form

$$\frac{1}{2\pi\mathbf{i}}\partial_p\overline{\partial}_p\big(g(p,q)\big) = \frac{1}{2\pi\mathbf{i}}\partial\overline{\partial}(\gamma)$$

is independent of q.

(3.2.3) The Genus 0 Case (With Empty Exceptional Set). In particular, the following is an intersection function:

$$g(p,q) = -\log\big|z(q)-z(p)\big| + \frac{1}{2}\log\Big(\big|z(p)\big|^2+1\Big) + \frac{1}{2}\log\Big(\big|z(q)\big|^2+1\Big).$$

I will leave it to the reader to show that this function is defined even when one of the arguments is ∞.

This example is interesting: it shows that there is an intersection function which is C^∞ and has an empty exceptional set (i.e., g is is an intersection function in the sense of the last section). In this case the Chern form is

$$-\frac{1}{\pi\mathbf{i}}\partial_p\overline{\partial}_p\big(g(p,q)\big) = -\frac{1}{\pi\mathbf{i}}\partial\overline{\partial}\left(\frac{1}{2}\log\Big(|z|^2+1\Big)\right) = -\frac{1}{2\pi\mathbf{i}}\frac{dz\wedge d\overline{z}}{\big(|z|^2+1\big)^2} = \frac{1}{\pi}\frac{dx\wedge dy}{(x^2+y^2+1)^2}.$$

A basic calculus calculation show that the integral over X of this real $(1,1)$-form is 1.

The discussion of (3.2.2) above implies that g' is an intersection function of X (with empty exceptional set) if and only if it is of the form

$$g'(p,q) = g(p,q) + \gamma(p) + \gamma(q)$$

where γ is a continuous function on all of X and g is as above.

(3.3) The Higher Genus Case. Let X be a nonsingular projective complex curve of positive genus. Then an intersection function exists on X and is unique up to a constant. This follows from Arakelov's theory of the canonical Green's function; in fact, when the canonical Green's function is suitably normalized, it is an intersection function, and all intersection functions differ from the canonical Green's function by a real constant. The order function $\Phi(f)$, defined in Property IV of (2.1), associated with an intersection function on X turns out to be the integral of $-\log|f|$ with respect to a certain positive real $(1,1)$-form called the *canonical* $(1,1)$-*form*.

In this section we will give a self-contained proof of the existence of intersection functions for nonsingular X. The construction here is different than those found in other treatments of Arakelov theory; the main novelty is the construction the intersection pairing via lifting the Neron pairing from divisor of degree zero to general divisors. In order to make this work we have to work for a little while with quasi-intersection functions with non-empty exceptional sets. In addition we will give a self-contained proof that the order function associated with an intersection function is integration with respect to the canonical $(1,1)$-form.

The Neron pairing λ is a pairing which associates to each pair of disjoint divisors D_1, D_2 of degree 0 a real number $\lambda(D_1, D_2)$, and it does so in such a way that

A. The pairing λ is C^∞ in the sense that, for fixed points q_1 and q_2, the function $\lambda(p_1 - q_1, p_2 - q_2)$ is C^∞ as the points p_1 and p_2 vary.
B. The pairing λ is bilinear and symmetric.
C. For any meromorphic function f on X, and divisor D not intersecting (f), $\lambda\big((f), D\big) = -\log\big|f(D)\big|$.

(3.3.1) Lemma. *Let e_1 and e_2 be two distinct points of X, and let Z_0 be any finite subset of X containing at least these two points. Define*

$$g_0(p,q) \stackrel{\text{def}}{=} \frac{1}{2}\lambda(p - e_1, q - e_2) + \frac{1}{2}\lambda(p - e_2, q - e_1).$$

Then the above function g_0, extended to pairs of divisors, is a quasi-intersection function with exceptional set Z_0. Furthermore, $g_0(p,q)$ is C^∞ where it is defined.

Proof. Property II of Definition 2.1 follows formally, and Property B of λ assures that Property I of (2.1) holds. Property A implies the first continuity statement of Property III of (2.1); i.e., the function $g_0(p,q)$ is continuous, actually C^∞, for distinct p and q outside Z_0.

Now

$$\begin{aligned}
-\log\big|f(p)\big| - g_0\big((f),p\big) &= -\log\big|f(p)\big| - \frac{1}{2}\lambda\big((f), p - e_2\big) - \frac{1}{2}\lambda\big((f), p - e_1\big) \\
&= -\log\big|f(p)\big| + \frac{1}{2}\log|f(p - e_2)| + \frac{1}{2}\log|f(p - e_1)| \\
&= -\log\big|f(p)\big| + \log|f(p)| - \frac{1}{2}\log|f(e_1)| - \frac{1}{2}\log|f(e_2)| \\
&= -\frac{1}{2}\Big(\log\big|f(e_1)\big| + \log\big|f(e_2)\big|\Big)
\end{aligned}$$

is independent of p. So we have established property IV of (2.1), where

$$\Phi(f) = -\frac{1}{2}\Big(\log\big|f(e_1)\big| + \log\big|f(e_2)\big|\Big).$$

We will now verify the rest of property III of (2.1). Let p_0 be any point of X outside Z_0. Let f be a Z_0 regular meromorphic function of X with a simple zero at p_0. So f is the coordinate of some analytic neighborhood U of p_0. For any q in U, let $D(q)$ be the divisor $\big(f - f(q)\big) - q$. Take U small enough so that $D(q)$ is outside Z_0 and U for all q in U. Property IV of (2.1), which we have just established, implies that the following is independent of p:

$$\begin{aligned}
\Phi(f - f(q)) &= -\log\big|f(p) - f(q)\big| - g_0\Big(\big(f - f(q)\big), p\Big) \\
&= -\log\big|f(p) - f(q)\big| - g_0(q,p) - g_0\big(D(q), p\big)
\end{aligned}$$

and so

$$\begin{aligned}\log\left|f(p)-f(q)\right|+g_0(q,p) &= -\Phi\big(f-f(q)\big)-g_0\big(D(q),p\big)\\ &= \frac{1}{2}\Big(\log\left|f(e_1)-f(q)\right|+\log\left|f(e_2)-f(q)\right|\Big)-g_0\big(D(q),p\big).\end{aligned}$$

The divisor $D(q)$ varies smoothly as q varies. So the right hand side of this equation is C^∞ as p and q vary, and is defined for all p and q in U, even when $p=q$. We can cover $X-Z_0$ by such U, hence property III of (2.1) follows. Note that we have shown that g_0 satisfies a stronger form of property III, namely the property resulting from replacing "continuous" by "C^∞". In particular g_0 is C^∞ where it is defined.

(3.3.2) Lemma. *Let g_0 and Z_0 be as above. The function g is a quasi-intersection function on X with exceptional set Z_0 if and only if on prime divisors it is of the form*

$$g(p,q)=g_0(p,q)+\gamma(p)+\gamma(q)$$

for some continuous function γ defined outside of Z_0.

Proof. If g is of this form then it is easy to check that it satisfies the various properties of Definition 2.1 with respect to Z_0, except perhaps Property V. Now suppose g satisfies all the defining properties of an intersection function with exceptional set Z_0 except perhaps Property V of (2.1). Consider the difference

$$h(p,q)=g(p,q)-g_0(p,q).$$

Property III of Definition 2.1 implies that $h(p,q)$ is defined and continuous even when $p=q$. Property IV of Defintion 2.1 implies that $h\big((f),D\big)=0$ for all divisors D of degree 0 disjoint from Z_0 and all Z_0-regular meromorphic functions f.

So for any fixed divisor D of degree 0, we have a continuous group homomorphism from the Jacobian of X to the additive group of real numbers sending each D' to $h(D',D)$. The Jacobian is compact, thus $h(D',D)=0$ for all D and D' of degree 0. Hence, fixing a point e outside Z_0,

$$\begin{aligned}h(p,q) &= h(p-e,q-e)+h(p,e)+h(q,e)-h(e,e)\\ &= h(p,e)+h(q,e)-h(e,e)=\gamma(p)+\gamma(q),\end{aligned}$$

Where $\gamma(p)\overset{\text{def}}{=}h(p,e)-h(e,e)/2$. The lemma follows.

(3.3.3) Proposition (Existence of Intersection Functions). *Let X be a non-singular projective complex curve of positive genus. There exists a intersection function on X, and it is unique up to addition by a constant. Furthermore, it is C^∞ where it is defined.*

Proof. Let Z_0 and g_0 be as in Lemma 3.3.1. As we mentioned before, g_0 is only a quasi-intersection function since Property V of Definition 2.1 is possibly not satisfied. What we will do now is determine which quasi-intersection functions satisfy, in addition, property V of Definition 2.1.

Let g be a quasi-intersection function also with exceptional set Z_0. So, by Lemma 3.3.2, any such g, restricted to prime divisors, is of the form

$$g(p,q)=g_0(p,q)+\gamma(p)+\gamma(q)$$

where γ is some continuous function on X defined outside Z_0.

Let ω be a Z_0-regular meromorphic section of the dualizing sheaf $\mathcal{K}_X$, so its divisor κ has support disjoint from Z_0. Let U be an analytic neighborhood of X, disjoint from Z_0 and κ, which is parameterized by a complex coordinate z. On U we can write $\omega = \psi\, dz$ for some nowhere vanishing holomorphic function ψ of U. Define

$$\Psi_g(\omega, p) \stackrel{\text{def}}{=} -\Big(g(\kappa, p) + \log\big|\psi(p)\big|\Big) - \lim_{q\to p}\Big(g(p,q) + \log\big|z(p) - z(q)\big|\Big).$$

This function is continuous outside of Z_0 and κ, and is independent of the coordinate z used to parameterize X around any given point p; Continuity follows from property III of (2.1), and the independence property follows from the definition of the derivative. If $g = g_0$ then this function is actually C^∞; this follows from the fact, demonstrated in the proof of Lemma 3.3.1, that g_0 satisfies a stronger form of Property III of (2.1), namely the property resulting from replacing the requirement of continuity with a C^∞ requirement.

Let P_a be the genus of X. A calculation shows that

$$\begin{aligned} \Psi_g(\omega, p) - \Psi_{g_0}(\omega, p) &= -\Big(g(\kappa, p) - g_0(\kappa, p)\Big) - \lim_{q\to p}\Big(g(p,q) - g_0(p,q)\Big) \\ &= -\Big(\gamma(\kappa) + (2P_a - 2)g_0(p) + 2\gamma(p)\Big) \\ &= -2p_a\gamma(p) + c \end{aligned}$$

for some constant c depending only on ω.

So g satisfies property V of (2.1) if and only if

$$\gamma(p) = \frac{1}{2P_a}\Psi_{g_0}(\omega, p) + c'$$

for some constant c'. This immediately implies that that there exists an intersection function on X with exceptional set Z_0, that it is C^∞ where it is defined, and that it is unique up to addition by a constant.

All we assumed about the finite set Z_0 is that it contained two different arbitrarily chosen points. Let Z_0' be another such set, and such that Z_0 and Z_0' are disjoint. Let g and g' be intersection functions with exceptional sets Z_0 and Z_0' respectively. Then both g and g' can be considered as intersection functions with exceptional set equal to the union of Z_0 and Z_0', and so, by the above discussion, they differ by a constant. Hence we can use g' to extend g to all of X. We easily check that, after being so extended, g is an intersection function with empty exceptional set.

(3.4) The Canonical (1,1)-form. Now we will define and study the canonical $(1,1)$-form associated with an intersection function g on a projective nonsingular complex curve X. Our goal is to prove that the order function $\Phi(f)$ associated with an intersection function g via Property IV of Definition 2.1 is integration of $-\log|f|$ with respect to the canonical $(1,1)$-form. In a later section, in Proposition 9.6, we will give an explicit formula for the canonical $(1,1)$-form in terms of global holomorphic differentials of X.

Let p be a point of X. We define the function g_p to be the function satisfying. $g_p(q) \stackrel{\text{def}}{=} g(p,q)$. We define the *canonical* $(1,1)$*-form* μ *on* X to be the $(1,1)$-form defined in the following lemma.

(3.4.1) Lemma. *Let g be an intersection function on X. The $(1,1)$-form*

$$\mu = -\frac{1}{\pi \mathbf{i}}\partial\overline{\partial}(g_p)$$

is independent of p and so is well-defined on all of X. Since g is unique up to a constant, μ only depends on X.

Proof. Fix a point e of X. For any divisor D of degree 0, define the $(1,1)$-form

$$\nu_D = -\frac{1}{\pi \mathbf{i}}\partial_p\partial_p\big(\lambda(D, p-e)\big).$$

If $D = (f)$ is a principal divisor, then $\lambda(D, p-e) = -\log|f(p)| + \log|f(e)|$, hence $\nu_D = 0$. Thus ν_D depends only on the class of D in the Jacobian of X.

For each point p, the map sending D to $\nu_D|_p$ is a continuous homomorphism from a compact group to the additive group of a one-dimensional vector space, hence is the zero map. Thus $\nu_D = 0$.

A consequence of this is that, for g_0 defined as in Lemma 3.3.1,

$$-\frac{1}{\pi \mathbf{i}}\partial_q\overline{\partial}_q\big(g_0(p,q)\big) = 0.$$

However, any intersection function g is of the form $g(p,q) = g_0(p,q) + \gamma(p) + \gamma(q)$ where γ is a C^∞ function on X. Hence

$$\mu \stackrel{\text{def}}{=} -\frac{1}{\pi \mathbf{i}}\partial_q\overline{\partial}_q\big(g(p,q)\big) = -\frac{1}{\pi \mathbf{i}}\partial\overline{\partial}(\gamma)$$

is independent of P.

(3.4.2) Lemma. *The canonical $(1,1)$-form μ satisfies*

$$\int_X \mu = 1.$$

Proof. Note that $\mu = d\big((1/2\pi)\, d^c g_p\big)$. By Green's theorem,

$$\int_X d\left(\frac{1}{2\pi} d^c(g_P)\right)$$

is the limit of a path integrals around p of $-d^c(g_p/2\pi)$ where the limit is taken as the path's radius goes to zero (in some fixed local coordinate system). But locally around p, $g_p(q) = -\log\big|z(q)\big| + h(q)$ for some C^∞ function h and some local coordinate z vanishing at p. So the line integral is the limit of a path integral around p of $\frac{1}{4\pi}d^c\log(z\overline{z})$ as the path's radius goes to zero which is easily computed to be 1.

(3.4.3) Lemma. *The integral $\int_X g_p\mu$ is independent of p.*

Proof. We can write

$$\mu = \frac{1}{2\pi}dd^c(g_P).$$

Let p, q be distinct points of X. By an elementary identity, $d(g_p) \cdot d^c(g_q) = d(g_q) \cdot d^c(g_p)$. So, using the product rule of calculus, we get

$$d(g_p \cdot d^c(g_q)) - g_p \cdot dd^c(g_q) = d(g_q \cdot d^c(g_p)) - g_q \cdot dd^c(g_p).$$

Multiplying by $\frac{1}{2\pi}$ gives us

$$d(\frac{1}{2\pi} g_P d^c(g_Q)) - g_p \cdot \mu = d(\frac{1}{2\pi} g_Q d^c(g_P)) - g_q \cdot \mu.$$

We integrate both sides. The singularities of g_p and g_q are very mild and μ has no singularities, so integrating with respect to μ is well defined.

By Green's theorem the integral

$$\int_X d\left(\frac{1}{2\pi} g_q d^c g_p\right)$$

equals the limit of a path integral around p of $-\frac{1}{2\pi} g_q d^c g_p$ as the path's radius goes to zero (in some fixed local coordinate system). So, by an argument similar to that found in the proof of the previous lemma, this limit is $g_q(p) = g(p, q)$. So

$$g(p, q) - \int_X g_p \mu = g(p, q) - \int_X g_q \mu.$$

The lemma follows.

(3.4.4) Proposition. *Let X be a projective nonsingular complex curve of positive genus, and let g be an intersection function on X. Let Φ be the a homomorphism associated to g in Property IV of Definition 2.1 which sends the multiplicative group of non-zero meromorphic functions on X to the additive group of real numbers. The map Φ is the following*

$$\Phi(f) = -\int_X \log|f|\mu.$$

Proof. By definition

$$\Phi(f) \stackrel{\text{def}}{=} -\log\big|f(p)\big| - g\big((f), p\big),$$

which is independent of p. The singularities of each term as p approaches (f) are very mild, hence we can integrate both sides with respect to μ:

$$\Phi(f) = \int \Phi(f)\mu = -\int \log|f|\mu - \int g\big((f), p\big)\mu$$

But, since (f) has degree 0, Lemma 3.4.3 implies that

$$\int g\big((f), p\big)\mu = 0.$$

The result follows.

(3.5) Arakelov's Canonical Green's Functions. Proposition 3.3.3 implies that the intersection function g is unique up to a constant, and Lemma 3.4.3 implies that the integral $\int_X g_p \mu$ is independent of p. The fact that $\int_X \mu = 1$ implies then that there is a *unique* intersection function g on X such that $\int_X g_p \mu = 0$ for all p in X. This intersection function is called the *Canonical Green's Function.*

Exercise. Show that the Canonical Green's Function defined here is, up to a constant multiple which depends on the conventions used, the same as the Canonical Green's Function defined in other accounts of Arakelov theory.

4. Intersection Functions on Singular Curves: The Existence Theorem

For this and the remaining sections of this chapter we turn our attention to singular curves. In this section we will demonstrate existence of intersection functions on singular curves; in the remaining sections of this chapter we will classify intersection functions on singular curvesand study their Chern forms.

Throughout this section let X be a projective complex curve, connected but not necessarily irreducible, and, as usual, let Z be the set of singular points of X. We assume that all the points of Z are ordinary double points.

A *C^∞-intersection function* on X is an intersection function on X whose restriction to pairs of points is a C^∞ function whenever it is defined.

Theorem (4.1). *There exists intersection functions on X. In fact, there are C^∞-intersection functions on X.*

Proof. We define a *nodal* curve to be any projective complex curve, connected but not necessarily irreducible, whose singular points are all ordinary double points. We will prove this theorem for the class of nodal curves, and will do so by induction on the number of singular points. For example, if X has no singular points, then the theorem follows from Proposition 3.3.3.

For any nodal curve X, and any two distinct non-singular points R and S of X, we can can form another nodal curve X' by identifying R and S in such a way that the resulting point is an ordinary double point of X'. Given any two nodal curves X_1 and X_2, a non-singular point R_1 on X_1, and a non-singular point R_2 on X_2, we can form a nodal curve by identifying R_1 and R_2 in such a way that the resulting point is an ordinary double point. It is clear that with the repeated use of these two operations, where we begin with the set of non-singular curves, we can produce all nodal curves. Therefore, to complete the induction argument, all we have to do is show that the theorem is true for any curve constructed in one of the above two ways from a curve, or curves, for which the theorem is assumed to be true.

Case 1. Let X be a nodal curve which has a C^∞-intersection function g. Let R and S be two non-singular points of X, and let X' be the new nodal curve resulting from identifying R and S into an ordinary double point. We will now show that there is a C^∞-intersection function on X'.

Let Z be the set of singularities of X and let Z' be the set of singularities of X'. Let P_a' be the arithmetic genus of X' which is one more than the arithmetic genus P_a of X. Let g be a C^∞-intersection function on X, and define g' as follows:

$$g'(p,q) \stackrel{\text{def}}{=} g(p,q) - \frac{1}{2P_a'}\, g(p+q, R+S)$$

where p and q are distinct points of $X' - Z'$; we will also consider p and q to be points of $X - Z$. Clearly g' is C^∞ for all pairs of distinct points outside of Z'. We extend the definition of g' bi-additively to pairs of divisors. It clear that g' satisfies Properties I, II,.and III of Definition 2.1.

Let f be any Z-regular meromorphic function of X'. Then

$$\begin{aligned}\Phi_{X'}(f) &\stackrel{\text{def}}{=} -\log\big|f(P)\big| - g'\big((f),p\big)\\ &= -\log\big|f(P)\big| - g\big((f),p\big) + \frac{1}{2P'_a}\, g\big((f),R+S\big)\\ &= \Phi_X(f) + \frac{1}{2P'_a}\, g\big((f),R+S\big),\end{aligned}$$

is independent of p. So g' satisfies Propery IV of Definition 2.1.

Now let ω be a Z'-regular meromorphic section of the dualizing sheaf $\mathcal{K}_{X'}$ of X', and let κ be the divisor of ω. Let U be an open neighborhood of X', disjoint from Z' and the support of κ, which is parameterized by a holomorphic coordinate z. On U we can write $\omega = \psi\, dz$ for some nowhere-zero holomorphic function ψ on U. Let p be a point of U, and consider the following:

$$\begin{aligned}\Psi'(\omega,p) &\stackrel{\text{def}}{=} -\Big(g'(\kappa,p) + \log\big|\psi(p)\big|\Big) - \lim_{q\to p}\Big(g'(p,q) + \log\big|z(p)-z(q)\big|\Big)\\ &= -\left(g(\kappa,p) - \frac{2P'_a-2}{2P'_a}\, g(p,R+S) - \frac{1}{2P'_a}\, g(\kappa,R+S) + \log\big|\psi(p)\big|\right)\\ &\qquad - \lim_{q\to p}\left(g(p,q) - \frac{1}{2P'_a}\, g(p+q,R+S) + \log\big|z(p)-z(q)\big|\right)\\ &= -g(\kappa,p) + g(p,R+S) + \frac{1}{2P'_a}\, g(\kappa,R+S) - \log\big|\psi(p)\big|\\ &\qquad - \lim_{q\to p}\Big(g(p,q) + \log\big|z(p)-z(q)\big|\Big).\end{aligned}$$

We can also consider ω as a Z-regular meromorphic section of the dualizing sheaf $\mathcal{K}_X$ of X. In this context it has divisor $\kappa - R - S$. So, by Property V of the intersection function,

$$\Psi(\omega) = -g\big(\kappa - R - S, p\big) - \log\big|\psi(p)\big| - \lim_{q\to p}\Big(g(p,q) + \log\big|z(p)-z(q)\big|\Big).$$

Therefore,

$$\Psi'(\omega,p) = \Psi(\omega) + \frac{1}{2P'_a}\, g(\kappa, R+S).$$

Since $\Psi'(\omega,p)$ is independent of p, the function g' satisfies the final propery, Property V of Definition 2.1. Therefore, g' is a C^∞-intersection function on X'.

Case 2. Let X_1 and X_2 be nodal curves with C^∞-intersection functions g_1 and g_2 respectively, and let Z_1 and Z_2 be the singularities of X_1 and X_2 respectively.

Let R_1 be a non-singular point of X_1, and let R_2 be a non-singular points of X_2. Let X' be the new nodal curve resulting from identifying R_1 and R_2 into an ordinary double point. Set Z' equal to the set of singularities of X'.

Let P_1 and P_2 be the arithmetic genera of X_1 and X_2 respectively. Then we have the formula

$$P'_a = P_1 + P_2$$

where P'_a is the arithmetic genus of X'.

We will now show that there is a C^∞-intersection function on X'.

Define a function g' for X' as follows. Let p and q be two distinct points of X' outside of Z'. So we can also consider p and q to be points of the disjoint union of X_1 with X_2. If p and q are both in X_1,

$$g'(p,q) \stackrel{\text{def}}{=} g_1(p,q) - \frac{P_2}{P'_a} g_1(p+q, R_1);$$

if p and q are both in X_2,

$$g'(p,q) \stackrel{\text{def}}{=} g_2(p,q) - \frac{P_1}{P'_a} g_2(p+q, R_2);$$

if p is in X_1 but q is in X_2,

$$g'(p,q) \stackrel{\text{def}}{=} \frac{P_1}{P'_a} g_1(p, R_1) + \frac{P_2}{P'_a} g_2(q, R_2);$$

finally, if p is in X_2 but q is in X_1, define

$$g'(p,q) \stackrel{\text{def}}{=} \frac{P_2}{P'_a} g_2(p, R_2) + \frac{P_1}{P'_a} g_1(q, R_1).$$

If $P_1 = P_2 = P'_a = 0$, we must modify the definition as follows. If p and q are both in X_i,

$$g'(p,q) \stackrel{\text{def}}{=} g_i(p,q) - \frac{1}{2} g_i(p+q, R_i);$$

if p is in X_i and q is in X_j where $i \neq j$,

$$g'(p,q) \stackrel{\text{def}}{=} \frac{1}{2} g_i(p, R_i) + \frac{1}{2} g_j(q, R_j).$$

Clearly g' is C^∞ for all pairs of distinct points outside of Z'. We extend g' bi-additively to pairs of divisors. It is clear that g' satisfies Properties I, II and III of Definition 2.1.

Case 2 (continued): Property IV. We now verify that g' satisfies Property IV of Definition 2.1.

Let f be any Z'-regular meromorphic function of X', and let f_i be the restriction of f to X_i. Let p be a point of X_1 not contained in Z' or the support of (f_1). Then

$$\begin{aligned}
-\log\bigl|f(p)\bigr| - g'\bigl((f),p\bigr) &= -\log\bigl|f_1(p)\bigr| - g'\bigl((f_1),p\bigr) - g'\bigl((f_2),p\bigr) \\
&= -\log\bigl|f_1(p)\bigr| - g_1\bigl((f_1),p\bigr) + \frac{P_2}{P'_a} g_1\bigl((f_1),R_1\bigr) \\
&\quad - \frac{P_2}{P'_a} g_2\bigl((f_2),R_2\bigr) \\
&= -\log\bigl|f_1(p)\bigr| + \Bigl(\log\bigl|f_1(p)\bigr| + \Phi_{X_1}(f_1)\Bigr) \\
&\quad - \frac{P_2}{P'_a}\Bigl(\log\bigl|f_1(R_1)\bigr| + \Phi_{X_1}(f_1)\Bigr) \\
&\quad + \frac{P_2}{P'_a}\Bigl(\log\bigl|f_2(R_2)\bigr| + \Phi_{X_2}(f_2)\Bigr)
\end{aligned}$$

$$
\begin{aligned}
&= \left(1-\frac{P_2}{P_a'}\right)\Phi_{X_1}(f_1) - \frac{P_2}{P_a'}\log\left|f_1(R_1)\right| + \frac{P_2}{P_a'}\log\left|f_2(R_2)\right| \\
&\quad + \frac{P_2}{P_a'}\Phi_{X_2}(f_2) \\
&= \frac{P_1}{P_a'}\Phi_{X_1}(f_1) + \frac{P_2}{P_a'}\Phi_{X_2}(f_2).
\end{aligned}
$$

The last equation uses the fact that $f_1(R_1) = f_2(R_2)$ which is the requirement that f is defined at the singularity formed from R_1 and R_2. We also use the fact that $P_a' = P_1 + P_2$.

The corresponding calculation for p in X_2 yields the same result. This implies that g' satisfies Property IV of the definition of the intersection function.

When $P_a' = 0$ we have to do a separate calculation, given here for a point p in X_1:

$$
\begin{aligned}
-\log\left|f(p)\right| - g'\big((f),p\big) &= -\log\left|f_1(p)\right| - g'\big((f_1),p\big) - g'\big((f_2),p\big) \\
&= -\log\left|f_1(p)\right| - g_1\big((f_1),p\big) + \frac{1}{2}g_1\big((f_1),R_1\big) - \frac{1}{2}g_2\big((f_2),R_2\big) \\
&= -\log\left|f_1(p)\right| + \Big(\log\left|f_1(p)\right| + \Phi_{X_1}(f_1)\Big) \\
&\quad -\frac{1}{2}\Big(\log\left|f_1(R_1)\right| + \Phi_{X_1}(f_1)\Big) + \frac{1}{2}\Big(\log\left|f_2(R_2)\right| + \Phi_{X_2}(f_2)\Big) \\
&= \frac{1}{2}\Phi_{X_1}(f_1) + \frac{1}{2}\Phi_{X_2}(f_2).
\end{aligned}
$$

Of course, we get the same result for p in X_2. Thus, g' satisfies Property IV of Definition 2.1.

Case 2 (conclusion): Property V. Finally, we verify that g' satisfies Property V of Definition 2.1.

First assume that $P_a' > 0$. Let ω be a Z'-regular meromorphic section of the dualizing sheaf $\mathcal{K}_{X'}$ of X', and let ω_1 and ω_2 be the restrictions of ω to X_1 and X_2 respectively. The fact that ω is defined at Z' implies that

- ω_1 considered as a meromorphic section of $\mathcal{K}_{X_1}$ has a simple pole at R_1,
- ω_2 considered as a meromorphic section of $\mathcal{K}_{X_2}$ has a simple pole at R_2, and
- $res_{R_1}(\omega_1) + res_{R_2}(\omega_2) = 0$.

Let κ be the divisor of ω, and let κ_1 and κ_2 be the restriction of κ to X_1 and X_2 respectively. Note that the divisor κ_i has degree $2P_i - 1$. Let U be an open neighborhood of X_1, disjoint from R_1, Z_1, and the support of κ_1, which is parameterized by a holomorphic coordinate z. So we can consider U as an open neighborhood of X' disjoint from Z' and the support of κ. On U we can write $\omega = \psi\, dz$ for some nowhere-zero holomorphic function ψ on U. Consider the following quantity, defined for all p in U, and consider the following series of equations:

$$
\begin{aligned}
\Psi'(\omega,p) &\stackrel{\text{def}}{=} -\Big(g'(\kappa,p) + \log\left|\psi(p)\right|\Big) - \lim_{q\to p}\Big(g'(p,q) + \log\left|z(p)-z(q)\right|\Big) \\
&= -g'(\kappa_1+\kappa_2,p) - \log\left|\psi(p)\right| - \lim_{q\to p}\Big(g'(p,q) + \log\left|z(p)-z(q)\right|\Big) \\
&= -g_1(\kappa_1,p) + \frac{(2P_1-1)P_2}{P_a'}g_1(p,R_1) + \frac{P_2}{P_a'}g_1(\kappa_1,R_1) \\
&\quad -\frac{(2P_2-1)P_1}{P_a'}g_1(p,R_1) - \frac{P_2}{P_a'}g_2(\kappa_2,R_2) - \log\left|\psi(p)\right|
\end{aligned}
$$

$$
\begin{aligned}
& - \lim_{q \to p} \left(g_1(p,q) - \frac{P_2}{P'_a}\, g_1(p+q, R_1) + \log\bigl|z(p) - z(q)\bigr| \right) \\
= \quad & -g_1(\kappa_1, p) + g_1(p, R_1) + \frac{P_2}{P'_a}\, g_1(\kappa_1, R_1) \\
& - \frac{P_2}{P'_a}\, g_2(\kappa_2, R_2) - \log\bigl|\psi(p)\bigr| - \lim_{q \to p} \left(g_1(p,q) + \log\bigl|z(p) - z(q)\bigr| \right).
\end{aligned}
$$

We can also consider ω_1 as a Z_1-regular meromorphic section of the dualizing sheaf $\mathcal{K}_{X_1}$ of X_1. In this context it has divisor $\kappa_1 - R_1$. So, the quantity

$$
\Psi_1(\omega_1) \stackrel{\text{def}}{=} -\Bigl(g_1\bigl(\kappa_1 - R_1, p\bigr) + \log\bigl|\psi(p)\bigr| \Bigr) - \lim_{q \to p} \Bigl(g_1(p,q) + \log\bigl|z(p) - z(q)\bigr| \Bigr)
$$

is independent of p. Combining this with the above equations gives us the expression

$$
\Psi'(\omega, p) \;=\; \frac{P_2}{P'_a}\, g_1(\kappa_1, R_1) \;-\; \frac{P_2}{P'_a}\, g_2(\kappa_2, R_2) \;+\; \Psi_1(\omega_1).
$$

This expression does not appear to be symmetric with respect to X_1 and X_2. However, we observe that if we write $\omega_1 = \phi\, dz/z$ where z is a holomorphic coordinate near R_1 vanishing to first order at R_1, and ϕ is a non-vanishing holomorphic function in a neighborhood of R_1, then

$$
\begin{aligned}
\Psi_1(\omega_1) \quad = \quad & \lim_{p \to R_1} \Bigl(-g_1\bigl(\kappa_1 - R_1, p\bigr) - \log\bigl|\phi(p)\bigr| \\
& \qquad + \log\bigl|z(p)\bigr| - \lim_{q \to p} \Bigl(g_1(p,q) + \log\bigl|z(p) - z(q)\bigr| \Bigr) \Bigr) \\
= \quad & -\log\bigl|\phi(R_1)\bigr| - g_1\bigl(\kappa_1, R_1\bigr) + \lim_{p \to R_1} \Bigl(g_1\bigl(R_1, p\bigr) + \log\bigl|z(p)\bigr| \Bigr) \\
& - \lim_{q \to R_1} \Bigl(g_1(R_1, q) + \log\bigl|z(q)\bigr| \Bigr) \\
= \quad & -\log\bigl|\phi(R_1)\bigr| - g_1\bigl(\kappa_1, R_1\bigr) \\
= \quad & -\log\bigl|\mathrm{res}_{R_1}(\omega_1)\bigr| - g_1\bigl(\kappa_1, R_1\bigr)
\end{aligned}
$$

We use this expression to continue the calculation of $\Psi'(\omega, P)$:

$$
\begin{aligned}
\Psi'(\omega, p) \quad = \quad & \frac{P_2}{P'_a}\, g_1(\kappa_1, R_1) \;-\; \frac{P_2}{P'_a}\, g_2(\kappa_2, R_2) \;+\; \Psi_1(\omega_1) \\
= \quad & \frac{P_2}{P'_a}\, g_1(\kappa_1, R_1) \;-\; \frac{P_2}{P'_a}\, g_2(\kappa_2, R_2) \;-\; \log\bigl|\mathrm{res}_{R_1}(\omega_1)\bigr| \;-\; g_1\bigl(\kappa_1, R_1\bigr) \\
= \quad & -\,\frac{P_1}{P'_a}\, g_1(\kappa_1, R_1) \;-\; \frac{P_2}{P'_a}\, g_2(\kappa_2, R_2) \;-\; \log\bigl|\mathrm{res}_{R_1}(\omega_1)\bigr|
\end{aligned}
$$

Since $\log\bigl|\mathrm{res}_{R_1}(\omega_1)\bigr| = \log\bigl|\mathrm{res}_{R_2}(\omega_2)\bigr|$, this expression is symmetric with respect to X_1 and X_2. So, if we preform the analogous calculation for U contained in $X_2 - Z_2$, we will get the same quantity. In other words, we get a formula for $\Psi'(\omega, p)$ which is independent of p. Hence g' satisfies property V of Definition 2.1.

The case $P'_a = 0$ is similar. I will leave this to the reader. This concludes the proof of the theorem.

5. Classification of Intersection Functions: Preliminaries

In this section we give a few results which will be useful for classifying intersection functions on singular curves. We begin by defining a map

$$\rho\colon \operatorname{Pic}(X) \rightarrow V$$

where V is a real vector space defined below. This map will play an important roll in the classification.

(5.1) Notation and Conventions. The following notation and conventions will be used in this and in the remaining sections of this chapter.

- Let X be a projective complex curves, connected but not necessarily irreducible. Let Z be the set of singularities of X. We assume that all the elements of Z are ordinary double points.
- Let P_a be the arithmetic genus of X.
- Let m be the number of components of X.
- Let $X_1, \ldots, X_m$ be the normalizations of the components of X.
- Let P_i be the genus of X_i.
- Let n be the number of nodes of X.
- Let $T_1, \ldots, T_n$ be the nodes of X.
- For each node T_j, let R_j and S_j be the two points of the disjoint union of the X_i's which map to T_j under the normalization map. Fix a labeling once and for all, i.e., we assume that one of these points will be consistently referred to as R_j and the other as S_j.
- For each node T_j, let r_j and s_j be the values of the index such that R_j is in X_{r_j}, and S_j is in X_{s_j}.
- Let $\mathcal{K}_X$ be the dualizing sheaf of X.
- A *Z-regular meromorphic function on* X is a meromorphic function on X defined and non-zero on a Zariski dense open subset of X containing Z. A *Z-regular meromorphic function on* X_i is an invertible meromorphic function on X_i defined and non-zero at any R_j or S_j contained in X_i.
- An *X-nonsingular point of* X_i is a point of X_i not equal to any R_j or S_j. Such points can be identified with non-singular points of X.
- A Cartier divisor of X is said to be *prime-to-Z* if is has support disjoint from Z. Since $X - Z$ is non-singular, such divisors can be thought of as Weil divisors. Note that every Cartier divisor class is represented, up to linear equivalence, by a prime-to-Z divisor. A divisor of X_i is said to be *prime-to-Z* if its support does not contain any R_j or S_j. Such divisors can be identified with prime-to-Z divisors on X.

(5.2) Divisor Groups on X. We will deal with four divisor groups listed here from largest to smallest:

1. The Picard group $\operatorname{Pic}(X)$ is defined to be the group of isomorphism classes of line bundles on X, or equivalently, the group of prime-to-Z divisors modulo linear equivalence.
2. $\operatorname{Pic}^0(X)$ is defined to be the group of isomorphism classes of line bundles on X of total degree 0, or equivalently, the divisor classes represented by prime-to-Z

divisors D whose restrictions $D_i = D|_{X_i}$ satisfy the relation $\sum_i \deg_{X_i} D_i = 0$. We say that such divisors have *total degree zero.*

3. $\mathrm{Pic}^{\mathrm{a}}(X)$ is defined to be the group of isomorphism classes of line bundles on X whose pullback to each X_i has degree 0. Equivalently, a prime-to-Z divisor D represents a divisor class in $\mathrm{Pic}^{\mathrm{a}}(X)$ if and only if its restrictions $D_i = D|_{X_i}$ have degree 0 for each normalized component X_i. The group $\mathrm{Pic}^{\mathrm{a}}(X)$ has an analytic topology under which it is a connected semi-abelian variety of dimension P_a.
4. $\mathrm{Pic}^{\mathrm{r}}(X)$ is defined to be the group of isomorphism classes of line bundles on X whose pullback to each normalized component X_i is trivial. Equivalently, it consists of the classes of prime-to-Z divisors D whose restriction D_i to each normalized component X_i is represented by the divisor of a meromorphic function f_i on X_i. In this situation, we say that $f_1, \ldots, f_m$ *represents* D, and that D is *representable.*

(5.3) The map ρ and the real vector space $\mathbf{V}$. Consider the multiplicative group of m-tuples of the form $(f_1, \ldots, f_m)$ where each f_i is a Z-regular meromorphic function on X_i. We define a group homomorphism from this group to the multiplicative group $(\mathbf{C}^\times)^n$ by the rule

$$(f_1, \ldots, f_m) \mapsto \left(\frac{f_{r_1}(R_1)}{f_{s_1}(S_1)}, \ldots, \frac{f_{r_n}(R_n)}{f_{s_n}(S_n)} \right).$$

This map is a surjection. This is a consequence of the following general principle: whenever we specify a finite set of points of X_i and then assign complex values to these points, we can always find a meromorphic function defined and taking on the assigned values at these points.

Let $N^\times$ be the image, under this map, of the subgroup of m-tuples of constants $(c_1, \ldots, c_m)$. The above map induces a surjective group homomorphism from the group of representable divisors to the group $(\mathbf{C}^\times)^n/N^\times$. The kernel is exactly the the subgroup of principal, prime-to-Z divisors. So this map induces a group isomorphism from $\mathrm{Pic}^{\mathrm{r}}(X)$ to $(\mathbf{C}^\times)^n/N^\times$. This map is, of course, bi-continuous with respect to the appropriate analytic topologies.

Consider the log-homomorphism from the multiplicative group $(\mathbf{C}^\times)^n$ to the additive group $\mathbf{R}^n$:

$$(c_1, \ldots, c_n) \mapsto \big(\log|c_1|, \ldots, \log|c_n| \big).$$

Note that this is a surjective homomorphism with compact kernel. Let N be the image of $N^\times$ under this map, i.e. N is the n-tuples of the form

$$\big(c_{r_1} - c_{s_1}, \ldots, c_{r_n} - c_{s_n}\big)$$

where $(c_1, \ldots, c_m)$ is a choice of real constants, one for each component of X. So N is the image of a linear transformation from $\mathbf{R}^m$ to $\mathbf{R}^n$. The kernel of this linear transformation is the constant m-tuples $(c, \ldots, c)$; this follows from connectivity. Thus N is a $m-1$ dimensional subspace of $\mathbf{R}^n$.

Define V to be the $n-m+1$ dimensional $\mathbf{R}$-vector space $\mathbf{R}^n/N$. The log-homomorphism induces a continuous group homomorphism ρ_0 from the multiplicative group $(\mathbf{C}^\times)^n/N^\times$ to the additive group of V. Note that, since the log-homomorphism is surjective with compact kernel, the homomorphism ρ_0 must also be surjective with compact kernel. When we compose the isomorphism from $\mathrm{Pic}^{\mathrm{r}}(X)$ to $(\mathbf{C}^{\times})^n/N^{\times}$ with ρ_0 we get a surjective, continuous group homomorphism with compact kernel from $\mathrm{Pic}^{\mathrm{r}}(X)$ to the additive group V. We will also call this homomorphism ρ_0.

Using intersection functions we can extend ρ_0 to a continuous group homomorphism ρ from $\mathrm{Pic}(X)$ to the additive group V. First, fix an intersection function g_i on each component of X_i. Given a prime-to-Z divisor D with restrictions $D_1, \ldots, D_m$ to the normalized components of X, define $\rho(D)$ as the class in V of

$$\Big(-g_{r_1}(D_{r_1}, R_1) + g_{s_1}(D_{s_1}, S_1),\ \ldots,\ -g_{r_n}(D_{r_n}, R_n) + g_{s_n}(D_{s_n}, S_n) \Big).$$

We now show that ρ is an extension of ρ_0, and, in particular, that it vanishes on principal divisors. For any representable divisor $D = (f_1) + \ldots + (f_m)$, we have that, for each i,

$$g_i\big((f_i), p\big) \;=\; -\log\big|f_i(p)\big| \;-\; \Phi_{X_i}(f_i)$$

by Property IV of the definition of the intersection function. So, under ρ, the divisor D goes to the class in V of

$$\left(\log\left|\frac{f_{r_1}(R_1)}{f_{s_1}(S_1)}\right| + \Phi(f_{r_1}) - \Phi(f_{s_1}),\ \ldots,\ \log\left|\frac{f_{r_n}(R_n)}{f_{s_n}(S_n)}\right| + \Phi(f_{r_n}) - \Phi(f_{s_n}) \right).$$

But

$$\Big(\Phi(f_{r_1}) - \Phi(f_{s_1}),\ \ldots,\ \Phi(f_{r_n}) - \Phi(f_{s_n}) \Big)$$

is an element of N, so D goes to the class in V of

$$\left(\log\left|\frac{f_{r_1}(R_1)}{f_{s_1}(S_1)}\right|,\ \ldots,\ \log\left|\frac{f_{r_n}(R_n)}{f_{s_n}(S_n)}\right| \right)$$

which agrees with the definition of ρ_0.

(5.3.1) Proposition. *Let ρ_1 be the restriction of ρ to $\mathrm{Pic}^{\mathrm{a}}(X)$.*
1. *The homomorphism ρ_1 is surjective with compact kernel.*
2. *The homomorphism ρ_1 is independent of the particular choice of intersection functions used to define ρ.*

Proof. We showed earlier that ρ_0, the restriction of ρ to $\mathrm{Pic}^{\mathrm{r}}(X)$, is surjective with compact kernel. Let J be the kernel of ρ_0, let A be the group $\mathrm{Pic}^{\mathrm{a}}(X)/J$, and let $\widetilde{\rho_1}$ be the homomorphism from A to V induced by the map ρ_1. The homomorphism ρ_0 induces an isomorphism between $\mathrm{Pic}^{\mathrm{r}}(X)/J$ and V. This allows us to identify V with a subgroup of A, and to view $\widetilde{\rho_1}$ as a projection of A onto V. Therefore, A is isomorphic to the sum of V and A/V, and we can regard $\widetilde{\rho_1}$ as the projection onto the V factor, and so it has kernel isomorphic to A/V. The group A/V can be identified with $\mathrm{Pic}^{\mathrm{a}}(X)/\mathrm{Pic}^{\mathrm{r}}(X)$, which in turn can be identified with the product of the Jacobians of the $X_i's$. Therefore, the kernel of $\widetilde{\rho_1}$ is compact. This, together with the compactness of J implies that the kernel of ρ_1 is compact.

Now we will prove that ρ_1 is independent of the choice of the intersection functions used to define it. In fact, we will prove something more general: we will show that we can replace each g_i by any function of the form

$$g_i'(p, q) \;=\; g_i(p, q) \;+\; \gamma_i(p) \;+\; \gamma_i(q)$$

where γ_i is a continuous function on X_i. We extend g_i' bi-additively to pairs of divisors on X_i. The remarks of (3.2.2) for genus 0, and Proposition 3.3.3 for higher genus, shows that the set of functions of this form include all the intersection functions on X_i. (In fact, Lemma 3.3.2 implies that all quasi-intersection functions are of this form) .

Let D be a prime-to-Z divisor whose equivalence class is in $\mathrm{Pic}^{\mathrm{a}}(X)$. Write $D = D_1 + \ldots + D_m$ where D_i has support in X_i. The resulting $\rho_1'(D)$, formed from the g_i''s in place of the g_i's, is

$$\begin{aligned}\rho_1'(D) &= \Big(-g_{r_1}'(D_{r_1}, R_1) + g_{s_1}'(D_{s_1}, S_1),\ \ldots,\ -g_{r_n}'(D_{r_n}, R_n) + g_{s_n}'(D_{s_n}, S_n) \Big) \\ &= \Big(-g_{r_1}(D_{r_1}, R_1) + g_{s_1}(D_{s_1}, S_1) - \gamma_{r_1}(D_{r_1}) + \gamma_{s_1}(D_{s_1}),\ \ldots, \\ &\qquad -g_{r_n}(D_{r_n}, R_n) + g_{s_n}(D_{s_n}, S_n) - \gamma_{r_n}(D_{r_n}) + \gamma_{s_n}(D_{s_n}) \Big) \\ &= \rho_1(D) + \Big(-\gamma_{r_1}(D_{r_1}) + \gamma_{s_1}(D_{s_1}), \ldots,\ -\gamma_{r_n}(D_{r_n}) + \gamma_{s_n}(D_{s_n}) \Big)\end{aligned}$$

where we define γ_i of a divisor by extending additively. Now

$$\Big(-\gamma_{r_1}(D_{r_1}) + \gamma_{s_1}(D_{s_1}), \ldots, -\gamma_{r_n}(D_{r_n}) + \gamma_{s_n}(D_{s_n}) \Big)$$

is in N, so its class in V is 0. We conclude that $\rho_1'(D) = \rho_1(D)$.

(5.3.2) Corollary. *Let ρ_1 be the restriction of ρ to* $\mathrm{Pic}^{\mathrm{a}}(X)$. *Then*

1. *any continuous homomorphism from* $\mathrm{Pic}^{\mathrm{a}}(X)$ *to* $\mathbf{R}$ *factors uniquely as ρ_1 followed by a linear map from V to* $\mathbf{R}$*;*
2. *all bi-additive, continuous maps h of* $\mathrm{Pic}^{\mathrm{a}}(X) \times \mathrm{Pic}^{\mathrm{a}}(X)$ *to* $\mathbf{R}$ *factor uniquely as $\rho_1 \times \rho_1$ followed by a bilinear transformation h_0 of $V \times V$ to* $\mathbf{R}$*, and h_0 is symmetric if and only if h is also.*

Proof. This is a consequence of the first statement of the previous proposition.

(5.4) Additive and Bi-additive Pairings of Pic(X).

(5.4.1) Lemma. *A map ϕ from* $\mathrm{Pic}(X)$ *to* $\mathbf{R}$ *is a continuous homomorphism if and only if it is of the form*

$$\phi(D) = \phi_V(\rho(D)) + \sum c_i \deg_i(D)$$

where ϕ_V is a linear functional on the real vector space V, where $c_1, \ldots, c_m$ are real constants, and where $\deg_i(D)$ is the degree of the restriction of D to the normalized component X_i. The functional ϕ_V and the constants $c_1, \ldots, c_m$ are uniquely determined by ϕ.

Proof. Any such ϕ is clearly a continuous homomorphism. Conversely, suppose ϕ a continuous homomorphism. By Corollary 5.3.2, there is a unique linear map ϕ_V from V to $\mathbf{R}$ such that the restriction of ϕ to $\mathrm{Pic}^{\mathrm{a}}(X)$ is equal to the restriction of $\phi_V \circ \rho$ to $\mathrm{Pic}^{\mathrm{a}}(X)$. So the difference $\beta \stackrel{\mathrm{def}}{=} \phi - \phi_V \circ \rho$ vanishes on $\mathrm{Pic}^{\mathrm{a}}(X)$. On each component X_i choose an X-nonsingular point e_i. Then

$$\beta(D) = \beta\Big(D - \sum \deg_i(D)\, e_i\Big) + \beta\Big(\sum \deg_i(D)\, e_i\Big) = \sum \deg_i(D) \beta(e_i)$$

thus ϕ is of the desired form. Note that $c_i = \beta(e_i)$ is independent of the choice of e_i, and so is uniquely determined by ϕ.

(5.4.2) Lemma. *A real-valued function h on* $\mathrm{Pic}(X) \times \mathrm{Pic}(X)$ *is a continuous, symmetric, bi-additive pairing if and only if it is of the form*

$$\begin{aligned} h(D,E) &= h_0\Big(\rho(D),\rho(E)\Big) + \sum \deg_i(E)\,\tau_i\big(\rho(D)\big) \\ &\qquad + \sum \deg_j(D)\,\tau_j\big(\rho(E)\big) + \sum c_{i,j}\,\deg_i(D)\deg_j(E) \end{aligned}$$

where h_0 is a real-valued, symmetric, bilinear pairing on V, where the $c_{i,j}$'s are constants such that $c_{i,j} = c_{j,i}$, where $\tau_1, \dots, \tau_m$ are linear maps from V to $\mathbf{R}$, *and where $\deg_i(D)$ is the degree of the restriction of D to X_i.*

The pairing h_0, the functions τ_i, and the constants $c_{i,j}$ are uniquely determined by h.

Proof. Any h of this form is clearly a continuous, symmetric, bi-additive pairing.

Now suppose h is a continuous, symmetric, bi-additive pairing. By Corollary 5.3.2 there is a unique real-valued, symmetric, bilinear pairing h_0 on V such that the restriction of h to $\mathrm{Pic}^{\mathrm{a}}(X) \times \mathrm{Pic}^{\mathrm{a}}(X)$ is equal to the composition of the restriction of $\rho\times\rho$ to $\mathrm{Pic}^{\mathrm{a}}(X)\times\mathrm{Pic}^{\mathrm{a}}(X)$ followed by h_0. Define the following pairing on $\mathrm{Pic}(X)$:

$$J(D,E) \stackrel{\text{def}}{=} h(D,E) - h_0\big(\rho(D),\rho(E)\big)$$

which vanishes when D and E are both in $\mathrm{Pic}^{\mathrm{a}}(X)$. On each normalized component X_i choose a X-nonsingular point p_i. Then

$$\begin{aligned} J(D,E) &= J\Big(D - \sum \deg_i(D)\,p_i,\ E - \sum \deg_j(E)\,p_j\Big) + J\Big(D,\ \sum \deg_j(E)\,p_j\Big) \\ &\qquad + J\Big(\sum \deg_i(D)\,p_i,\ E\Big) - J\Big(\sum \deg_i(D)\,p_i,\ \sum \deg_j(E)\,p_j\Big) \\ &= \sum_j \deg_j(E)\,J(D,p_j) + \sum_i \deg_i(D)\,J(E,p_i) - \sum_{i,j} \deg_i(D)\deg_j(E)\,J(p_i,p_j). \end{aligned}$$

Now $J(D,p_j)$, for each j, is a continuous homomorphism from $\mathrm{Pic}(X)$ to $\mathbf{R}$, so, by Lemma 5.4.1, it is of the form

$$J(D,p_j) = \tau_j\big(\rho(D)\big) + \sum b_{k,j}\deg_k(D)$$

for some linear functional τ_j on V and some constants $b_{k,j}$. By substitution and by combining constants we get

$$J(D,E) = \sum_j \deg_j(E)\tau_j\big(\rho(D)\big) + \sum_i \deg_i(D)\tau_i\big(\rho(E)\big) + \sum_{i,j} c_{i,j}\,\deg_i(D)\deg_j(E)$$

which shows that h is of the required form.

The uniqueness of the pairing h_0, the functions τ_i, and the constants $c_{i,j}$ follows from restricting h to certain types of divisors and by using Corollary 5.3.2.

(5.4.3) Lemma.
1. *Any continuous, symmetric, bi-additive pairing* $h_0\colon \mathrm{Pic}^0(X) \times \mathrm{Pic}^0(X) \to \mathbf{R}$ *extends to a continuous, symmetric, bi-additive pairing from* $\mathrm{Pic}(X) \times \mathrm{Pic}(X)$ *to* $\mathbf{R}$.

2. *Let h_1 and h_2 be continuous, symmetric, bi-additive pairings*

$$\mathrm{Pic}(X)\times\mathrm{Pic}(X) \longrightarrow \mathbf{R}.$$

Then h_1 and h_2 restrict to the same pairing of $\mathrm{Pic}^0(X)$ *if and only if*

$$h_2(D,E) \;=\; h_1(D,E) \,+\, \deg(E)\,\tau(D) \,+\, \deg(D)\,\tau(E)$$

for some continuous homomorphism τ from $\mathrm{Pic}(X)$ *to* $\mathbf{R}$.

Proof.

1. Fix a non-singular point p_0 of X. Define

$$h(D_1,D_2) \;\stackrel{\text{def}}{=}\; h_0\Big(\, D_1 - \deg(D_1)\,p_0,\; D_2 - \deg(D_2)\,p_0 \,\Big).$$

Clearly, h is such an extension.

2. Suppose that h_1 and h_2 restrict to the same pairing on $\mathrm{Pic}^0(X)$. Consider the difference

$$J(D,E) \;\stackrel{\text{def}}{=}\; h_1(D,E) \,-\, h_2(D,E).$$

Fix a non-singular point p_0 of X. Then for D and E prime-to-Z divisors of total degree d and e respectively,

$$\begin{aligned}
J(D,E) &= J\big(D - d\,p_0,\; E - e\,p_0\big) + e\,J(D,p_0) + d\,J(E,p_0) - d\,e\,J(p_0,p_0)\\
&= e\Big(J(D,p_0) - \frac{d}{2}J(p_0,p_0)\Big) + d\Big(J(E,p_0) - \frac{e}{2}J(p_0,p_0)\Big)\\
&= e\,\tau(D) + d\,\tau(E)
\end{aligned}$$

where $\tau(D) \;\stackrel{\text{def}}{=}\; J(D,p_0) - \frac{d}{2}J(p_0,p_0)$.

(5.5) Adjusted Pairings in the Positive Genus Situation. Now we assume that the arithmetic genus P_a of X is strictly positive. In this case there is a class of pairings which will prove useful in the classification of intersection functions.

Given a continuous, symmetric, bi-additive pairing $h\colon \mathrm{Pic}(X) \times \mathrm{Pic}(X) \to \mathbf{R}$, define the *adjustment of h by the canonical divisor*, written $\widehat{h}$, as follows: if $P_a > 1$, then

$$\begin{aligned}
\widehat{h}(D,E) \;\stackrel{\text{def}}{=}\; h(D,E) &- \frac{\deg(E)}{2P_a}h(D,\kappa) - \frac{\deg(D)}{2P_a}h(E,\kappa) - \frac{\deg(E)}{2P_a}\sum d_i\,h(D_i,D_i)\\
&- \frac{\deg(D)}{2P_a}\sum e_j\,h(E_j,E_j) + \frac{\deg(D)\deg(E)}{2P_a(2P_a-2)}\,h(\kappa,\kappa),
\end{aligned}$$

but if $P_a = 1$, then

$$\begin{aligned}
\widehat{h}(D,E) \;\stackrel{\text{def}}{=}\; h(D,E) &- \frac{\deg(E)}{2}\,h(D,\kappa) - \frac{\deg(D)}{2}\,h(E,\kappa)\\
&- \frac{\deg(E)}{2}\sum d_i\,h(D_i,D_i) - \frac{\deg(D)}{2}\sum e_j\,h(E_j,E_j),
\end{aligned}$$

where here κ is a prime-to-Z divisor of the dualizing sheaf $\mathcal{K}_X$, D and E are any prime-to-Z divisors, and $D = \sum d_i\,D_i$ and $E = \sum e_j\,E_j$ are expressions of D and E in terms of prime divisors (i.e., points) D_i and E_j. Note that $\widehat{h}$ does not necessarily vanish when one of the divisors is principal, so it is a pairing of divisors, not of divisor classes.

This adjusted pairing $\widehat{h}$ has properties which make it useful in the classification of intersection functions. The following lemma states some of these properties.

(5.5.1)Lemma. *Let h be a continuous, symmetric, bi-additive pairing from the product* $\mathrm{Pic}(X) \times \mathrm{Pic}(X)$ *to* $\mathbf{R}$. *Then*

1. *For any pair of divisors D and E, each of total degree 0, $\widehat{h}(D,E) = h(D,E)$.*
2. *For any prime-to-Z divisor κ of the dualizing sheaf $\mathcal{K}_X$, the sum $\widehat{h}(p,p)+\widehat{h}(p,\kappa)$ is a constant independent of p, where p varies among the non-singular point of X.*
3. *If $P_a > 1$ then, for any pair of prime-to-Z divisors κ and κ' of the dualizing sheaf,*

$$\widehat{h}(\kappa,\kappa') = (2P_a - 2)\Big(\widehat{h}(p,p) + \widehat{h}(p,\kappa)\Big) + (2P_a - 2)\Big(\widehat{h}(p,p) + \widehat{h}(p,\kappa')\Big)$$

for all non-singular points p in X.

Proof. The first claim is evident from the definition.
Let $\kappa = \sum d_i \kappa_i$ be a prime-to-Z divisor of the dualizing sheaf, where each κ_i is a prime divisor, i.e., a point. If $P_a > 1$,

$$\begin{aligned}
\widehat{h}(p,p) + \widehat{h}(p,\kappa) \quad &= \quad h(p,p) - \frac{1}{P_a} h(p,\kappa) - \frac{1}{P_a} h(p,p) + \frac{1}{2P_a(2P_a-2)} h(\kappa,\kappa) + h(p,\kappa) \\
&\quad - \frac{2P_a - 2}{2P_a} h(p,\kappa) - \frac{1}{2P_a} h(\kappa,\kappa) - \frac{2P_a-2}{2P_a} h(p,p) \\
&\quad - \frac{1}{2P_a} \sum d_i\, h(\kappa_i,\kappa_i) + \frac{1}{2P_a} h(\kappa,\kappa) \\
&= \quad \frac{1}{2P_a(2P_a-2)} h(\kappa,\kappa) - \frac{1}{2P_a} \sum d_i\, h(\kappa_i,\kappa_i)
\end{aligned}$$

which is, as claimed, a constant independent of p.

For $P_a = 1$ we have instead

$$\widehat{h}(p,p) + \widehat{h}(p,\kappa) = -\frac{1}{2} h(\kappa,\kappa) - \frac{1}{2} \sum d_i\, h(\kappa_i,\kappa_i).$$

This proves the second claim.

Now suppose $P_a > 1$. Let $\kappa = \sum d_i \kappa_i$ and $\kappa' = \sum e_j \kappa'_j$ be two prime-to-Z divisors of the dualizing sheaf $\mathcal{K}_X$, where each κ_i and κ'_j is a prime Weil divisor (i.e., a point). Then

$$\begin{aligned}
\widehat{h}(\kappa,\kappa') \quad &\stackrel{\text{def}}{=} \quad h(\kappa,\kappa) - \frac{2P_a-2}{2P_a} h(\kappa,\kappa) - \frac{2P_a-2}{2P_a} h(\kappa,\kappa) - \frac{2P_a-2}{2P_a} \sum d_i\, h(\kappa_i,\kappa_i) \\
&\quad - \frac{2P_a-2}{2P_a} \sum e_j\, h(\kappa'_j,\kappa'_j) + \frac{(2P_a-2)^2}{2P_a(2P_a-2)} h(\kappa,\kappa) \\
&= \quad \frac{1}{P_a} h(\kappa,\kappa) - \left(1 - \frac{1}{P_a}\right) \sum d_i\, h(\kappa_i,\kappa_i) - \left(1 - \frac{1}{P_a}\right) \sum e_j\, h(\kappa'_j,\kappa'_j).
\end{aligned}$$

Using the formula developed earlier in this proof, we get

$$(2P_a - 2)\Big(\widehat{h}(p,p) + \widehat{h}(p,\kappa)\Big) \;=\; \frac{1}{2P_a} h(\kappa,\kappa) - \left(1 - \frac{1}{P_a}\right) \sum d_i\, h(\kappa_i,\kappa_i).$$

Comparing these, we see that

$$\widehat{h}\big(\kappa,\kappa'\big) \;=\; (2P_a - 2)\Big(\widehat{h}(p,p) + \widehat{h}(p,\kappa)\Big) \;+\; (2P_a - 2)\Big(\widehat{h}(p,p) + \widehat{h}(p,\kappa')\Big).$$

This proves the third claim.

(5.5.2) Lemma. *Let h be a continuous, symmetric, bi-additive pairing from the product $\mathrm{Pic}(X) \times \mathrm{Pic}(X)$ to $\mathbf{R}$. Suppose that θ is a symmetric, bi-additive pairing on the group of prime-to-Z divisors of X such that θ and h restrict to the same pairing on $\mathrm{Pic}^0(X) \times \mathrm{Pic}^0(X)$, and suppose that, for any prime-to-Z divisor κ of the dualizing sheaf $\mathcal{K}_X$, $\theta(p,p) + \theta(p,\kappa)$ is a constant independent of the non-singular point p. Then θ is of the form*

$$\theta(D,E) \;=\; \widehat{h}(D,E) \;+\; c\,\deg(D)\deg(E)$$

for some real constant c.

Proof. Consider the difference $J(D,E) \overset{\text{def}}{=} \theta(D,E) - \widehat{h}(D,E)$. Then J vanishes on pairs of divisors of total degree 0. Fix a non-singular point p_0 on X. Then

$$\begin{aligned}
J(D,E) &= J\Big(D - \deg(D)\,p_0,\ E - \deg(E)\,p_0\Big) + J\Big(D, \deg(E)\,p_0\Big) + J\Big(\deg(D)\,p_0, E\Big) \\
&\qquad - J\Big(\deg(D)\,p_0, \deg(E)\,p_0\Big) \\
&= \deg(D')\,J(D,p_0) \;+\; \deg(D)\,J(E,p_0) \;-\; \deg(D)\deg(E)\,J(p_0,p_0) \\
&= \deg(E)\,\Big(J(D,p_0) - \frac{1}{2}\deg(D)\,J(p_0,p_0)\Big) \\
&\qquad + \deg(D)\,\Big(J(E,p_0) - \frac{1}{2}\deg(E)\,J(p_0,p_0)\Big).
\end{aligned}$$

So we can write

$$\theta(D,E) \;=\; \widehat{h}(D,E) \;+\; \deg(E)\,\gamma(D) \;+\; \deg(D)\,\gamma(E)$$

for some additive, real-valued map γ defined on the group of prime-to-Z divisors. For any prime-to-Z divisor κ of the dualizing sheaf,

$$\begin{aligned}
\theta(p,p) + \theta(p,\kappa) &= \widehat{h}(p,p) + 2\gamma(p) + \widehat{h}(p,\kappa) + (2P_a - 2)\gamma(p) + \gamma(\kappa) \\
&= \widehat{h}(p,p) + 2P_a\gamma(p) + \widehat{h}(p,\kappa) + \gamma(\kappa).
\end{aligned}$$

Since $\theta(p,p) + \theta(p,\kappa)$ and $\widehat{h}(p,p) + \widehat{h}(p,\kappa)$ are both constants, $\gamma(p)$ must be a constant. The lemma follows.

(5.5.3) Lemma. *Suppose that $P_a > 1$. Let h_1 and h_2 be continuous, symmetric, bi-additive pairings from $\mathrm{Pic}(X) \times \mathrm{Pic}(X)$ to $\mathbf{R}$. The following three conditions are equivalent:*

1. *$\widehat{h}_1 = \widehat{h}_2$.*
2. *$h_1(D,E) = h_2(D,E)$ for all divisor classes D and E in $\mathrm{Pic}^0(X)$.*
3. *There is a continuous homomorphism τ from $\mathrm{Pic}(X)$ to $\mathbf{R}$ such that*

$$h_2(D,E) \;=\; h_1(D,E) \;+\; \deg(E)\,\tau(D) \;+\; \deg(D)\,\tau(E).$$

Proof. Lemma 5.4.3 shows that (2) and (3) are equivalent. Lemma 5.5.1 show that (1) implies (2). So it is enough to show that (3) implies (1).

Consider the difference, $J(D,E) \overset{\text{def}}{=} h_2(D,E) - h_1(D,E)$. Assume (3), so

$$J(D,E) \;=\; \deg(E)\,\tau(D) \;+\; \deg(D)\,\tau(E)$$

for some continuous homomorphism τ from $\mathrm{Pic}(X)$ to $\mathbf{R}$.

Let κ be a prime-to-Z divisor of the dualizing sheaf, and let D and E be any prime-to-Z divisors. Write out $D = \sum d_i D_i$ and $E = \sum e_j E_j$ where each D_i and E_j is a prime divisor, i.e., a point. Then

$$\begin{aligned}
\widehat{h}_1(D,E) - \widehat{h}_2(D,E) &= J(D,E) - \frac{\deg(E)}{2P_a}J(D,\kappa) - \frac{\deg(D)}{2P_a}J(E,\kappa) \\
&\quad -\frac{\deg(E)}{2P_a}\sum d_i J(D_i,D_i) - \frac{\deg(D)}{2P_a}\sum e_j J(E_j,E_j) \\
&\quad +\frac{\deg(D)\deg(E)}{2P_a(2P_a-2)}J(\kappa,\kappa) \\
&= \Big(\deg(E)\tau(D) + \deg(D)\tau(E)\Big) \\
&\quad -\frac{\deg(E)}{2P_a}\Big((2P_a-2)\tau(D) + \deg(D)\tau(\kappa)\Big) \\
&\quad -\frac{\deg(D)}{2P_a}\Big((2P_a-2)\tau(E) + \deg(E)\tau(\kappa)\Big) \\
&\quad -\frac{\deg(E)}{2P_a}\sum 2\, d_i\, \tau(D_i) - \frac{\deg(D)}{2P_a}\sum 2\, e_j\, \tau(E_j) \\
&\quad +\frac{\deg(D)\deg(E)}{2P_a(2P_a-2)}\Big(2(2P_a-2)\tau(\kappa)\Big).
\end{aligned}$$

All the terms of the last expression cancel, so $\widehat{h}_1 = \widehat{h}_2$.

(5.5.4) Lemma. *Suppose $P_a = 1$, and let h_1 and h_2 be continuous, symmetric, bi-additive pairings from $\mathrm{Pic}(X) \times \mathrm{Pic}(X)$ to $\mathbf{R}$. Let c be a real constant. Let κ be a prime-to-Z divisor of the dualizing sheaf $\mathcal{K}_X$. The following are equivalent:*

1. *For all prime-to-Z divisors D and E*

$$\widehat{h}_2(D,E) \;=\; \widehat{h}_1(D,E) \;-\; c\,\deg(D)\deg(E).$$

2. *For all divisors D and E of total degree 0 and all non-singular points p of X*

$$h_2(D,E) \;=\; h_1(D,E) \quad \text{and} \quad h_2(\kappa,p) \;=\; h_1(\kappa,p) \;+\; c.$$

3. *There is a continuous homomorphism τ from $\mathrm{Pic}(X)$ to $\mathbf{R}$ such that, for all-prime-Z divisors D and E,*

$$h_2(D,E) \;=\; h_1(D,E) \;+\; \deg(E)\,\tau(D) \;+\; \deg(D)\,\tau(E) \quad \text{and} \quad \tau(\kappa) \;=\; c.$$

Proof. Lemma 5.4.3 together with a quick calculation shows that 2 and 3 are equivalent.

Suppose 1. Then Lemma 5.5.1 (1.) implies the first equation, and the following calculation implies the second equation:

$$\begin{aligned}
-c &= \widehat{h}_2(p,p) - \widehat{h}_1(p,p) \\
&= h_2(p,p) - \frac{1}{2}h_2(p,\kappa) - \frac{1}{2}h_2(p,\kappa) - \frac{1}{2}h_2(p,p) - \frac{1}{2}h_2(p,p) \\
&\qquad -h_1(p,p) + \frac{1}{2}h_1(p,\kappa) + \frac{1}{2}h_1(p,\kappa) + \frac{1}{2}h_1(p,p) + \frac{1}{2}h_1(p,p) \\
&= -h_2(p,\kappa) + h_1(p,\kappa).
\end{aligned}$$

Finally, we show that 3 implies 1. Let κ be a prime-to-Z divisor of the dualizing sheaf $\mathcal{K}_X$, let D and E be prime-to-Z divisors, and let $D = \sum d_i D_i$ and $E = \sum e_j E_j$ be expressions of D and E in terms of prime divisors (i.e., points) D_i and E_j. Then, assuming 3 holds,

$$\begin{aligned}
\widehat{h}_2(D,E) - \widehat{h}_1(D,E) &= \Big(h_2(D,E) - h_1(D,E)\Big) - \frac{\deg(E)}{2}\Big(h_2(D,\kappa) - h_1(D,\kappa)\Big) \\
&\quad - \frac{\deg(D)}{2}\Big(h_2(E,\kappa) - h_1(E,\kappa)\Big) \\
&\quad - \frac{\deg(E)}{2}\sum_i d_i\Big(h_2(D_i,D_i) - h_1(D_i,D_i)\Big) \\
&\quad - \frac{\deg(D)}{2}\sum_j e_j\Big(h_2(E_j,E_j) - h_1(E_j,E_j)\Big) \\
&= \Big(\deg(E)\tau(D) + \deg(D)\tau(E)\Big) \\
&\quad - \frac{\deg(E)}{2}\Big(\deg(D)\tau(\kappa)\Big) - \frac{\deg(D)}{2}\Big(\deg(E)\tau(\kappa)\Big) \\
&\quad - \frac{\deg(E)}{2}\sum d_i\Big(2\,\tau(D_i)\Big) - \frac{\deg(D)}{2}\sum e_j\Big(2\,\tau(E_j)\Big) \\
&= \deg(E)\tau(D) + \deg(D)\tau(E) - \deg(D)\deg(E)\tau(\kappa) - \deg(E)\tau(D) \\
&\quad - \deg(D)\tau(E) \\
&= -\deg(D)\deg(E)\,\tau(\kappa) \\
&= -c\deg(D)\deg(E).
\end{aligned}$$

6. The Classification of Intersection Functions

(6.1) Theorem. *Suppose the arithmetic genus P_a of X is strictly positive.*

1. *Let g be an intersection function on X. If g' is any other intersection function of X, then there is a continuous, symmetric, bi-additive pairing h from the product $\mathrm{Pic}(X) \times \mathrm{Pic}(X)$ to $\mathbf{R}$ and a real constant c such that*

$$g'(D,E) = g(D,E) - \widehat{h}(D,E) + c\deg(D)\deg(E)$$

for all D and E prime-to-Z divisors with disjoint support. Conversely, any function of this form is an intersection function. (The pairing $\widehat{h}$ is as defined in (5.5).)

2. *The set of intersection functions on X can be described explicitly as follows. Let g_0 be a fixed intersection function; we will usually take it to be of the type explicitly constructed in Section 4. Let $\rho\colon \mathrm{Pic}(X) \to V$ be as in Section 5. Recall that ρ depends on a choice of intersection function on each normalized component; fix such a choice. Let g be any real-valued, bi-additive pairing on pairs of prime-to-Z divisors with disjoint supports. Then g is an intersection function if and only if, for all pairs of disjoint non-singular points p and q,*

$$\begin{aligned}
g(p,q) &= g_0(p,q) - h(p,q) + \frac{1}{2P_a}h(p,\kappa) + \frac{1}{2P_a}h(q,\kappa) \\
&\quad + \frac{1}{2P_a}h(p,p) + \frac{1}{2P_a}h(q,q) - \frac{1}{2P_a(2P_a-2)}h(\kappa,\kappa) + c
\end{aligned}$$

when $P_a > 1$, or

$$g(p,q) = g_0(p,q) - h(p,q) + \frac{1}{2}h(p,\kappa) + \frac{1}{2}h(q,\kappa) + \frac{1}{2}h(p,p) + \frac{1}{2}h(q,q) + c$$

when $P_a = 1$, where c is a real constant, κ is a prime-to-Z divisor of the dualizing sheaf $\mathcal{K}_X$, and h is a function of the form

$$\begin{aligned} h(p,q) &= h_0\big(\rho(p),\rho(q)\big) + \sum_i \deg_i(q)\,\tau_i\big(\rho(p)\big) + \sum_i \deg_i(p)\,\tau_i\big(\rho(q)\big) \\ &\quad + \sum_{i,j} c_{i,j}\,\deg_i(p)\deg_j(q) \end{aligned}$$

extended bi-additively to divisors. Here h_0 is a symmetric, bilinear pairing of the real vector space V, the τ_i's are linear functionals of V, one for each component of X, and the $c_{i,j}$'s are real constants symmetric in i and j. The expression $\deg_i(p)$ is defined to be 1 if p is in X_i, otherwise 0. An intersection function written in the above form will be said to be represented by $\big(h_0,\{\tau_i\},\{c_{i,j}\},c\big)$.

3. *The uniqueness of the above representation can be described as follows: if g is an intersection function represented by $\big(h_0,\{\tau_i\},\{c_{i,j}\},c\big)$, and g' is an intersection function represented by $\big(h_0',\{\tau_i'\},\{c_{i,j}'\},c'\big)$, then $g = g'$ if and only if, when $P_a > 1$,*
   i. *$h_0 = h_0'$,*
   ii. *$c = c'$,*
   iii. *there is a linear functional τ on V such that $\tau_i = \tau_i' + \tau$ for each i, and*
   iv. *for each i and j*

$$c_{i,j} - \frac{1}{2}\big(c_{i,i} + c_{j,j}\big) = c_{i,j}' - \frac{1}{2}\big(c_{i,i}' + c_{j,j}'\big),$$

   or, if $P_a = 1$,
   i. *$h_0 = h_0'$,*
   ii. *for each i,*

$$c - \tau_i\big(\rho(\kappa)\big) - \frac{1}{2}\sum c_{j,j}\,\deg_j(\kappa) = c' - \tau_i'\big(\rho(\kappa)\big) - \frac{1}{2}\sum c_{j,j}'\,\deg_j(\kappa)$$

   where κ is a prime-to-Z divisor of the dualizing sheaf $\mathcal{K}_X$,
   iii. *there is a linear functional τ on V such that $\tau_i = \tau_i' + \tau$ for each i,*
   iv. *for each i and j*

$$c_{i,j} - \frac{1}{2}\big(c_{i,i} + c_{j,j}\big) = c_{i,j}' - \frac{1}{2}\big(c_{i,i}' + c_{j,j}'\big).$$

(6.1.1) Corollary. *All intersection functions on X are C^∞. The set of intersection functions on X is parameterized by a real vector space of dimension $1 + (n^2+n)/2$.*

Proof of Corollary 6.1.1. By Theorem 4.1 there is a C^∞ intersection function g_0 on X. The function ρ depends on a choice of intersection functions on each normalized component of g; by Theorem 4.1 we can choose these functions to be C^∞. The definition of ρ implies that the resulting function $p \mapsto \rho(p)$ is a C^∞

function on $X - Z$. So the formula for the general intersection function g given in statement 2 of Theorem 6.1 implies that the function $(p,q) \mapsto g(p,q)$ is C^∞.

It follows from statement 2 and 3 of Theorem 6.1 that the intersection functions on X are parameterized by a finite dimensional real vector space. The dimension is easily calculated from the fact that V is an $n-m+1$ dimensional real vector space.

Proof of Theorem 6.1. Let g' be an intersection function. Consider the difference $\theta \stackrel{\text{def}}{=} g - g'$. The properties of Definition 2.1 have the following implications for θ:

- Property III implies that θ is a pairing of prime-to-Z divisors, even if the two divisors do not have disjoint support.
- Properties I and II imply that θ is a symmetric, bi-additive pairing.
- Property IV implies that θ, restricted to divisors of total degree 0, is a pairing $\mathrm{Pic}^0(X)\times\mathrm{Pic}^0(X) \to \mathbf{R}$, and Property III implies that this pairing is continuous.
- So Lemma 5.4.3 implies that we can extend θ to a continuous, real, symmetric, bi-additive pairing h on all of $\mathrm{Pic}(X)$.
- Finally, Property V implies that, for any prime-to-Z divisor κ of the dualizing sheaf $\mathcal{K}_X$, the quantity $\theta(p,p) + \theta(p,\kappa)$ is a constant independent of the non-singular point p.

By Lemma 5.5.2, $\theta(D,E) = \widehat{h}(D,E) + c\,\deg(D)\deg(E)$ for some real constant c. Therefore, g' is of the desired form.

Conversely suppose that

$$g'(D,E) = g(D,E) - \widehat{h}(D,E) + c\,\deg(D)\deg(E)$$

for some h, a continuous, symmetric, bi-additive pairing $\mathrm{Pic}(X)\times\mathrm{Pic}(X) \to \mathbf{R}$, and some real constant c. Properties I, II, and IV of the definition of the intersection function follow from the definition of $\widehat{h}$.

The pairing $\widehat{h}$ is defined in terms of the pairing h which is continuous on pairs of divisors. In particular, the function $(p,q) \mapsto h(p,q)$ and consequently the function $(p,q) \mapsto \widehat{h}(p,q)$ are defined and continuous for all pairs of non-singular points. Thus Property III is satisfied.

Property V is equivalent to the requirement that, for any prime-to-Z divisor of the dualizing sheaf of X, $\widehat{h}(p,p) + \widehat{h}(p,\kappa)$ is a constant independent of the non-singular point p. So Lemma 5.5.1, statement 2 implies Property V.

Therefore, g' is an intersection function of X.

The second statement follows from the first statement by using Lemma 5.4.2 together with the definition of $\widehat{h}$. Finally, the third statement follows easily from the following lemma.

(6.1.2) Lemma. *Consider the following real, continuous, symmetric, bi-additive pairing of* $\mathrm{Pic}^{\mathrm{a}}(X)$*:*

$$\begin{aligned} h(D,E) &= h_0\Big(\rho(D),\rho(E)\Big) + \sum \deg_i(E)\,\tau_i(\rho(D)) \\ &\qquad + \sum \deg_j(D)\,\tau_j(\rho(E)) + \sum c_{i,j}\,\deg_i(D)\deg_j(E) \end{aligned}$$

where h_0 is a symmetric, bilinear pairing of the real vector space V, each τ_i is a linear function from V to $\mathbf{R}$, and the $c_{i,j}$'s are real constants symmetric in i and j. Let c be a real constant.

Suppose $P_a > 1$. Then $\widehat{h}(D,E) + c\deg(D)\deg(E) = 0$ for all pairs of prime-to-Z divisors D and E if and only if
1. *h_0 is the zero pairing,*
2. *$c = 0$,*
3. *τ_i is independent of i, and*
4. *$c_{i,j} = \frac{1}{2}\big(c_{i,i} + c_{j,j}\big)$ for every i and j.*

Suppose $P_a = 1$. Then $\widehat{h}(D,E) + c\deg(D)\deg(E) = 0$ for all pairs of prime-to-Z divisors D and E if and only if
1. *h_0 is the zero pairing,*
2. *$c = \tau_i\big(\rho(\kappa)\big) + \frac{1}{2}\sum \deg_j(\kappa)\, c_{j,j}$ for each i, where κ is a prime-to-Z divisor of the dualizing sheaf $\mathcal{K}_X$,*
3. *τ_i is independent of i, and*
4. *$c_{i,j} = \frac{1}{2}\big(c_{i,i} + c_{j,j}\big)$ for every i and j.*

Proof. First assume that $P_a > 1$. Suppose that $\widehat{h}(D,E) = -c\deg(D)\deg(E)$ for all prime-to-Z divisors D and E. By Lemma 5.5.1

$$\widehat{h}(\kappa,\kappa) - 2(2P_a - 2)\left(\widehat{h}(p,p) + \widehat{h}(p,\kappa)\right) = 0$$

for any non-singular point p of X and any prime-to-Z divisor κ of the dualizing sheaf $\mathcal{K}_X$. So

$$\begin{aligned} 0 &= \widehat{h}(\kappa,\kappa) - 2(2P_a - 2)\left(\widehat{h}(p,p) + \widehat{h}(p,\kappa)\right) \\ &= -(2P_a-2)^2 c - 2(2P_a - 2)\Big(-c - (2P_a - 2)\, c\Big) \\ &= -(2P_a - 2)(-2P_a)\, c. \end{aligned}$$

Hence $c = 0$ and, consequently, $\widehat{h} = 0$.

Since $\widehat{h} = 0$, it follows from Lemmas 5.4.1 and 5.5.3 that

$$\begin{aligned} h(D,E) &= \deg(E)\left(\tau\big(\rho(D)\big) + \sum \deg_i(D)\, b_i\right) + \deg(D)\left(\tau\big(\rho(E)\big) + \sum \deg_i(E)\, b_i\right) \\ &= \deg(E)\,\tau\big(\rho(D)\big) + \deg(D)\,\tau\big(\rho(E)\big) + \sum_{i,j} \deg_i(D)\deg_j(E)(b_i + b_j) \end{aligned}$$

for some linear functional τ on V and set of real constants $\{b_1, \ldots, b_m\}$.

By Lemma 5.4.2, the pairing h_0, the functions τ_i, and the constants $c_{i,j}$ are uniquely determined by h. So the above expression for h implies that $h_0 = 0$ and that $\tau_i = \tau$. The uniqueness of $c_{i,j}$ implies that

$$c_{i,j} = b_i + b_j.$$

This implies that

$$b_i = \frac{1}{2} c_{i,i}$$

and so

$$c_{i,j} = \frac{1}{2}(c_{i,i} + c_{j,j}).$$

The converse, in the case $P_a > 1$, follows simply by reversing the above arguments. We leave it to the reader to check the details.

Now assume $P_a = 1$, and suppose that $\widehat{h}(D,E) = -c\deg(D)\deg(E)$ for all prime-to-Z divisors D and E. So, by Lemmas 5.5.4 and 5.3.2,

$$\begin{aligned} h(D,E) &= \deg(E)\Big(\tau\big(\rho(D)\big) + \sum \deg_i(D)\, b_i\Big) + \deg(D)\Big(\tau\big(\rho(E)\big) + \sum \deg_i(E)\, b_i\Big) \\ &= \deg(E)\,\tau\big(\rho(D)\big) + \deg(D)\,\tau\big(\rho(E)\big) + \sum_{i,j}(b_i + b_j)\,\deg_i(D)\deg_j(E) \end{aligned}$$

for some linear functional τ on V and set of real constants $\{b_1, \ldots, b_m\}$. Again, by Lemma 5.5.4, if κ is a prime-to-Z divisor of the dualizing sheaf, then

$$\tau\big(\rho(\kappa)\big) + \sum_i b_i \deg_i(\kappa) = c.$$

By Lemma 5.3.2, the pairing h_0, the functions τ_i, and the constants $c_{i,j}$ are uniquely determined by h. So the above expression for h implies that $h_0 = 0$ and that $\tau_i = \tau$. The uniqueness of $c_{i,j}$ implies that

$$c_{i,j} = b_i + b_j.$$

This implies that

$$b_i = \frac{1}{2}c_{i,i} \qquad \text{and} \qquad c_{i,j} = \frac{1}{2}(c_{i,i} + c_{j,j}).$$

Finally, the equation

$$\tau\big(\rho(\kappa)\big) + \sum b_j \deg_j(\kappa) = c$$

becomes

$$\tau_i\big(\rho(\kappa)\big) + \sum_j \frac{1}{2}c_{j,j}\deg_j(\kappa) = c$$

for each i.

The converse, in the case $P_a = 1$, follows simply by reversing the above arguments. We leave it to the reader to check the details.

(6.2) Definition of the Associated Bilinear Pairing. Assume that P_a is strictly positive, and let V be the real vector space defined in Section 5. For any intersection function g on X, we associate to g a real, symmetric, bilinear pairing h_g on V as follows: express g as in Theorem 6.1, statement 2 where g_0 is an intersection function on X of the type constructed in the proof of Theorem 4.1, and define h_g to be the real, symmetric, bilinear pairing h_0 occurring in this expression.

Call this pairing the *bilinear pairing associated to* g. This pairing is characterized by the property that $g - g_0$ restricted to pairs of divisors whose classes are in $\mathrm{Pic}^{\mathrm{a}}(X)$ is equal to $-h_g\big(\rho(D), \rho(E)\big)$. Although ρ depends on a choice of intersection function on each normalized component of X, Proposition 5.3.1 and Corollary 5.3.2 imply that h_g is independent of this choice. The following proposition shows that h_g is, in fact, independent of g_0.

Theorem 6.1, statement 2, shows that all real, symmetric, bilinear pairings of V occur as the bilinear pairing associated to some intersection function. Since V is canonically a quotient of $\mathbf{R}^n$, we can also think of h_g as an n by n matrix.

(6.2.1) Proposition. *Assume $P_a > 0$, and let g be an intersection function on Z. The bilinear pairing h_g, defined above, is independent of the choice of g_0.*

Proof. It is enough to show that if g_0 and g_0' are two intersection functions on X of the type constructed in the proof of Theorem 4.1, then $g_0 - g_0'$ is zero for pairs of divisors whose classes are in $\mathrm{Pic}^{\mathrm{a}}(X)$.

For any two divisors classes of $\mathrm{Pic}^{\mathrm{a}}(X)$, we can choose a divisor from each class, D and E say, such that D and E are prime-to-Z divisors with disjoint supports. For each normalized component X_i, let D_i and E_i be the restrictions to X_i of D and E respectively. I claim that

$$g_0(D, E) = \sum_i g_i(D_i, E_i)$$

where each g_i is any choice of an intersection function on the normalized component X_i. We prove this claim by induction on the number of singularities of X. For X non-singular this equality follows from Propostion 3.3.3 for X of positive genus, and the remarks at the end of (3.2.3) for X of genus zero. I leave it to the reader to verify the induction step by looking at the constructions given in the proof of Theorem 4.1.

The same thing is true when we replace g_0 by g_0'. Hence $g_0' - g_0$ is zero on pairs of divisors whose classes are in $\mathrm{Pic}^{\mathrm{a}}(X)$.

(6.2.2) Corollary. *Assume that X is irreducible and that $P_a > 0$. Then h_g determines g up to a constant. In other words, if g and g' are two intersection functions such that $h_g = h_{g'}$, then there is a constant c such that*

$$g'(D, E) = g(D, E) + c \deg(D) \deg(E)$$

for all pairs of prime-to-Z divisors D and E with disjoint support.

Proof. This follows from Theorem 6.1, statement 3.

(6.3) Theorem (Classification - Arithmetic Genus 0). *Suppose $P_a = 0$, and let g be an intersection function on X. Let g' be a real-valued, bi-additive pairing defined on pairs of prime-to-Z divisors with disjoint supports. Then g' is an intersection function if and only if its restriction to disjoint non-singular points p and q is of the form*

$$g'(p, q) = g(p, q) + \gamma(p) + \gamma(q) + \sum c_{i,j} \deg_i(p) \deg_j(q)$$

where γ is a continuous function on the set of non-singular points of X, and $c_{i,j}$ are real constants, symmetric in i and j, satisfying

$$c_{i,i} + \sum_j l_j c_{i,j} - 2 \sum_j c_{i,j} = 0$$

for all i. Here l_j is the number of R_i's and S_i's contained in the normalized component X_j.

Proof. Suppose that g' is an intersection function, and consider the difference $\theta \stackrel{\text{def}}{=} g' - g$. As in the proof of Theorem 6.1 above, the restriction of θ to divisors of total degree 0 is a real, symmetric, continuous, bi-additive pairing on $\mathrm{Pic}^0(X)$.

Let p_0 be a fixed non-singular point of X. Note that

$$h(D,E) \stackrel{\text{def}}{=} \theta\Big(D - \deg(D)\,p_0,\ E - \deg(E)\,p_0\Big)$$

is a real, symmetric, continuous, bi-additive pairing h on $\operatorname{Pic}(X)$.

Let p and q be distinct non-singular points of X. Then

$$\begin{aligned} \theta(p,q) &= \theta(p-p_0, q-p_0) + \theta(p,p_0) + \theta(q,p_0) - \theta(p_0,p_0) \\ &= h(p,q) + \gamma(p) + \gamma(q) \end{aligned}$$

where γ is a continuous function on the non-singular points of X.

Since $P_a = 0$, the space V has dimension 0. This follows from the fact that V has dimension $n - m + 1$. So Lemma 4.3.2 implies that

$$h(D,E) = \sum_{i,j} c_{i,j} \deg_i(D) \deg_j(E)$$

where $c_{i,j}$ are real constants such that $c_{i,j} = c_{j,i}$. Therefore,

$$\theta(p,q) = \sum_{i,j} c_{i,j} \deg_i(p) \deg_j(q) + \gamma(p) + \gamma(q).$$

Property V of Definition 2.1 implies that, for any prime-to-Z divisor κ of the dualizing sheaf $\mathcal{K}_X$, the quantity $\theta(p,p) + \theta(p,\kappa)$ is a constant independent of the non-singular point p. Extend γ bi-additively to divisors, and observe the following

$$\begin{aligned} \theta(p,p) + \theta(p,\kappa) &= c_{i,i} + 2\gamma(p) + \sum_j c_{i,j} \deg_j(\kappa) - 2\gamma(p) + \gamma(\kappa) \\ &= c_{i,i} + \sum_j c_{i,j} \deg_j(\kappa) + \gamma(\kappa) \end{aligned}$$

where i is such that p is in X_i. In particular,

$$c_{i,i} + \sum_j c_{i,j} \deg_j(\kappa)$$

is a constant independent of i.

Now, $\deg_j(\kappa)$ is equal to $l_j - 2$ where l_j is the number of R_i's and S_i's contained in X_j. Thus

$$c_{i,i} + \sum_j l_i\, c_{i,j} - 2\sum_j c_{i,j}$$

is a constant c, independent of i.

By adding c to each $c_{i,j}$ and subtracting $c/2$ from γ, we can assume that

$$c_{i,i} + \sum_j l_i\, c_{i,j} - 2\sum_j c_{i,j} = 0$$

for each i.

Hence

$$g'(p,q) = g(p,q) + \gamma(p) + \gamma(q) + \sum c_{i,j} \deg_i(p) \deg_j(q)$$

where the constants satisfy the above equations.

We leave the converse to the reader.

7. Chern Forms of Intersection Functions

We assume throughout this section that the arithmetic genus P_a of X is strictly positive. We continue to use the notation of Section 5.

If g is an intersection function on X, and if D is a prime-to-Z divisor of X, then we define the function g_D by the rule $p \mapsto g(D,p)$. So g_D is a C^∞-function defined on points of X outside of Z and outside of the support of D. We can use g_D to define a C^∞-metric ν on the line bundle $\mathcal{O}(D)$ by the rule

$$\log\big\|\mathbf{1}|_p\big\|_\nu = -2\,g_D(p)$$

where $\mathbf{1}$ is the canonical meromorphic section of $\mathcal{O}(D)$. The asymptotic behavior of g is such that this metric extends to all of $X - Z$.

Recall that the Chern form associated to a C^∞-metrized line bundle (L,ν) is the $(1,1)$-form given locally around a non-singular point p by the formula

$$\mathrm{chern}(L,\nu) \stackrel{\mathrm{def}}{=} \frac{1}{2\pi\mathbf{i}}\partial\overline{\partial}\log\|s\|_\nu$$

where s is a holomorphic section of L defined and nowhere vanishing in an analytic neighborhood of p. This is well-defined since the ratio of any two holomorphic section of L on such a neighborhood is a holomorphic nowhere vanishing function f on the neighborhood and

$$\partial\overline{\partial}\log\|f\| = 0.$$

Therefore, the Chern form of the metric on $\mathcal{O}(D)$ defined by the function g_D is a $(1,1)$-form on $X - Z$ given, outside the support of D, by the formula

$$\frac{1}{2\pi\mathbf{i}}\partial\overline{\partial}(-2g_D) = \frac{\mathbf{i}}{\pi}\partial\overline{\partial}(g_D).$$

We call this $(1,1)$-form the *Chern form of* g_D and we write it *chern*(g_D).

Since the Chern forms of canonical Green's functions play such a central role in the standard Arakelov theory of non-singular curves (see [L] Chapter II), it is natural to expect that the Chern forms of intersection functions will be of some interest. In this section and the next we calculate the Chern forms associated with intersection functions. We use this in Section 9 to make a connection between Chern forms of intersection functions and certain symmetric, hermitian pairings associated with the P_a-dimensional complex vector space of global sections of the dualizing sheaf. This association forms a pleasing generalization of the relationship on non-singular curves between the Chern form of the canonical Green's function, usually called the canonical or the Arakelov $(1,1)$-form, and the natural inner product structure on the space of global holomorphic differentials.

One corollary of the main theorem of this section, Theorem 7.3, is that the Chern form of g_p is independent of p where p is any non-singular point regarded as a prime divisor of X. So this $(1,1)$-form gives an invariant of the intersection function. In the previous section we defined, for each intersection function g, an invariant h_g which was a real, symmetric, bilinear pairing on the space V. Recall that in Section 5 the real vector space V was defined as the quotient of $\mathbf{R}^n$; thus h_g can be represented as a real, symmetric, bilinear pairing of $\mathbf{R}^n$, i.e., a real, symmetric, n by n matrix. One of the things we will do in this section is show the connection between this matrix and the Chern form associated to g.

(7.1) The Canonical (1,1)-form. Let X_i be a normalized component of X of positive genus. Recall that in (3.4) the canonical $(1,1)$-form of X_i, written μ_i, was defined to be *chern* (g_p) where g is an intersection function on X_i and p is any point of X_i regarded as a prime divisor. By Lemma 3.4.1, μ_i is independent of p.

We will also consider μ_i as a $(1,1)$-form on the whole non-singular set of X by defining it to be zero outside of X_i.

We do not define the canonical $(1,1)$-form of X_i if X_i is of genus 0. However, from (3.2.2) it is clear that if g is any C^∞-intersection function of X_i then chern g_p is independent of p.

(7.2) Definition. For each pair of distinct points p and q on a fixed normalized component X_i of X, let $\omega_i(p,q)$ be the meromorphic differential, holomorphic outside p and q, defined by

$$\omega_i(p,q) \stackrel{\text{def}}{=} -2\,\partial_x\big(g_i(p,x) - g_i(q,x)\big)$$

where g_i is a C^∞-intersection function on X_i.

(7.2.1) Lemma.

1. *The above definition indeed results in a meromorphic differential on X_i, holomorphic outside p and q.*
2. *This differential is independent of the choice of C^∞-intersection function g_i on X_i.*
3. *There is an analytic neighborhood U of p with local coordinate z vanishing to first order at p such that on U*

$$\omega_i(p,q) = \frac{dz}{z} + \beta$$

for some holomorphic differential β defined on U. So $\omega_i(p,q)$ has a simple pole at p with residue $+1$. Likewise, $\omega_i(p,q)$ has a simple pole at q but with residue -1.

Proof. As we mentioned above in (7.1), if g is any intersection function of X_i, the Chern form of g_p is independent of the point p, or equivalently $\partial_x\overline{\partial}_x\big(g_i(p,x)\big)$ is independent of p. So

$$\overline{\partial}_x\Big(\partial_x\big(g_i(p,x) - g_i(q,x)\big)\Big) = -\partial_x\overline{\partial}_x\big(g_i(p,x)\big) + \partial_x\overline{\partial}_x\big(g_i(q,x)\big) = 0.$$

Therefore,

$$\omega_i(p,q) \stackrel{\text{def}}{=} -2\,\partial_x\Big(g_i(p,x) - g_i(q,x)\Big)$$

is a holomorphic differential outside p and q.

Let U be an analytic neighborhood of p, not containing q, with local coordinate z vanishing to first order at p. Property III of Definition 2.1 tells us that, on U,

$$g_i(p,x) - g_i(q,x) = -\log\big|z(x)\big| + f(x)$$

for some function f of U which is C^∞ even at $x = p$. An elementary calculation yields

$$\partial \log \|z\| = \frac{dz}{z}$$

Therefore,

$$\omega_i(p,q) = \frac{dz}{z} + \beta$$

for some holomorphic differential β defined on U. We obtain a similar result (with opposite sign) for q.

Now suppose g_i' is another choice of C^∞ intersection function, or more generally that

$$g_i'(p,q) = g_i(p,q) + \gamma(p) + \gamma(q)$$

for some C^∞ function γ. (By Propositions 3.3.3, for positive genus, and the discussion in (3.2.3), for genus 0, this includes at least all intersection functions on X.) Then

$$g_i'(p,x) - g_i'(q,x) = g_i(p,x) - g_i(q,x) + \gamma(p) - \gamma(q).$$

So, since the two expressions differ by a quantity independent of x,

$$\partial\Big(g_i'(p,x) - g_i'(q,x)\Big) = \partial\Big(g_i(p,x) - g_i(q,x)\Big).$$

(7.2.3) Definition. For each pair of points p and q on a fixed normalized component X_{i_0} of X, let $\omega(p,q)$ be the meromorphic differential on the non-singular points of X defined by

$$\omega(p,q) \stackrel{\text{def}}{=} \omega_{i_0}(p,q)$$

on X_{i_0} and

$$\omega(p,q) \stackrel{\text{def}}{=} 0$$

elsewhere.

Let $T_1,\ldots,T_n$ be the nodes of X and let R_j, r_j, S_j, and s_j be as in Section 5. From each normalized component X_i, choose, once and for all, a non-singular point e_i. For each node T_j define

$$\alpha_j \stackrel{\text{def}}{=} \omega(R_j, e_{r_j}) - \omega(S_j, e_{s_j}).$$

By the above lemma,

1. α_j is a holomorphic differential on the disjoint union of the X_i's except at R_j, e_{r_j}, S_j, and e_{s_j}. This means that α_j is a section of $\mathcal{K}_X$ outside the node T_j and the non-singular points e_{r_j} and e_{s_j}.
2. α_j is independent of the choice of g_i's, but does depend on the choice of the e_i's.
3. α_j has simple poles at R_j and S_j. The residue of α_i at R_j is $+1$ and at S_j it is -1. If $s_j = r_j$ then α_j is holomorphic at e_{r_j} and e_{s_j}. Otherwise, α_j has simple poles at e_{r_j} and e_{s_j} where the residue of α_i at e_{s_j} is $+1$ and at e_{r_j} it is -1. In any case, α_j is a global section of $\mathcal{K}_X\big(e_{r_j} + e_{s_j}\big)$.

Now we can state the main theorem of this section:

(7.3) Theorem. *Let g be an intersection function on X and let h_g be the associated real, symmetric, bilinear pairing of V. Let $H = [h_{i,j}]$ be the n by n matrix representation of h_g. Then, for any non-singular point p of X considered as a prime divisor,*

$$\operatorname{chern}(g_p) = \frac{1}{P_a}\sum_i P_i\,\mu_i + \frac{\mathbf{i}}{4\pi P_a}\sum_{i,j} h_{i,j}\,\alpha_i \wedge \overline{\alpha_j}$$

where μ_i is the canonical $(1,1)$-form of the normalized component X_i, extended by zero to all the non-singular points of X, and P_i is the genus of X_i. (if X_i has genus 0, μ_i is not well-defined but $P_i\,\mu_i$ is always well-defined). This formula is defined on the non-singular points of X outside $\{e_1,\ldots,e_m\}$, but, by choosing $\{e_1,\ldots,e_m\}$ differently, we see that this gives an expression for all the non-singular points of X.

The proof of this theorem will be postponed until the next section.

(7.3.1) Corollary. *For any intersection function g on X, $\mathrm{chern}(g_p)$ is independent of the non-singular point p. We define this real $(1,1)$-form to be the Chern form of g and we write it $\mathrm{chern}(g)$.*

(7.3.2) Corollary. *A real (1,1)-form μ is the Chern form of an intersection function of X if and only if μ is of the form*

$$\frac{1}{P_a}\sum_i P_i\,\mu_i + \frac{\mathbf{i}}{4\pi P_a}\sum_{i,j} h_{i,j}\,\alpha_i\wedge\overline{\alpha_j}$$

where μ_i is the canonical $(1,1)$-form of the normalized component X_i, P_i is its genus, and where $H=[h_{i,j}]$ is a matrix representing some real, symmetric, bilinear pairing h of V.

Proof. Theorem 6.1, statement 2 shows that every real, symmetric, bilinear pairing of V occurs as the associated pairing h_g of some intersection function g.

8. Chern Forms of Intersection Functions: Proofs.

This entire section is dedicated to the proof of Theorem 7.3. The notation will be, of course, that of the previous section.

(8.1) Lemma. *Let g be a C^∞-intersection function on X of the type constructed in Section 4. Then $\mathrm{chern}(g_p)$ is independent of p where p is any non-singular point of X. Furthermore, if $P_a>0$, then*

$$\mathrm{chern}(g_p)=\frac{1}{P_a}\sum P_i\,\mu_i.$$

Note that if $P_i=0$ then μ_i is not well-defined, but the product $P_i\,\mu_i$ is always well-defined.

Proof. For non-singular curves this follows from Lemma 3.4.1 for positive genus and from (3.2.2) for genus zero. For singular curves we proceed by an induction parallel to that of Section 4. Thus we can restrict ourselves to two cases.

Case 1. Let X be a nodal curve which has a C^∞-intersection function g constructed with the techniques of Section 4. Let R and S be two non-singular points of X, and let X' be the new nodal curve resulting from identifying R and S into an ordinary double point. Let P_a be the arithmetic genus of X and let $P_a'=P_a+1$ be the arithmetic genus of X'. The constructions of Section 4 defines the following intersection function on X':

$$g'(p,q)\overset{\mathrm{def}}{=}g(p,q)-\frac{1}{2P_a'}\,g(p+q,R+S).$$

First consider the case $P_a > 0$. The normalized components of X' are the same as those of X; fix one such X_i. By induction, we can assume that

$$\operatorname{chern}(g_p)\Big|_{X_i} \stackrel{\text{def}}{=} \frac{\mathbf{i}}{\pi}\,\partial_q\overline{\partial}_q\big(g(p,q)\big)\Big|_{X_i} = \frac{P_i}{P_a}\,\mu_i$$

where P_i is the genus of X_i. For all non-singular points p of X

$$\begin{aligned}\operatorname{chern}\big(g_p'\big)\Big|_{X_i} &= \frac{\mathbf{i}}{\pi}\,\partial_q\overline{\partial}_q\big(g(p,q)\big)\Big|_{X_i} - \frac{1}{2P_a'}\frac{\mathbf{i}}{\pi}\,\partial_q\overline{\partial}_q\big(g(q,R)+g(q,S)\big)\Big|_{X_i}\\ &= \frac{P_i}{P_a}\mu_i - \frac{1}{2P_a'}\left(2\frac{P_i}{P_a}\mu_i\right) = \frac{P_i}{P_a}\,\frac{P_a'-1}{P_a'}\mu_i = \frac{P_i}{P_a}\,\frac{P_a}{P_a'}\,\mu_i = \frac{P_i}{P_a'}\,\mu_i\end{aligned}$$

which is the desired expression.

Now consider the case $P_a = 0$. Hence $P_a' = 1$, and by induction $\mathit{chern}(g_p)$ is independent of p. Therefore,

$$\begin{aligned}\operatorname{chern}\big(g_p'\big) &= \frac{\mathbf{i}}{\pi}\,\partial_q\overline{\partial}_q\big(g_p(p,q)\big) - \frac{1}{2}\frac{\mathbf{i}}{\pi}\,\partial_q\overline{\partial}_q\big(g(q,R)+g(q,S)\big)\\ &= \operatorname{chern}(g_p) - \frac{1}{2}\big(\operatorname{chern}(g_R)+\operatorname{chern}(g_S)\big) = 0.\end{aligned}$$

In this case the geometric genus P_i of each component X_i is also 0, so we also get the desired expression.

Case 2. Let X and Y be nodal curves with intersection functions g_X and g_Y, respectively, each constructed as in Section 4. Let R_X and R_Y be non-singular points on X and Y respectively, and let X' be the nodal curve formed from identifying R_X and R_Y into an ordinary double point. Let P_X, P_Y, and P_a' be the arithmetic genera of X, Y, and X' respectively. Recall that $P_a' = P_X + P_Y$.

Case 2a. First we assume that $P_a' > 0$.

The construction of Section 4 gives an intersection function g' on X' of the following form. If p and q are both in X then

$$g'(p,q) \stackrel{\text{def}}{=} g_X(p,q) - \frac{P_Y}{P_a'}\,g_X(p+q,R_X),$$

and if p is in Y, but q is in X, then

$$g'(p,q) \stackrel{\text{def}}{=} \frac{P_Y}{P_a'}\,g_Y(p,R_Y) + \frac{P_X}{P_a'}\,g_X(q,R_X).$$

Fix a normalized component X_i of X with genus P_i.

First suppose $P_X > 0$. If p is a point of X, then

$$\begin{aligned}\operatorname{chern}\big(g_p'\big)\Big|_{X_i} &= \frac{\mathbf{i}}{\pi}\,\partial_q\overline{\partial}_q\big(g_X(p,q)\big) - \frac{P_Y}{P_a'}\frac{\mathbf{i}}{\pi}\,\partial_q\overline{\partial}_q\big(g_X(q,R_X)\big) = \frac{P_i}{P_X}\,\mu_i - \frac{P_Y}{P_a'}\,\frac{P_i}{P_X}\,\mu_i\\ &= \frac{P_i}{P_X}\left(\frac{P_a'-P_Y}{P_a'}\right)\mu_i = \frac{P_i}{P_X}\,\frac{P_X}{P_a'}\,\mu_i = \frac{P_i}{P_a'}\,\mu_i,\end{aligned}$$

and, if p is a point of Y, then

$$\operatorname{chern}\big(g_p'\big)\Big|_{X_i} = \frac{P_X}{P_a'}\frac{\mathbf{i}}{\pi}\,\partial_q\overline{\partial}_q\big(g_X(q,R_X)\big) = \frac{P_X}{P_a'}\,\frac{P_i}{P_X}\,\mu_i = \frac{P_i}{P_a'}\,\mu_i$$

So we get the desired result.
Now suppose $P_X = 0$, hence $P_Y = P'_a$ and $P_i = 0$. By induction, we can assume that $\frac{\mathbf{i}}{\pi}\,\partial_q\overline{\partial}_q\big(g_X(p,q)\big)$ is independent of p. If p is in X, then

$$\begin{aligned}\operatorname{chern}\big(g'_p\big)\Big|_{X_i} &= \frac{\mathbf{i}}{\pi}\,\partial_q\overline{\partial}_q\big(g_X(p,q)\big) - \frac{P_Y}{P'_a}\frac{\mathbf{i}}{\pi}\,\partial_q\overline{\partial}_q\big(g_X(q,R_X)\big)\\ &= \frac{\mathbf{i}}{\pi}\,\partial_q\overline{\partial}_q\big(g_X(p,q)\big) - \frac{\mathbf{i}}{\pi}\,\partial_q\overline{\partial}_q\big(g_X(R_X,q)\big) = 0,\end{aligned}$$

and, if p is in Y, then

$$\operatorname{chern}\big(g'_p\big)\Big|_{X_i} = \frac{0}{P'_a}\frac{\mathbf{i}}{\pi}\,\partial_q\overline{\partial}_q\big(g_X(q,R_X)\big) = 0.$$

So the Chern form of g_p restricted to X_i is 0, which is of the desired form since $P_i = 0$.
A similar result follows for each normalized component of Y.

Case 2b. We continue the previous case, but now we assume that $P'_a = P_X = P_Y = 0$.
If p and q are both in X then

$$g'(p,q) = g_X(p,q) - \frac{1}{2}\,g_X(p+q,R_X),$$

and if p is in Y but q is in X then

$$g'(p,q) = \frac{1}{2}\,g_X(q,R_X) + \frac{1}{2}\,g_Y(p,R_Y).$$

We can assume, by induction, that $\frac{\mathbf{i}}{\pi}\,\partial_q\overline{\partial}_q\big(g_X(p,q)\big)$ independent of p. So, if p is in X,

$$\operatorname{chern}\big(g'_p\big)\Big|_{X} = \frac{\mathbf{i}}{\pi}\,\partial_q\overline{\partial}_q\big(g_X(p,q)\big) - \frac{1}{2}\frac{\mathbf{i}}{\pi}\,\partial_q\overline{\partial}_q\big(g_X(q,R_X)\big) = \frac{1}{2}\frac{\mathbf{i}}{\pi}\,\partial_q\overline{\partial}_q\big(g_X(p,q)\big)$$

and, if p is in Y, then

$$\operatorname{chern}\big(g'_p\big)\Big|_{X} = \frac{1}{2}\frac{\mathbf{i}}{\pi}\,\partial_q\overline{\partial}_q\big(g_X(q,R_X)\big) = \frac{1}{2}\frac{\mathbf{i}}{\pi}\,\partial_q\overline{\partial}_q\big(g_X(q,p)\big).$$

Since, by induction, $\frac{\mathbf{i}}{\pi}\,\partial_q\overline{\partial}_q\big(g_X(p,q)\big)$ is independent of p, the above calculation shows that $chern\big(g'_p\big)\Big|_{X}$ is independent of p. By the same sort of argument we can show that $chern\big(g'_p\big)\Big|_{Y}$ is also independent of p.

(8.2) Lemma. *Let ϕ be a continuous homomorphism from $\operatorname{Pic}(X)$ to $\mathbf{R}$. Then ϕ restricted to non-singular points of X is a C^∞-function and $\partial\overline{\partial}(\phi) = 0$.*

Proof. By Lemma 5.4.1,

$$\phi(q) = \phi_V\big(\rho(q)\big) + c(q)$$

where ϕ_V is a linear functional on V, and $c(q)$ is a locally constant function defined on the non-singular points of X. Thus, $q \mapsto \phi(q)$ is a C^∞-function, and

$$\partial\overline{\partial}(\phi) = \partial_q\overline{\partial}_q\Big(\phi_V(\rho(q))\Big).$$

Since V is defined as a quotient of $\mathbf{R}^n$, the functional ϕ_V pulls back to a linear functional on $\mathbf{R}^n$ where n is the number of nodes of X. Let $[a_1, \ldots, a_n]$ be the matrix representation of this functional with respect to the standard basis on $\mathbf{R}^n$. Let R_i, S_i, r_i, and s_i be as in Section 5. The definition of ρ in (5.3) implies that

$$\phi_V(\rho(q)) = \sum_i a_i\Big(-g_{r_i}(q_{r_i}, R_i) + g_{s_i}(q_{s_i}, S_i)\Big)$$

where q_i is defined to be q if q is a point of X_i, otherwise it is defined to be the zero divisor. Thus

$$\begin{aligned}\frac{\mathbf{i}}{\pi}\partial\overline{\partial}(\phi) &= \frac{\mathbf{i}}{\pi}\partial_q\overline{\partial}_q\Big(\phi_V(\rho(q))\Big)\\ &= \sum_i a_i\Big(-\frac{\mathbf{i}}{\pi}\partial_q\overline{\partial}_q\big(g_{r_i}(q_{r_i}, R_i)\big) + \frac{\mathbf{i}}{\pi}\partial_q\overline{\partial}_q\big(g_{s_i}(q_{s_i}, S_i)\big)\Big)\\ &= \sum_i a_i\left(-\mu_{r_i} + \mu_{s_i}\right)\end{aligned}$$

where μ_i is the canonical $(1,1)$-form of X_i extended by zero to all of X. (Or, if X_i has genus 0, then it is some $(1,1)$-form independent of p).

The fact that ϕ_V is defined on V implies that

$$\sum_i a_i\left(c_{r_i} - c_{s_i}\right) = 0$$

for all real m-tuples $(c_1, \ldots, c_m)$. This in turn implies that

$$\frac{\mathbf{i}}{\pi}\partial\overline{\partial}(\phi) = \sum_i a_i\left(-\mu_{r_i} + \mu_{s_i}\right) = 0,$$

so $\partial\overline{\partial}(\phi) = 0$.

(8.3) Lemma. *Suppose $P_a > 0$. Let h be a real, continuous, symmetric, bi-additive pairing on* Pic(X), *and let $\widehat{h}$ its associated pairing defined in Section 7. By restricting $\widehat{h}$ to non-singular points of X, regarded as prime Weil divisors of X, we get a C^∞-function of two variables on X. Considered as such,*

$$\partial_q\overline{\partial}_q\left(\widehat{h}(p,q)\right) = -\frac{1}{2P_a}\partial_q\overline{\partial}_q\Big(h(q,q)\Big).$$

Proof. Fix, once and for all, a non-singular point p of X. By the definition of (5.5)

$$\widehat{h}(p,q) = h(p,q) - \frac{1}{2P_a}h(q,\kappa) - \frac{1}{2P_a}h(q,q) + c$$

where κ is a prime-to-Z divisor of the dualizing sheaf $\mathcal{K}_X$, and c is a term which depends only on p. The previous lemma implies that we can ignore all but the third term. So we get the desired result:

$$\partial_q\overline{\partial}_q\left(\widehat{h}(p,q)\right) = \partial_q\overline{\partial}_q\left(-\frac{1}{2P_a}h(q,q)\right)$$

(8.4) Recall the definition, in the previous section, of the differentials $\alpha_1, \ldots, \alpha_n$ one for each node of X. This definition depends on the choice of fixed points $e_1, \ldots, e_m$, one for each component of X. We fix such a choice once and for all.

(8.4.1) Lemma. *Let h be a real-valued, continuous, symmetric, bi-additive pairing on* Pic(X). *Let h_0 be the real, symmetric, bilinear pairing on V which occurs when we write h in the form given in Lemma 5.4.2. Consider the function $q \mapsto h(q,q)$ defined on all the non-singular points of X. Outside $\{e_1, \ldots, e_m\}$,*

$$\partial_q \overline{\partial}_q \big(h(q,q)\big) \;=\; \frac{1}{2} \sum_{i,j} h_{i,j}\, \alpha_i \wedge \overline{\alpha_j}$$

where $H = [h_{i,j}]$ is the n by n matrix representing h_0.

Proof. Lemma 5.4.2 implies that

$$h(q,q) \;=\; h_0\big(\rho(q), \rho(q)\big) \;+\; 2\, \tau_{i(q)}\big(\rho(q)\big) \;+\; c(q)$$

where $c(q)$ a locally constant function, $\{\tau_i\}$ a set of linear functionals on V, and $i(q)$ the integer such that q is in $X_{i(q)}$. By Lemma 8.2 only the first term is significant:

$$\partial_q \overline{\partial}_q \big(h(q,q)\big) \;=\; \partial_q \overline{\partial}_q \Big(h_0\big(\rho(q), \rho(q)\big)\Big).$$

The definition of ρ given in (5.3) implies that

$$h_0\big(\rho(p), \rho(p)\big) \;=\; \sum_{i,j} h_{i,j}\, \rho_i(p)\, \rho_j(p)$$

where, for any prime-to-Z divisor D,

$$\rho_i(D) \;\stackrel{\text{def}}{=}\; -\,g_{r_i}(D_{r_i}, R_i) \;+\; g_{s_i}(D_{s_i}, S_i)$$

where D_i is the restriction of D to X_i. Therefore,

$$\overline{\partial}\Big(h_0\big(\rho(q), \rho(q)\big)\Big) = \sum_{i,j} h_{i,j}\Big(\rho_i(q)\, \overline{\partial}\big(\rho_j(q)\big) + \overline{\partial}\big(\rho_i(q)\big)\, \rho_j(q)\Big) = 2 \sum_{i,j} h_{i,j}\, \rho_i(q)\, \overline{\partial}\big(\rho_j(q)\big).$$

This last step follows by symmetry of H. So we have

$$\begin{aligned} \partial_q \overline{\partial}_q \Big(h(q,q)\Big) \;&=\; \partial_q \overline{\partial}_q \Big(h_0\big(\rho(q), \rho(q)\big)\Big) \\ &=\; 2 \sum_{i,j} h_{i,j} \partial\big(\rho_i(q)\big) \wedge \overline{\partial}\big(\rho_j(q)\big) + 2 \sum_{i,j} h_{i,j}\, \rho_i(q)\, \partial_q \overline{\partial}_q \big(\rho_j(q)\big). \end{aligned}$$

Let $\mu_i = \frac{\mathbf{i}}{\pi}\, \partial_q \overline{\partial}_q \big(g_i(p,q)\big)$ be the canonical $(1,1)$-form on X_i extended by zero to all the other non-singular points of X. So

$$\frac{\mathbf{i}}{\pi}\, \partial \overline{\partial}\big(\rho_i(q)\big) \;=\; -\mu_{r_i} \;+\; \mu_{s_i}.$$

Recall that in Section 5 the vector space V was defined to be the quotient of $\mathbf{R}^n$ by the space N. Since h_0 is a pairing of V,

$$\sum_j h_{i,j}(c_{r_j} - c_{s_j}) \;=\; 0$$

for each i and each real m-tuple $(c_1, \ldots, c_m)$. This implies that, as $(1,1)$-forms,

$$\sum_j h_{i,j}\, \partial_q\overline{\partial}_q\big(\rho_j(q)\big) \;=\; \frac{\pi}{\mathbf{i}}\sum_j h_{i,j}(-\mu_{r_j} + \mu_{s_j}) \;=\; 0.$$

So, in particular,

$$\sum_{i,j} h_{i,j}\, \rho_i(q)\;\partial_q\overline{\partial}_q\big(\rho_j(q)\big) \;=\; 0.$$

Our computation becomes

$$\partial_q\overline{\partial}_q\big(h(q,q)\big) \;=\; 2\sum_{i,j} h_{i,j}\, \partial\rho_i(q) \wedge \overline{\partial}\rho_j(q).$$

For each component X_i, we have chosen a non-singular point e_i of X_i. We did this when we defined $\alpha_1, \ldots, \alpha_n$. Define

$$\sigma_i(q) \;\stackrel{\text{def}}{=}\; g_{r_i}(q_{r_i}, e_{r_i}) \;-\; g_{s_i}(q_{s_i}, e_{s_i})$$

where $q_i = q$ if q is in X_i, otherwise $q_i = 0$.

The equation

$$\sum_j h_{i,j}(c_{r_j} - c_{s_j}) = 0,$$

true for any fixed i and m-tuple $(c_1, \ldots, c_m)$, implies that, as 1-forms,

$$\sum_i h_{i,j}\big(\partial\sigma_i(q)\big) \;=\; \sum_i h_{i,j}\Big(\partial_q\, g_{r_i}(q_{r_i}, e_{r_i}) \;-\; \partial_q\, g_{s_i}(q_{s_i}, e_{s_i})\Big) \;=\; 0$$

for each j, and

$$\sum_j h_{i,j}\,\big(\overline{\partial}\sigma_j(q)\big) \;=\; \sum_j h_{i,j}\Big(\overline{\partial}_q\, g_{r_j}(q_{r_j}, e_{r_j}) \;-\; \overline{\partial}_q\, g_{s_j}(q_{s_j}, e_{s_j})\Big) \;=\; 0$$

for each i. Hence, outside $e_1, \ldots, e_m$,

$$\begin{aligned}\partial_q\overline{\partial}_q\big(h(q,q)\big) \;&=\; 2\sum_{i,j} h_{i,j}\, \partial\rho_i(q) \wedge \overline{\partial}\rho_j(q)\\ &=\; 2\sum_{i,j} h_{i,j}\, \partial\Big(\rho_i(q) + \sigma_i(q)\Big) \wedge \overline{\partial}\Big(\rho_j(q) + \sigma_j(q)\Big).\end{aligned}$$

By definition,

$$\alpha_i \;=\; 2\,\partial\big(\rho_i(p) + \sigma_i(p)\big) \qquad \text{and so} \quad \overline{\alpha_i} \;=\; 2\,\overline{\partial}\big(\rho_i(p) + \sigma_i(p)\big).$$

Therefore, outside $e_1, \ldots, e_m$,

$$\partial_q\overline{\partial}_q\big(h(q,q)\big) \;=\; \frac{1}{2}\sum_{i,j} h_{i,j}\, \alpha_i \wedge \overline{\alpha_j}.$$

Proof of Theorem 7.3. Let g_0 be an intersection function of X constructed by the techniques of Section 4. The classification theorem, Theorem 6.1, states that any intersection functions g is of the form

$$g(p,q) \;=\; g_0(p,q) \;-\; \widehat{h}(p,q) \;+\; c$$

where h is a continuous, bi-additive, symmetric pairing from $\mathrm{Pic}(X)\times\mathrm{Pic}(X)$ into $\mathbf{R}$, and where c is a real constant. Let $H = [h_{i,j}]$ be the n by n matrix representation of the real, symmetric, bilinear form h_g.

Then, using Lemmas 8.1, 8.3, and 8.4.1, we get

$$\begin{aligned}
\mathrm{chern}(g_p) &= \frac{\mathbf{i}}{\pi}\,\partial_q\overline{\partial}_q\Big(g_0(p,q)\Big) \;-\; \frac{\mathbf{i}}{\pi}\,\partial_q\overline{\partial}_q\Big(\widehat{h}(p,q)\Big)\\
&= \frac{1}{P_a}\sum_i P_i\,\mu_i \;+\; \frac{\mathbf{i}}{\pi}\frac{1}{2P_a}\partial_q\overline{\partial}_q\big(h(q,q)\big)\\
&= \frac{1}{P_a}\sum_i P_i\,\mu_i + \frac{\mathbf{i}}{4\pi P_a}\sum_{i,j} h_{i,j}\,\alpha_i\wedge\overline{\alpha_j}
\end{aligned}$$

outside $\{e_1,\ldots,e_m\}$.

9. Chern Forms of Intersection Function: Another Interpretation

Throughout this section we suppose that X has positive arithmetic genus: $P_a > 0$. We will continue to use the notation introduced in Section 5.

(9.1) Overview. In the non-singular case of this theory, studied first by Arakelov, the Chern form of the intersection function (the canonical Green's function) turns out to be closely related to the natural inner product structure of $H^0(\mathcal{K}_X)$. We will generalize this relationship to the singular case. In this more general case there is no unique natural inner product structure of $H^0(\mathcal{K}_X)$, but, there again, we don't have a unique natural intersection function on X either. This parallel ambiguity is exactly what we will investigate here.

Since this section ties together several different concepts, it will be convenient to give a brief description of the individual topics which will be covered:

(1) We begin with the relationship between V and $H^0(\mathcal{K}_X)$, where V is the real $n-m+1$ dimensional vector space defined in (5.3). From the relationship we obtain, we show how to construct a hermitian pairing θ_h on the dual of $H^0(\mathcal{K}_X)$ from any symmetric, bilinear pairing h on V.

(2) Then we will explain how any hermitian pairing θ on the dual of $H^0(\mathcal{K}_X)$ determines naturally a real $(1,1)$-form μ_θ. So, using (1), this gives a way of associating to any symmetric pairing h of V a real $(1,1)$-form μ_{θ_h}.

(3) Next we will derive an explicit formula for the real $(1,1)$-form μ_{θ_h} determined, via (2), by a given symmetric pairing h of V.

(4) As we showed in Definition 6.2, we can associate a real symmetric pairing $h = h_g$ of V to any given intersection function g on X. So this associates to g, via (1) and (2), a real $(1,1)$-form μ_{θ_h}. What is interesting is that this $(1,1)$-form is, up to a multiplicative constant, the Chern form of g. We end this chapter by proving this relationship.

(9.2) The Relationship between V and $H^0(\mathcal{K}_X)$. Let U be the subspace of $H^0(\mathcal{K}_X)$ consisting of the differentials ω whose pull-back $\omega|_{X_i}$ to each normalized component X_i is a holomorphic differential of X_i. (The general element of $H^0(\mathcal{K}_X)$ when pulled back to X_i will have simple poles at the points corresponding to singularities of X.) We note that U is a $\sum P_i$ dimensional subspace of $H^0(\mathcal{K}_X)$. The dimension of $H^0(\mathcal{K}_X)$ as a whole is, of course, P_a. We have the basic genus formula

$$P_a = n - m + 1 + \sum_i P_i.$$

So the codimension of U in $H^0(\mathcal{K}_X)$ is $n - m + 1$, which is exactly the dimension of the real vector space V defined in (5.3). What is interesting is that *there is a natural complement U' to the subspace U, and U' is canonically dual to $V \otimes_{\mathbf{R}} \mathbf{C}$.*

Let $\{T_1, \dots, T_n\}$ be the singularities of X, and let R_j, r_j, S_j, and s_j be as in Section 5. As in Definition 7.2.3, fix a non-singular point e_i on each normalized component X_i of X, and define, for each node T_j, a differential α_j. We state here some of the properties of each α_j.

1. α_j is a holomorphic differential on the disjoint union of the X_i's except at R_j, e_{r_j}, S_j, and e_{s_j}.
2. α_j depends on the choice of e_{r_j} and e_{s_j}, but is otherwise canonically defined.
3. α_j has simple poles at R_j and S_j. The residue of α_i at R_j is $+1$ and at S_j it is -1. If $s_j = r_j$ then α_j is holomorphic at e_{r_j} and e_{s_j}. Otherwise, α_j has simple poles at e_{r_j} and e_{s_j} where the residue of α_i at e_{s_j} is $+1$ and at e_{r_j} it is -1. In any case, α_j is an element of $H^0\big(\mathcal{K}_X(e_{r_j} + e_{s_j})\big)$.

Each α_j is an element of $H^0\big(\mathcal{K}_X(e_1 + \ldots + e_m)\big)$. Let W be the *real* span of $\{\alpha_j\}$ in this space, i.e, linear combinations of $\{\alpha_j\}$ with real coefficients.

We can identify W with the dual of $\mathbf{R}^n$ as follows: Let ω be an element of W, and $(a_1, \dots, a_n)$ an element of $\mathbf{R}^n$. Define the pairing

$$\Big\langle \omega,\, (a_1, \dots, a_n) \Big\rangle \stackrel{\text{def}}{=} \sum_i a_i \operatorname{res}_{R_i}(\omega).$$

Under this pairing, $\alpha_1, \dots, \alpha_n$ is the dual basis to the standard basis of $\mathbf{R}^n$.

In (5.3) the real vector space V was defined as the quotient of $\mathbf{R}^n$ by the subspace N where N is the image of the linear transformation of $\mathbf{R}^m$ to $\mathbf{R}^n$ defined by

$$(c_1, \dots, c_m) \;\mapsto\; \big(c_{r_1} - c_{s_1}, \dots, c_{r_n} - c_{s_n}\big).$$

Let W_0 be the subspace of W which acts trivially on N, i.e., W_0 is the subspace which is dual to V. So W_0 has real dimension $n - m + 1$.

(9.2.1) Lemma. *The space W_0 is the set of ω of W such that $\operatorname{res}_{e_i} \omega = 0$ for each e_i. In other words, W_0 is the intersection of W with $H^0(\mathcal{K}_X)$. Although W was defined relative to a choice of the points $e_1, \dots, e_m$, the space W_0, considered as a real subspace of $H^0(\mathcal{K}_X)$, is completely independent of this choice.*

Proof. For any normalized component X_t, consider the m-tuple

$$\mathbf{c_t} \stackrel{\text{def}}{=} (c_{t,1}, \dots, c_{t,m})$$

where $c_{t,t} = 1$, but $c_{t,i} = 0$ for $i \neq t$. So $\mathbf{c_1}, \ldots, \mathbf{c_m}$ is the standard basis of $\mathbf{R}^m$. Therefore, N is spanned by vectors of the form

$$(c_{t,r_1} - c_{t,s_1}, \ldots, c_{t,r_n} - c_{t,s_n}).$$

Let ω be an element of W. It follows that ω is in W_0 if and only if for every normalized component X_t

$$\sum_i \Big(c_{t,r_i} - c_{t,s_i}\Big) \operatorname{res}_{R_i}(\omega) \;=\; 0.$$

But, if $\{P_{t,j}\}$ is the set of R_i and S_i contained in X_t, then, using the fact that $\operatorname{res}_{S_i} \omega = -\operatorname{res}_{R_i} \omega$, we get

$$\sum_i \Big(c_{t,r_i} - c_{t,s_i}\Big) \operatorname{res}_{R_i}(\omega) \;=\; \sum_j \operatorname{res}_{P_{t,j}} \omega.$$

The sum of the residues of ω restricted to X_t must be zero:

$$\operatorname{res}_{e_t} \omega + \sum_j \operatorname{res}_{P_{t,j}} \omega = 0.$$

Putting this all together, we get that ω is in W_0 if and only if

$$\operatorname{res}_{e_t}(\omega) = 0$$

for all t. This proves the first statement.

Now suppose for some normalized component, X_1 say, that instead of e_1 we choose another non-singular point e_1'. So, for each node T_i of X we have another differential α_i' resulting from this alternate choice. Let W' be the real span of $\{\alpha_i'\}$ in $H^0(\mathcal{K}_X(e_1 + \ldots + e_m))$, and W_0' the intersection of W' with $H^0(\mathcal{K}_X)$.

We look up the definition of α_i and α_i' in (7.2.3) and conclude that

$$\alpha_i' \;=\; \alpha_i \;+\; (\operatorname{res}_{e_1} \alpha_i)\,\beta$$

where β is the meromorphic differential defined, on X_1, by

$$\beta \;\stackrel{\text{def}}{=}\; -2\,\partial_x\Big(g_1(e_1', x) - g_1(e_1, x)\Big)$$

and extended by zero to the whole disjoint union of the normalized components.

This relation holds for linear combinations, i.e., if

$$\omega = \sum_i b_i \alpha_i \qquad \text{and} \qquad \omega' = \sum_i b_i \alpha_i'$$

then

$$\omega' = \omega + (\operatorname{res}_{e_1} \omega)\,\beta.$$

In particular, if ω is in W_0, then $\omega' = \omega$. So ω' is in $H^0(\mathcal{K}_X)$, hence in W_0'. Likewise, if ω' is in W_0', then it is in W_0. Therefore, considered as real subspaces of $H^0(\mathcal{K}_X)$, $W_0 = W_0'$.

Let U be subspace of $H^0(\mathcal{K}_X)$ defined above. We can also write

$$U = \bigoplus_i H^0(\mathcal{K}_{X_i})$$

where each $H^0(\mathcal{K}_{X_i})$ is considered naturally a subspace of $H^0(\mathcal{K}_X)$ by extending by zero any global differential of X_i to the rest of the disjoint union of the normalized components.

We are ready to provide the promised complement to the subspace U:

(9.2.2) Proposition. *Let U' to the complex subspace of $H^0(\mathcal{K}_X)$ generated by W_0. Then U' is naturally isomorphic to $V^* \otimes_{\mathbf{R}} \mathbf{C}$ where V^* is the dual vector space to V. Hence U has dimension $n-m+1$. In addition, we have the following natural decomposition:*

$$H^0(\mathcal{K}_X) = U' \oplus U \xrightarrow{\sim} \left(W_0 \otimes_{\mathbf{R}} \mathbf{C}\right) \oplus \left(\bigoplus_i H^0(\mathcal{K}_{X_i})\right) \xrightarrow{\sim} \left(V^* \otimes_{\mathbf{R}} \mathbf{C}\right) \oplus \left(\bigoplus_i H^0(\mathcal{K}_{X_i})\right).$$

Proof. Consider the map from $H^0(\mathcal{K}_X)$ to $\mathbf{C}^n$ sending ω to $(\operatorname{res}_{R_1} \omega, \ldots, \operatorname{res}_{R_n} \omega)$. By definition, U is in the kernel of this map.

Note that W_0 does not contain any non-zero elements in the kernel of this map, and its image is a real subspace of dimension $n - m + 1$ in $\mathbf{R}^n$. It follows that U', the $\mathbf{C}$-span of W_0, cannot contain any elements of the kernel and has image of complex dimension $n - m + 1$.

This implies that U and U' intersect in dimension zero, and that we have the following natural isomorphisms:

$$U' \xrightarrow{\sim} W_0 \otimes_{\mathbf{R}} \mathbf{C} \xrightarrow{\sim} V^* \otimes_{\mathbf{R}} \mathbf{C}.$$

By a dimension count, we see that U and U' together span $H^0(\mathcal{K}_X)$, so U' is the complement of U.

(9.3) Definitions. Let E be a complex vector space. We define the *conjugate vector space* $\overline{E}$ as follows. As an abelian group $\overline{E}$ is defined to be equal to E; if x is an element of E, $\overline{x}$ will signify this same element, but considered as an element of $\overline{E}$. Define $c \cdot \overline{x}$ to be $\overline{\overline{c} \cdot x}$ for any complex number c.

In the case of $E = H^0(\mathcal{K}_X)$ with element ω we can identify $\overline{\omega}$ with the $(0,1)$-form conjugate to ω (which we also denote by $\overline{\omega}$). Thus the conjugate space $\overline{H^0(\mathcal{K}_X)}$ can be identified with the space of such $(0,1)$-forms.

A *sesquilinear pairing* on a complex vector space E is a map θ from $E \times E$ to $\mathbf{C}$ which is linear in the first variable, and conjugate-linear in the second, i.e.

$$\theta(ax+by, z) = a\,\theta(x,z) + b\,\theta(y,z) \qquad \text{and} \qquad \theta(x, ay+bz) = \overline{a}\,\theta(x,y) + \overline{b}\,\theta(x,z).$$

Equivalently, a sesquilinear pairing on E is a linear map from $E \otimes \overline{E}$ to $\mathbf{C}$, or, alternately, an element of $E^* \otimes \overline{E}^*$.

For example, sesquilinear pairings on the dual of $H^0(\mathcal{K}_X)$ correspond to elements of $H^0(\mathcal{K}_X) \otimes \overline{H^0(\mathcal{K}_X)}$.

A *hermitian pairing* on a complex vector space E is a sesquilinear pairing θ such that $\theta(x,y) = \overline{\theta(y,x)}$. A *hermitian inner product* on E is a hermitian pairing θ such that $\theta(x,x) > 0$ for all $x \neq 0$.

If X is any non-singular curve over $\mathbf{C}$, the *standard hermitian inner product* on $H^0(\mathcal{K}_X)$ is that defined by the formula

$$\frac{\mathbf{i}}{4\pi}\int \omega \wedge \overline{\omega'}$$

where ω and ω' are any two global sections of $\mathcal{K}_X$.

Under the standard hermitian inner product, $H^0(\mathcal{K}_{X_i})$, for each X_i, is naturally isomorphic to its dual. The dual of U' is naturally isomorphic to $V \otimes_{\mathbf{R}} \mathbf{C}$. Thus, by the decomposition of Proposition 9.2.2, the dual of $H^0(\mathcal{K}_X)$ is naturally isomorphic to

$$\left(V \otimes_{\mathbf{R}} \mathbf{C}\right) \oplus \left(\bigoplus H^0(\mathcal{K}_{X_i})\right).$$

An *admissible pairing* on the dual of $H^0(\mathcal{K}_X)$ is a hermitian pairing such that

1. the above decomposition an orthogonal decomposition,
2. the restriction to each $H^0(\mathcal{K}_{X_i})$ is the standard hermitian inner product, and
3. the restriction to $V \otimes_{\mathbf{R}} \mathbf{C}$ is induced from a real symmetric bilinear map on V.

An *admissible inner product* on the dual of $H^0(\mathcal{K}_X)$ is an admissible pairing which is a hermitian inner product.

This definition gives a one-to-one correspondence between symmetric bilinear pairings on V and admissible hermitian pairings on the dual of $H^0(\mathcal{K}_X)$. If h is a bilinear pairing of V, then the corresponding admissible pairing on the dual of $H^0(\mathcal{K}_X)$ will be called the *admissible pairing induced by* h, and will be written θ_h.

Consider the $\mathbf{C}$-linear map

$$\omega \otimes \overline{\omega'} \quad \mapsto \quad \frac{\mathbf{i}}{2}\omega \wedge \overline{\omega}'$$

sending $H^0(\mathcal{K}_X) \otimes \overline{H^0(\mathcal{K}_X)}$ to the space of global $(1,1)$-forms on the non-singular points of X. Any sesquilinear pairing θ of the dual of $H^0(\mathcal{K}_X)$ can be identified with an an element of $H^0(\mathcal{K}_X) \otimes \overline{H^0(\mathcal{K}_X)}$. So, by the above map, θ determines a global $(1,1)$-form of the non-singular set of X. This form, written μ_θ, is called the $(1,1)$*-form determined by* θ.

(9.4) Proposition. *If θ is hermitian pairing on the dual of $H^0(\mathcal{K}_X)$, then μ_θ, the $(1,1)$-form determined by θ, is a real $(1,1)$-form. If, in addition, θ is hermitian inner product, then μ_θ is a semi-positive $(1,1)$-form on the non-singular points of X; on the components with positive geometric genus, μ_θ is strictly positive.*

Proof. Consider the switching isomorphism σ on the underlying real vector space of $H^0(\mathcal{K}_X) \otimes \overline{H^0(\mathcal{K}_X)}$ taking $\omega \otimes \overline{\omega}'$ to $\omega' \otimes \overline{\omega}$.

As we mentioned earlier, any sesquilinear pairing θ on the dual of $H^0(\mathcal{K}_X)$ corresponds to an element θ' of $H^0(\mathcal{K}_X) \otimes \overline{H^0(\mathcal{K}_X)}$. Such a sesquilinear pairing is hermitian if and only if θ' is invariant under σ. Under the above defined linear map Ψ from $H^0(\mathcal{K}_X) \otimes \overline{H^0(\mathcal{K}_X)}$ to the space of global $(1,1)$-forms on the non-singular points of X,

$$\Psi\big(\sigma(\theta')\big) \;=\; \overline{\Psi(\theta')}.$$

Hence, if θ is hermitian, $\Phi(\theta')$, the $(1,1)$-form determined by θ, is invariant under complex conjugation, i.e, it is a real $(1,1)$-form.

Now suppose that θ is an hermitian inner product on the dual of $H^0(\mathcal{K}_X)$. Let $\delta_1, \ldots, \delta_{P_a}$ be an orthonormal basis, with respect to θ, of the dual of $H^0(\mathcal{K}_X)$, and let $\omega_1, \ldots, \omega_{P_a}$ be its dual basis in $H^0(\mathcal{K}_X)$. Then the $(1,1)$-form determined by θ is

$$\sum_i \frac{\mathbf{i}}{2} \omega_i \wedge \overline{\omega_i}.$$

Outside the vanishing set of ω_i, $\frac{\mathbf{i}}{2}\omega_i \wedge \overline{\omega_i}$ is a positive $(1,1)$-form. For any X-nonsingular point p of a component X_i of positive genus, there is an element of $H^0(\mathcal{K}_{X_i})$, hence of $H^0(\mathcal{K}_X)$, not vanishing at p. Hence the above sum is positive at such p.

(9.5) For each normalized component X_i, let $\omega_1^i, \ldots, \omega_{P_i}^i$ be an orthonormal basis of $H^0(\mathcal{K}_{X_i})$ under the standard hermitian inner product. As usual, we consider $H^0(\mathcal{K}_{X_i})$ as a subspace of $H^0(\mathcal{K}_X)$.

Let h be a real, symmetric, bilinear pairing on V, and let $H = [h_{i,j}]$ be the n by n matrix representing h. As above, the pairing h induces an admissible hermitian pairing θ_h on the dual of $H^0(\mathcal{K}_X)$. Similarly, θ_h determines a $(1,1)$-form μ_{θ_h}.

Let α_j be as defined in Definition 7.2.3. Recall that α_j is a section of $\mathcal{K}_X$ outside e_{r_j} and e_{s_j} where it can have simple poles.

(9.5.1) Proposition. *Using the notation above,*

$$\mu_{\theta_h} = \mu' + \mu_1 + \ldots + \mu_m$$

where

$$\mu' = \frac{\mathbf{i}}{2}\sum_{i,j} h_{i,j}\, \alpha_i \wedge \overline{\alpha_j} \qquad \text{and} \qquad \mu_i = \frac{\mathbf{i}}{2}\sum_j \omega_j^i \wedge \overline{\omega_j^i} \quad \text{for each } i.$$

Proof. Recall that the real span of $\alpha_1, \ldots, \alpha_n$ is a real subspace W of the space $H^0\big(\mathcal{K}_X(e_1 + \ldots + e_m)\big)$ which is dual to $\mathbf{R}^n$, and $\alpha_1, \ldots, \alpha_n$ is the dual basis to the standard basis of $\mathbf{R}^n$. Under this duality, W_0, the intersection of W with $H^0(\mathcal{K}_X)$, is dual to the quotient space V of $\mathbf{R}^n$.

The $\mathbf{C}$ span of W_0 is a complex subspace of $H^0(\mathcal{K}_X)$ isomorphic to $W_0 \otimes \mathbf{C}$. The above duality induces a duality between $W_0 \otimes \mathbf{C}$ and $V \otimes \mathbf{C}$.

Recall that the dual of $H^0(\mathcal{K}_X)$ decomposes as follows:

$$\Big(V \otimes_{\mathbf{R}} \mathbf{C}\Big) \oplus \Big(\bigoplus_i H^0(\mathcal{K}_{X_i})\Big).$$

Let $\delta_1, \ldots, \delta_{n-m+1}$ be a basis of V and $\omega'_1, \ldots, \omega'_{n-m+1}$ its dual basis in W_0. Then $H^0(\mathcal{K}_X)$ has basis

$$\omega'_1, \ldots, \omega'_{n-m+1};\ \omega_1^1, \ldots, \omega_{p_i}^1;\ \ldots;\ \omega_1^m, \ldots, \omega_{p_i}^m$$

with dual basis

$$\delta_1, \ldots, \delta_{n-m+1};\ \omega_1^1, \ldots, \omega_{p_i}^1;\ \ldots;\ \omega_1^m, \ldots, \omega_{p_i}^m.$$

The induced admissible hermitian pairing θ_h on the dual of $H^0(\mathcal{K}_X)$ is such that, for all i, j, and k,

$$\theta_h(\delta_i, \delta_j) = h(\delta_i, \delta_j), \qquad \text{and} \qquad \theta_h(\omega_j^i, \delta_k) = 0.$$

In addition,

$$\theta_h(\omega^i_j, \omega^k_l) = \begin{cases} 1 & \text{if } i = k \text{ and } j = l, \\ 0 & \text{otherwise.} \end{cases}$$

Hence, the real $(1,1)$-form determined by θ_h is

$$\mu_{\theta_h} = \mu' + \mu_1 + \ldots + \mu_m$$

where

$$\mu' = \frac{\mathbf{i}}{2} \sum_{i,j} h(\delta_i, \delta_j)\, \omega'_i \wedge \overline{\omega'_j}. \qquad \text{and} \qquad \mu_i = \frac{\mathbf{i}}{2} \sum_j \omega^i_j \wedge \overline{\omega^i_j}, \quad \text{for each } i.$$

Now we will show that μ' has the desired form.

Recall that in (5.3) the vector space V is defined to be $\mathbf{R}^n/N$. So each δ_i is an element modulo N of $\mathbf{R}^n$. Fix an element $(\delta_{i,1}, \ldots, \delta_{i,n})$ of this modulo class. Likewise, each ω'_i is of the form

$$\omega'_i = b_{i,1}\alpha_1 + \ldots + b_{i,n}\alpha_n.$$

The fact that $\{\delta_i\}$ is dual to $\{\omega'_i\}$ implies that the matrices $D = [\delta_{i,j}]$ and $B = [b_{i,j}]$ satisfy

$$BD^t = I$$

where I is the $n - m + 1$ identity matrix. Since W_0 is dual to $V = \mathbf{R}^n/N$, the kernel of the matrix B is exactly N. Both these facts together imply that the image of matrix $D^tB - I_n$, i.e., the space generated by the columns of this matrix, is contained in N.

Since N is contained in the null space of h (considered as a pairing on $\mathbf{R}^n$),

$$HD^tB = H$$

where $H = [h_{i,j}]$ is the matrix representation of h.

We compute

$$h(\delta_i, \delta_j) = \sum_{k,l} \delta_{i,k}\, \delta_{j,l}\, h_{k,l}$$

or in terms of matrices, the matrix $R = [h(\delta_i, \delta_j)]$ is equal toDHD^t.

We have

$$\mu' = \frac{\mathbf{i}}{2} \sum_{i,j} h(\delta_i, \delta_j)\, \omega'_i \wedge \overline{\omega'_j} = \frac{\mathbf{i}}{2} \sum_{k,l} \left(\sum_{i,j} h(\delta_i, \delta_j) b_{i,k} b_{j,l} \right) \alpha_k \wedge \overline{\alpha_l}.$$

In the above sum, the coefficient for the $\alpha_k \wedge \overline{\alpha_l}$ term is the (k,l)-th entry of the matrix B^tRB. But by our earlier formulas

$$B^tRB = B^t(DHD^t)B = B^tD(HD^tB) = B^tDH = (HD^tB)^t = H^t = H.$$

Thus

$$\mu' = \frac{\mathbf{i}}{2} \sum_{k,l} h_{k,l}\, \alpha_k \wedge \overline{\alpha_l}.$$

(9.6) Proposition. *Suppose X is non-singular. Let $\omega_1, \ldots, \omega_{P_a}$ be an orthonormal basis of $H^0(\mathcal{K}_X)$ with respect to the standard hermitian inner product. The canonical $(1,1)$-form μ of X is*

$$\mu = \frac{\mathbf{i}}{4\pi P_a} \sum_i \omega_i \wedge \overline{\omega_i}.$$

Proof. This is a standard result of Arakelov theory. See [L] Chapter II, Sections 2 and 3; keep in mind that Lang's definition of the standard hermitian inner product on $H^0(\mathcal{K}_X)$ is 2π times mine. In this I have strayed from the norm.

(9.7) Theorem. *Let g be an intersection function of X, and let $h = h_g$ be its associated bilinear pairing on V. Let θ_h be the hermitian pairing on the dual of $H^0(\mathcal{K}_X)$ induced by h, and let μ_{θ_h} be its associated real $(1,1)$-form. Then we have the following expression for the chern form of g:*

$$\mathrm{chern}(g) = \frac{1}{2\pi P_a} \mu_{\theta_h}$$

Proof. Compare the expressions of Theorem 7.3, Proposition 9.5.1, and Proposition 9.6.

Chapter 5

The Arithmetic Riemann-Roch Isomorphism

In this chapter we gather together the elements considered in the earlier chapters to form the arithmetic Riemann-Roch isomorphism. The only piece missing is a norm for the determinant of cohomology which is compatible with the norm on the intersection pairing defined in the previous chapter. Such a norm is defined in the first section of this chapter. After this we are ready to define the arithmetic Riemann-Roch isomorphism; this is done in the second and final section.

1. Norms for Determinants of Cohomology

Throughout this section, X will be a connected one-dimensional scheme projective over the complex numbers $\mathbf{C}$; we will also assume that X is non-singular except for possibly a finite number of ordinary double points. We choose, once and for all, an intersection function g on X in the sense of Chapter 4, Section 2. As in that section, let Φ be the order function on X associated with g.

Recall that an admissible line bundle on X is a line bundle L together with an order function Φ_L on L which is compatible with the order function Φ. Recall also that $\mathbf{D}(L)$, the determinant of cohomology of L, is a one-dimensional complex vector space. In this section we define a norm for the one-dimensional complex vector space $\mathbf{D}(L)$ which depends on the order function Φ_L as well as the underlying line bundle L. Then we show that this norm has the property that the various canonical isomorphisms involving determinants of cohomology are also isometries.

To do so we will use a basic idea of B. Mazur: he used a definition similar to the following one in order to give an alternate proof for the existence of Faltings volumes (in the case when X is non-singular).

(1.1) Definition. Fix a non-trivial norm on the complex vector space $\mathbf{D}(\mathcal{O}_X)$; this can be chosen arbitrarily, but choose it once and for all. Let L be a line bundle on X with order function Φ_L. Consider the Riemann-Roch isomorphism of Chapter 3, Section 4:

$$\frac{\mathbf{D}(L)^{\otimes 2}}{\mathbf{D}(\mathcal{O}_X)^{\otimes 2}} \xrightarrow{\sim} \left\langle L, \frac{L}{\mathcal{K}_X} \right\rangle$$

The right hand side of this equation has a norm defined in Section 2 of Chapter 4. We simply define the norm on $\mathbf{D}(L)$ to be the unique norm which makes this isomorphism an isometry. (Note that there are actually two Riemann-Roch isomorphisms, but that they differ only by a sign, so the above norm is well-defined.)

(1.2) Definition. Let L be a line bundle on X with order function Φ_L, and let D be an effective Cartier divisor on X whose support does not contain any singular points of X. We define the norm on $\mathbf{D}(L|_D)$ to be the unique norm which makes the isomorphism

$$\mathbf{D}(L|_D)^{\otimes 2} \xrightarrow{\sim} \frac{\langle L, \mathcal{O}_X(D)\rangle}{\langle \mathcal{K}_X(D)\otimes L^{-1}, \mathcal{O}_X(D)\rangle}$$

an isometry, where the norm on the right hand side is as defined in Section 2 of Chapter 4, and where this isomorphism is defined to be the composition of the following three isomorphisms.

To form the first isomorphism consider the duality isomorphism of Definition 3.2, Chapter 3 which takes $\mathbf{D}(\mathcal{K}_D \otimes L^{-1}|_D)$ to $\mathbf{D}(L|_D)^{-1}$. By taking the dual of this isomorphism we get an isomorphism from $\mathbf{D}(L|_D)$ to $\mathbf{D}(\mathcal{K}_D \otimes L^{-1}|_D)^{-1}$. The desired isomorphism is obtained by tensoring this isomorphism with the identity isomorphism on $\mathbf{D}(L|_D)$:

$$\alpha\colon \mathbf{D}(L|_D)^{\otimes 2} \xrightarrow{\sim} \mathbf{D}(L|_D)\otimes \mathbf{D}(\mathcal{K}_D\otimes L^{-1}|_D)^{-1} = \frac{\mathbf{D}(L|_D)}{\mathbf{D}(\mathcal{K}_X(D)\otimes L^{-1}|_D)}.$$

Next, the definitions of (3.2) and (5.1) in Chapter 1 gives us naturally the isomorphism

$$\beta\colon \frac{\mathbf{D}(L|_D)}{\mathbf{D}(\mathcal{K}_X(D)\otimes L^{-1}|_D)} \xrightarrow{\sim} \frac{\mathbf{D}(L|_D)}{\mathbf{D}(\mathcal{O}_D)} \otimes \frac{\mathbf{D}(\mathcal{O}_D)}{\mathbf{D}(\mathcal{K}_X(D)\otimes L^{-1}|_D)} = \frac{\mathbf{N}_D(L)}{\mathbf{N}_D(\mathcal{K}_X(D)\otimes L^{-1})};$$

Finally, Proposition 6.2, Chapter 1 gives the following isomorphism:

$$\gamma\colon \frac{\mathbf{N}_D(L)}{\mathbf{N}_D(\mathcal{K}_X(D)\otimes L^{-1})} \xrightarrow{\sim} \frac{\langle L, \mathcal{O}_X(D)\rangle}{\langle \mathcal{K}_X(D)\otimes L^{-1}, \mathcal{O}_X(D)\rangle}.$$

The isomorphisms α, β, and γ compose to give the desired isomorphism.

"For the record," we reformulate these two definitions in the following proposition:

(1.3) Proposition. *Let L be a line bundle on X with order function Φ_L.*

1. *The Riemann-Roch isomorphism*

$$\frac{\mathbf{D}(L)^{\otimes 2}}{\mathbf{D}(\mathcal{O}_X)^{\otimes 2}} \xrightarrow{\sim} \left\langle L, \frac{L}{\mathcal{K}_X}\right\rangle$$

of Chapter 3, Section 4 is actually an isometry, where the norm on the right hand side is as in Section 2 of Chapter 4, and the norm on the left hand side is as in Definition 1.1 above.

2. *Let D be an effective Cartier divisor on X whose support does not contain any singular points of X. Then the following isomorphism, defined in (1.2) above, is an isometry:*

$$\mathbf{D}(L|_D)^{\otimes 2} \xrightarrow{\sim} \frac{\langle L, \mathcal{O}_X(D)\rangle}{\langle \mathcal{K}_X(D)\otimes L^{-1}, \mathcal{O}_X(D)\rangle}.$$

Here the norm of the right hand side is as in Section 2 of Chapter 4, the norm of the left hand side is as in Definition 1.2 above.

Proof. This is just a formal restatement of Definitions 1.1 and 1.2.

(1.4) Proposition. *Let L be a line bundle on X with order function Φ_L, and let D be an effective Cartier divisor on X whose support does not contain any singular points of X. Then canonical isomorphism*

$$\mathbf{D}(L) \xrightarrow{\sim} \mathbf{D}\big(L(-D)\big) \otimes \mathbf{D}\big(L|_D\big)$$

induced by the exact sequence

$$0 \to L(-D) \to L \to L|_D \to 0$$

is an isometry with respect to the norms given to the various vector spaces.

Proof. From the isomorphism

$$\mathbf{D}(L) \xrightarrow{\sim} \mathbf{D}\big(L(-D)\big) \otimes \mathbf{D}\big(L|_D\big)$$

we form, in the obvious way, the isomorphism

$$\frac{\mathbf{D}(\mathcal{O}_X)^{\otimes 2}}{\mathbf{D}(L)^{\otimes 2}} \xrightarrow{\sim} \frac{\mathbf{D}(\mathcal{O}_X)^{\otimes 2}}{\mathbf{D}\big(L(-D)\big)^{\otimes 2}} \otimes \mathbf{D}\big(L|_D\big)^{\otimes -2}.$$

The statement of the proposition is equivalent to proving that this second isomorphism is an isometry. Our strategy will be to show that the following diagram commutes up to sign:

$$\begin{array}{ccc}
\dfrac{\mathbf{D}(\mathcal{O}_X)^{\otimes 2}}{\mathbf{D}(L)^{\otimes 2}} & \longrightarrow & \left\langle L, \frac{\mathcal{K}_X}{L} \right\rangle \\
\downarrow & & \downarrow \\
\dfrac{\mathbf{D}(\mathcal{O}_X)^{\otimes 2}}{\mathbf{D}\big(L(-D)\big)^{\otimes 2}} \otimes \mathbf{D}\big(L|_D\big)^{\otimes -2} & \longrightarrow & \left\langle L(-D), \frac{\mathcal{K}_X}{L(-D)} \right\rangle \otimes \dfrac{\left\langle \mathcal{O}_X(D), \frac{\mathcal{K}_X}{L(-D)} \right\rangle}{\left\langle L, \mathcal{O}_X(D) \right\rangle}
\end{array}$$

where the left vertical map is the isomorphism discussed above; the right vertical map is the natural isomorphism induced by the bilinearity of the intersection pairing; the top horizontal map is constructed from the Riemann-Roch isomorphism of Chapter 3, Section 4; the bottom horizontal map is constructed from the tensor product of the Riemann-Roch isomorphism with the the isomorphism of Definition 1.2.

Once we have shown that this diagram commutes, then, since the horizontal maps are isometries by Proposition 1.3, and the right vertical map is an isometry by Proposition 2.6.1 of Chapter 4, it follows that the left vertical map must be an isometry.

We will show that the above diagram commutes up to sign in two steps:

Step 1. First we show that the following diagram commutes:

$$\begin{array}{ccc}
\dfrac{\mathbf{D}(\mathcal{O}_X)^{\otimes 2}}{\mathbf{D}(L)^{\otimes 2}} & \longrightarrow & \dfrac{\mathbf{D}(\mathcal{K}_X) \otimes \mathbf{D}(\mathcal{O}_X)}{\mathbf{D}(L) \otimes \mathbf{D}(\mathcal{K}_X \otimes L^{-1})} \\
\downarrow & & \downarrow \\
\dfrac{\mathbf{D}(\mathcal{O}_X)^{\otimes 2}}{\mathbf{D}\big(L(-D)\big)^{\otimes 2}} \otimes \mathbf{D}\big(L|_D\big)^{\otimes -2} & \longrightarrow & \dfrac{\mathbf{D}(\mathcal{K}_X) \otimes \mathbf{D}(\mathcal{O}_X)}{\mathbf{D}\big(L(-D)\big) \otimes \mathbf{D}\big(\mathcal{K}_X \otimes L^{-1}(D)\big)} \otimes \dfrac{\mathbf{D}\big(\mathcal{K}_D \otimes L^{-1}|_D\big)}{\mathbf{D}\big(L|_D\big)}
\end{array}$$

where the horizontal maps are constructed by taking (inverses) of the various duality isomorphisms of Chapter 3, Section 3 and tensoring them with identity maps; the vertical maps are formed from the isomorphisms

$$\mathbf{D}\big(L\big) \to \mathbf{D}\big(L(-D)\big) \otimes \mathbf{D}\big(L|_D\big)$$

and

$$\mathbf{D}\big(\mathcal{K}_X \otimes L^{-1}\big) \to \mathbf{D}\big(\mathcal{K}_X \otimes L^{-1}(D)\big) \otimes \mathbf{D}\big(\mathcal{K}_D \otimes L^{-1}|_D\big)^{-1}$$

which arise from the sequences

$$0 \longrightarrow L(-D) \longrightarrow L \longrightarrow L|_D \longrightarrow 0$$

and

$$0 \longrightarrow \mathcal{K}_X \otimes L^{-1} \longrightarrow \mathcal{K}_X \otimes L^{-1}(D) \longrightarrow \mathcal{K}_D \otimes L^{-1}|_D \longrightarrow 0.$$

The fact that this diagram commutes is a corollary of Proposition 3.3 of Chapter 3. So, by the definition of the Riemann-Roch isomorphism given in (4.1) of Chapter 3 and the definition of the isomorphism constructed in Definition 1.2, we have reduced the problem to the following:

Step 2. Now we show that the following diagram commutes up to sign:

$$\begin{array}{ccc}
\dfrac{\mathbf{D}(\mathcal{K}_X) \otimes \mathbf{D}(\mathcal{O}_X)}{\mathbf{D}(L) \otimes \mathbf{D}(\mathcal{K}_X \otimes L^{-1})} & \longrightarrow & \left\langle L\,,\ \frac{\mathcal{K}_X}{L} \right\rangle \\
\downarrow & & \downarrow \\
\dfrac{\mathbf{D}(\mathcal{K}_X)\otimes\mathbf{D}(\mathcal{O}_X)}{\mathbf{D}(L(-D))\otimes\mathbf{D}(\mathcal{K}_X\otimes L^{-1}(D))} \otimes \dfrac{\mathbf{D}(\mathcal{K}_D\otimes L^{-1}|_D)}{\mathbf{D}(L|_D)} & \longrightarrow & \left\langle L(-D), \frac{\mathcal{K}_X}{L(-D)} \right\rangle \otimes \dfrac{\left\langle \mathcal{O}_X(D), \frac{\mathcal{K}_X}{L(-D)} \right\rangle}{\left\langle L, \mathcal{O}_X(D) \right\rangle}
\end{array}$$

where the left vertical map is just the right vertical map of the diagram in step 1 above; the right vertical map is the natural isomorphism induced by the bilinearity of the intersection pairing; the top horizontal map is the isomorphism of Proposition 5.7, Chapter 2 (where we always take $L_2 = \mathcal{O}_X$, $M_2 = \mathcal{O}_X$, and ρ and ρ' any acceptable values); and the bottom horizontal map is the tensor product of two isomorphisms: first, the isomorphism of Proposition 5.7, Chapter 2, and second, the isomorphism constructed in the obvious way from the composition of the maps β and γ of Definition 1.2.

The fact that this commutes up to sign is a corollary to Lemma 1.4.1 below. To see this, let $M = \mathcal{K}_X \otimes L^{-1}(D)$, and observe that Lemma 1.4.1 gives the following diagram, commutative up to sign:

$$\begin{array}{ccc}
\dfrac{\mathbf{D}(\mathcal{K}_X)\otimes\mathbf{D}(\mathcal{O}_X)}{\mathbf{D}(L)\otimes\mathbf{D}(\mathcal{K}_X\otimes L^{-1})} \otimes \dfrac{\mathbf{D}(\mathcal{K}_X(D)|_D)}{\mathbf{D}(\mathcal{K}_X\otimes L^{-1}(D)|_D)} & \longrightarrow & \left\langle L\,,\ \frac{\mathcal{K}_X}{L} \right\rangle \otimes \left\langle L\,,\ \mathcal{O}_X(D) \right\rangle \\
\uparrow & & \uparrow \\
\dfrac{\mathbf{D}(\mathcal{K}_X(D))\otimes\mathbf{D}(\mathcal{O}_X)}{\mathbf{D}(L)\otimes\mathbf{D}(\mathcal{K}_X\otimes L^{-1}(D))} & \longrightarrow & \left\langle L\,,\ \frac{\mathcal{K}_X}{L(-D)} \right\rangle \\
\downarrow & & \downarrow \\
\dfrac{\mathbf{D}(\mathcal{K}_X)\otimes\mathbf{D}(\mathcal{O}_X)}{\mathbf{D}(L(-D))\otimes\mathbf{D}(\mathcal{K}_X\otimes L^{-1}(D))} \otimes \dfrac{\mathbf{D}(\mathcal{K}_X(D)|_D)}{\mathbf{D}(L|_D)} & \longrightarrow & \left\langle L(-D), \frac{\mathcal{K}_X}{L(-D)} \right\rangle \otimes \left\langle \mathcal{O}_X(D), \frac{\mathcal{K}_X}{L(-D)} \right\rangle
\end{array}$$

Eliminating the middle row, we get the following diagram, commutative up to sign:

$$\begin{array}{ccc}
\dfrac{\mathbf{D}\big(\mathcal{K}_X\big)\otimes\mathbf{D}\big(\mathcal{O}_X\big)}{\mathbf{D}\big(L\big)\otimes\mathbf{D}\big(\mathcal{K}_X\otimes L^{-1}\big)}\otimes\dfrac{\mathbf{D}\big(\mathcal{K}_X(D)|_D\big)}{\mathbf{D}\big(\mathcal{K}_X\otimes L^{-1}(D)|_D\big)} & \longrightarrow & \Big\langle L,\frac{\mathcal{K}_X}{L}\Big\rangle\otimes\Big\langle L,\mathcal{O}_X(D)\Big\rangle \\
\downarrow & & \downarrow \\
\dfrac{\mathbf{D}\big(\mathcal{K}_X\big)\otimes\mathbf{D}\big(\mathcal{O}_X\big)}{\mathbf{D}\big(L(-D)\big)\otimes\mathbf{D}\big(\mathcal{K}_X\otimes L^{-1}(D)\big)}\otimes\dfrac{\mathbf{D}\big(\mathcal{K}_X(D)|_D\big)}{\mathbf{D}\big(L|_D\big)} & \longrightarrow & \Big\langle L(-D),\frac{\mathcal{K}_X}{L(-D)}\Big\rangle\otimes\Big\langle\mathcal{O}_X(D),\frac{\mathcal{K}_X}{L(-D)}\Big\rangle
\end{array}$$

We modify this diagram using the following isomorphism (described in Lemma 1.4.1 below):

$$\frac{\mathbf{D}\big(\mathcal{K}_X(D)|_D\big)}{\mathbf{D}\big(\mathcal{K}_X\otimes L^{-1}(D)|_D\big)}\ \tilde{\to}\ \Big\langle L,\ \mathcal{O}_X(D)\Big\rangle;$$

essentially what we do is "divide" the two leftmost terms of the above diagram by $\frac{\mathbf{D}(\mathcal{K}_X(D)|_D)}{\mathbf{D}(\mathcal{K}_X\otimes L^{-1}(D)|_D)}$ and the two rightmost terms by $\big\langle L,\mathcal{O}_X(D)\big\rangle$. We leave it to the reader to show that the resulting diagram, commutative up to sign, is the desired one.

(1.4.1) Lemma. *Let L and M be line bundles on X, and let D be an effective Cartier divisor on X.*

1. *The following diagram commutes up to sign:*

$$\begin{array}{ccc}
\dfrac{\mathbf{D}\big(L\otimes M\big)\otimes\mathbf{D}\big(\mathcal{O}_X\big)}{\mathbf{D}\big(L\big)\otimes\mathbf{D}\big(M\big)} & \longrightarrow & \big\langle L,M\big\rangle \\
\downarrow & & \downarrow \\
\dfrac{\mathbf{D}\big(L\otimes M(-D)\big)\otimes\mathbf{D}\big(\mathcal{O}_X\big)}{\mathbf{D}\big(L\big)\otimes\mathbf{D}\big(M(-D)\big)}\otimes\dfrac{\mathbf{D}\big(L\otimes M|_D\big)}{\mathbf{D}\big(M|_D\big)} & \longrightarrow & \big\langle L,M(-D)\big\rangle\otimes\big\langle L,\mathcal{O}_X(D)\big\rangle
\end{array}$$

where the maps are as follows:

• The top horizontal map is the isomorphism of Proposition 5.7, Chapter 2, where we take the values $L_1 = L$, $M_1 = M$ and $L_2 = M_2 = \mathcal{O}_X$. We take ρ and ρ' to be any permissible set of values: a different choice will only change the isomorphism by a sign.

• The bottom horizontal map is the tensor product of two isomorphisms. The first is the isomorphism of Proposition 5.7, Chapter 2, where we take $L_1 = L$, $M_1 = M(-D)$ and $L_2 = M_2 = \mathcal{O}_X$; we take ρ and ρ' to be any permissible set of values – a different choice will only change the isomorphism by a sign. The second is the isomorphism formed by the following composition

$$\begin{aligned}
\frac{\mathbf{D}\big(L\otimes M|_D\big)}{\mathbf{D}\big(M|_D\big)} &\ \tilde{\to}\ \frac{\mathbf{D}\big(L\otimes M|_D\big)}{\mathbf{D}\big(\mathcal{O}_D\big)}\otimes\frac{\mathbf{D}\big(\mathcal{O}_D\big)}{\mathbf{D}\big(M|_D\big)}\ \tilde{\to}\ \mathbf{N}_D\big(L\otimes M\big)\otimes\mathbf{N}_D\big(M\big)^{-1} \\
&\ \tilde{\to}\ \mathbf{N}_D\big(L\big)\ \tilde{\to}\ \big\langle L,\mathcal{O}_X(D)\big\rangle.
\end{aligned}$$

• The rightmost vertical map is the natural "bilinearity" isomorphism of Proposition 7.4 of Chapter 1.

• The leftmost vertical map is formed from the isomorphisms

$$\mathbf{D}(L\otimes M)\ \tilde{\to}\ \mathbf{D}\big(L\otimes M(-D)\big)\otimes\mathbf{D}\big(L\otimes M|_D\big) \quad\text{and}\quad \mathbf{D}\big(M\big)\ \tilde{\to}\ \mathbf{D}\big(M(-D)\big)\otimes\mathbf{D}\big(M|_D\big)$$

which come from the exact sequences

$$0 \longrightarrow L\otimes M(-D) \longrightarrow L\otimes M \longrightarrow L\otimes M|_D \longrightarrow 0$$

and

$$0 \longrightarrow M(-D) \longrightarrow M \longrightarrow M|_D \longrightarrow 0.$$

2. Likewise, the following diagram commutes up to sign:

$$\begin{array}{ccc} \dfrac{\mathbf{D}(L\otimes M)\otimes\mathbf{D}(\mathcal{O}_X)}{\mathbf{D}(L)\otimes\mathbf{D}(M)} & \longrightarrow & \langle L, M\rangle \\ \downarrow & & \downarrow \\ \dfrac{\mathbf{D}(L\otimes M(-D))\otimes\mathbf{D}(\mathcal{O}_X)}{\mathbf{D}(L(-D))\otimes\mathbf{D}(M)} \otimes \dfrac{\mathbf{D}(L\otimes M|_D)}{\mathbf{D}(L|_D)} & \longrightarrow & \langle L(-D), M\rangle \otimes \langle \mathcal{O}_X(D), M\rangle \end{array}$$

where the maps are analogous to those of statement 1.

Proof. We only need to prove that the first diagram commutes up to sign; the proof for the second diagram is, of course, completely analogous. Choose meromorphic sections, l of L and m of M, such that their associated Cartier divisors (l) and (m) have disjoint supports which do not intersect D. Find effective Cartier divisors D_1 and D_2 such that $(l) = D_1 - D_2$ and such that D_1 and D_2 are disjoint from the supports of (m) and D. Let $\mathbf{1}$ be the canonical meromorphic section of $\mathcal{O}_X(-D)$.

Consider the line bundle $L_0 = L(-D_1)$; as in (4.3) of Chapter 1, we construct from L_0 and l the exact sequences

$$0 \longrightarrow L_0 \longrightarrow L \longrightarrow L|_{D_1} \longrightarrow 0$$

and

$$0 \longrightarrow L_0 \longrightarrow \mathcal{O}_X \longrightarrow \mathcal{O}_{D_2} \longrightarrow 0.$$

As in the proof of Proposition 5.7 of Chapter 2, the isomorphism

$$\frac{\mathbf{D}(L\otimes M)\otimes\mathbf{D}(\mathcal{O}_X)}{\mathbf{D}(L)\otimes\mathbf{D}(M)} \longrightarrow \langle L, M\rangle$$

can be factored as the composition

$$\frac{\mathbf{D}(L\otimes M)\otimes\mathbf{D}(\mathcal{O}_X)}{\mathbf{D}(L)\otimes\mathbf{D}(M)} \xrightarrow{\Psi_{(l,D_1,D_2)}} \frac{\mathbf{D}(L\otimes M|_{D_1})\otimes\mathbf{D}(\mathcal{O}_{D_2})}{\mathbf{D}(L|_{D_1})\otimes\mathbf{D}(M|_{D_2})} \xrightarrow{\Theta_{(D_1,D_2;m)}} \mathbf{C} \xrightarrow{\bullet\langle l,m\rangle} \langle L, M\rangle$$

where $\Psi_{(l,D_1,D_2)}$ and $\Theta_{(D_1,D_2;m)}$ are as defined in (4.7) of Chapter 1, and the last map is multiplication by $\langle l, m\rangle$.

Likewise, the isomorphism

$$\frac{\mathbf{D}(L\otimes M(-D))\otimes\mathbf{D}(\mathcal{O}_X)}{\mathbf{D}(L)\otimes\mathbf{D}(M(-D))} \longrightarrow \langle L, M(-D)\rangle$$

can be factored as the composition

$$\frac{\mathbf{D}(L\otimes M(-D))\otimes\mathbf{D}(\mathcal{O}_X)}{\mathbf{D}(L)\otimes\mathbf{D}(M(-D))} \xrightarrow{\widetilde{\Psi}_{(l,D_1,D_2)}} \frac{\mathbf{D}(L\otimes M(-D)|_{D_1})\otimes\mathbf{D}(\mathcal{O}_{D_2})}{\mathbf{D}(L|_{D_1})\otimes\mathbf{D}(M(-D)|_{D_2})} \xrightarrow{\Theta_{(D_1,D_2;m\otimes\mathbf{1})}} \mathbf{C} \xrightarrow{\bullet\langle l,m\otimes\mathbf{1}\rangle} \langle L, M(-D)\rangle$$

where $\widetilde{\Psi}_{(l,D_1,D_2)}$ and $\Theta_{(D_1,D_2;\, m\otimes\mathbf{1})}$ are as defined in (4.7) of Chapter 1; the tilde on $\widetilde{\Psi}_{(l,D_1,D_2)}$ is there just to distinguish it from $\Psi_{(l,D_1,D_2)}$ above.

Now consider the two maps $L_0 \to L$ and $L_0 \to \mathcal{O}_X$ mentioned above. When we tensor these maps with M and then restrict to D, we get isomorphisms

$$L_0 \otimes M|_D \xrightarrow{\sim} L \otimes M|_D \qquad \text{and} \qquad L_0 \otimes M|_D \xrightarrow{\sim} M|_D.$$

Using the determinants of these two isomorphisms,

$$\mathbf{D}(L_0 \otimes M|_D) \xrightarrow{\sim} \mathbf{D}(L \otimes M|_D) \qquad \text{and} \qquad \mathbf{D}(L_0 \otimes M|_D) \xrightarrow{\sim} \mathbf{D}(M|_D),$$

we can form a map δ as the following composition

$$\delta\colon\ \frac{\mathbf{D}\big(L\otimes M|_D\big)}{\mathbf{D}\big(M|_D\big)} \longrightarrow \frac{\mathbf{D}\big(L_0\otimes M|_D\big)}{\mathbf{D}\big(L_0\otimes M|_D\big)} \longrightarrow \mathbf{C}$$

where the last map is the natural "cancellation" isomorphism. The reader can verify that

$$\frac{\mathbf{D}\big(L\otimes M|_D\big)}{\mathbf{D}\big(M|_D\big)} \xrightarrow{\ \delta\ } \mathbf{C} \xrightarrow{\ \bullet\langle l,\mathbf{1}\rangle\ } \big\langle L, \mathcal{O}_X(D)\big\rangle$$

is a factorization of the isomorphism defined in the statement of the lemma.

We break the remainder of the proof into stages:

Stage 1. First, we claim that the following commutes up to sign:

$$\begin{array}{ccc}
\dfrac{\mathbf{D}(L\otimes M)\otimes\mathbf{D}(\mathcal{O}_X)}{\mathbf{D}(L)\otimes\mathbf{D}(M)} & \xrightarrow{\ \Psi_{(l,D_1,D_2)}\ } & \dfrac{\mathbf{D}\big(L\otimes M|_{D_1}\big)\otimes\mathbf{D}\big(\mathcal{O}_{D_2}\big)}{\mathbf{D}\big(L|_{D_1}\big)\otimes\mathbf{D}\big(M|_{D_2}\big)} \\
\downarrow & & \downarrow \\
\dfrac{\mathbf{D}\big(L\otimes M(-D)\big)\otimes\mathbf{D}\big(\mathcal{O}_X\big)}{\mathbf{D}\big(L\big)\otimes\mathbf{D}\big(M(-D)\big)} \otimes \dfrac{\mathbf{D}\big(L\otimes M|_D\big)}{\mathbf{D}\big(M|_D\big)} & \xrightarrow{\ \widetilde{\Psi}_{(l,D_1,D_2)}\otimes\delta\ } & \dfrac{\mathbf{D}\big(L\otimes M(-D)|_{D_1}\big)\otimes\mathbf{D}\big(\mathcal{O}_{D_2}\big)}{\mathbf{D}\big(L|_{D_1}\big)\otimes\mathbf{D}\big(M(-D)|_{D_2}\big)}\otimes\mathbf{C}
\end{array}$$

where the leftmost vertical map is as in the statement of the lemma, and the rightmost vertical map is formed from the determinants of the isomorphisms

$$L\otimes M|_{D_1} \to L\otimes M(-D)|_{D_1} \qquad \text{and} \qquad M|_{D_2} \to M(-D)|_{D_2}$$

which are, in turn, formed from tensoring with the isomorphisms

$$\mathcal{O}_{D_i} \to \mathcal{O}_X(-D)|_{D_i} \qquad \text{defined by} \qquad \mathbf{1} \mapsto \mathbf{1}|_{D_i}.$$

To prove this claim, consider the diagram of exact sequences

$$\begin{array}{ccccccccc}
 & & (\zeta_1) & & (\zeta_2) & & (\zeta_3) & & \\
 & & 0 & & 0 & & 0 & & \\
 & & \downarrow & & \downarrow & & \downarrow & & \\
(\eta_1) & 0 \longrightarrow & L_0 \otimes M(-D) & \longrightarrow & L \otimes M(-D) & \longrightarrow & L \otimes M(-D)|_{D_1} & \longrightarrow & 0 \\
 & & \downarrow & & \downarrow & & \downarrow & & \\
(\eta_2) & 0 \longrightarrow & L_0 \otimes M & \longrightarrow & L \otimes M & \longrightarrow & L \otimes M|_{D_1} & \longrightarrow & 0 \\
 & & \downarrow & & \downarrow & & \downarrow & & \\
(\eta_3) & 0 \longrightarrow & L_0 \otimes M|_D & \longrightarrow & L \otimes M|_D & \longrightarrow & 0 & & \\
 & & \downarrow & & \downarrow & & & & \\
 & & 0 & & 0 & & & &
\end{array}$$

which gives, by Theorem 2.5.1 of Chapter 1, the following diagram, commutative up to sign:

$$\begin{array}{ccc}
\mathbf{D}\big(L \otimes M\big) & \xrightarrow{\mathbf{D}(\eta_2)} & \mathbf{D}\big(L_0 \otimes M\big) \otimes \mathbf{D}\big(L \otimes M|_{D_1}\big) \\
\Big\downarrow {\scriptstyle \mathbf{D}(\zeta_2)} & & \Big\downarrow {\scriptstyle \mathbf{D}(\zeta_1)\otimes \mathbf{D}(\zeta_3)} \\
\mathbf{D}\big(L \otimes M(-D)\big) \otimes \mathbf{D}\big(L \otimes M|_D\big) & \xrightarrow{\mathbf{D}(\eta_1)\otimes \mathbf{D}(\eta_3)} & \mathbf{D}\big(L_0 \otimes M(-D)\big) \otimes \mathbf{D}\big(L \otimes M(-D)|_{D_1}\big) \otimes \mathbf{D}\big(L_0 \otimes M|_D\big).
\end{array}$$

Likewise, the diagram of exact sequences

$$\begin{array}{ccccccccc}
 & & (\zeta_1') & & (\zeta_2') & & (\zeta_3') & & \\
 & & 0 & & 0 & & 0 & & \\
 & & \downarrow & & \downarrow & & \downarrow & & \\
(\eta_1') & 0 \longrightarrow & L_0 \otimes M(-D) & \longrightarrow & M(-D) & \longrightarrow & M(-D)|_{D_2} & \longrightarrow & 0 \\
 & & \downarrow & & \downarrow & & \downarrow & & \\
(\eta_2') & 0 \longrightarrow & L_0 \otimes M & \longrightarrow & M & \longrightarrow & M|_{D_2} & \longrightarrow & 0 \\
 & & \downarrow & & \downarrow & & \downarrow & & \\
(\eta_3') & 0 \longrightarrow & L_0 \otimes M|_D & \longrightarrow & M|_D & \longrightarrow & 0 & & \\
 & & \downarrow & & \downarrow & & & & \\
 & & 0 & & 0 & & & &
\end{array}$$

gives the following diagram, commutative up to sign:

$$\begin{array}{ccc}
\mathbf{D}\big(M\big) & \xrightarrow{\mathbf{D}(\eta_2')} & \mathbf{D}\big(L_0 \otimes M\big) \otimes \mathbf{D}\big(M|_{D_2}\big) \\
\Big\downarrow {\scriptstyle \mathbf{D}(\zeta_2')} & & \Big\downarrow {\scriptstyle \mathbf{D}(\zeta_1')\otimes \mathbf{D}(\zeta_3')} \\
\mathbf{D}\big(M(-D)\big) \otimes \mathbf{D}\big(M|_D\big) & \xrightarrow{\mathbf{D}(\eta_1')\otimes \mathbf{D}(\eta_3')} & \mathbf{D}\big(L_0 \otimes M(-D)\big) \otimes \mathbf{D}\big(M(-D)|_{D_2}\big) \otimes \mathbf{D}\big(L_0 \otimes M|_D\big)
\end{array}$$

These allow us to form the following diagram, commutative up to sign:

$$\begin{array}{ccc}
\dfrac{\mathbf{D}(L\otimes M)\otimes\mathbf{D}(\mathcal{O}_X)}{\mathbf{D}(L)\otimes\mathbf{D}(M)} & \longrightarrow & \dfrac{\mathbf{D}(L_0\otimes M)\otimes\mathbf{D}(L\otimes M|_{D_1})\otimes\mathbf{D}(L_0)\otimes\mathbf{D}(\mathcal{O}_{D_2})}{\mathbf{D}(L_0)\otimes\mathbf{D}(L|_{D_1})\otimes\mathbf{D}(L_0\otimes M)\otimes\mathbf{D}(M|_{D_2})} \\
\downarrow & & \downarrow \\
\dfrac{\mathbf{D}(L\otimes M(-D))\otimes\mathbf{D}(\mathcal{O}_X)}{\mathbf{D}(L)\otimes\mathbf{D}(M(-D))}\otimes\dfrac{\mathbf{D}(L\otimes M|_D)}{\mathbf{D}(M|_D)} & \longrightarrow & \dfrac{\mathbf{D}(L_0\otimes M(-D))\otimes\mathbf{D}(L\otimes M(-D)|_{D_1})\otimes\mathbf{D}(L_0)\otimes\mathbf{D}(\mathcal{O}_{D_2})}{\mathbf{D}(L_0)\otimes\mathbf{D}(L|_{D_1})\otimes\mathbf{D}(L_0\otimes M(-D))\otimes\mathbf{D}(M(-D)|_{D_2})}.\otimes\dfrac{\mathbf{D}(L_0\otimes M|_D)}{\mathbf{D}(L_0\otimes M|_D)}
\end{array}$$

where the maps are formed from the isomorphisms of the previous two diagrams, together with the isomorphisms

$$\mathbf{D}(\mathcal{O})\to\mathbf{D}(L_0)\otimes\mathbf{D}(\mathcal{O}_{D_2})\qquad\text{and}\qquad\mathbf{D}(L)\to\mathbf{D}(L_0)\otimes\mathbf{D}(L|_{D_1})$$

resulting from the exact sequences

$$0\longrightarrow L_0\longrightarrow\mathcal{O}_X\longrightarrow\mathcal{O}_{D_2}\longrightarrow 0$$

and

$$0\longrightarrow L_0\longrightarrow L\longrightarrow L|_{D_1}\longrightarrow 0.$$

We also have the following commutative diagram, where the horizontal maps are the natural "cancellation" isomorphisms:

$$\begin{array}{ccc}
\dfrac{\mathbf{D}(L_0\otimes M)\otimes\mathbf{D}(L\otimes M|_{D_1})\otimes\mathbf{D}(L_0)\otimes\mathbf{D}(\mathcal{O}_{D_2})}{\mathbf{D}(L_0)\otimes\mathbf{D}(L|_{D_1})\otimes\mathbf{D}(L_0\otimes M)\otimes\mathbf{D}(M|_{D_2})} & \longrightarrow & \dfrac{\mathbf{D}(L\otimes M|_{D_1})\otimes\mathbf{D}(\mathcal{O}_{D_2})}{\mathbf{D}(L|_{D_1})\otimes\mathbf{D}(M|_{D_2})} \\
\downarrow & & \downarrow \\
\dfrac{\mathbf{D}(L_0\otimes M(-D))\otimes\mathbf{D}(L\otimes M(-D)|_{D_1})\otimes\mathbf{D}(L_0)\otimes\mathbf{D}(\mathcal{O}_{D_2})}{\mathbf{D}(L_0)\otimes\mathbf{D}(L|_{D_1})\otimes\mathbf{D}(L_0\otimes M(-D))\otimes\mathbf{D}(M(-D)|_{D_2})}\otimes\dfrac{\mathbf{D}(L_0\otimes M|_D)}{\mathbf{D}(L_0\otimes M|_D)} & \longrightarrow & \dfrac{\mathbf{D}(L\otimes M(-D)|_{D_1})\otimes\mathbf{D}(\mathcal{O}_{D_2})}{\mathbf{D}(L|_{D_1})\otimes\mathbf{D}(M(-D)|_{D_2})}\otimes\mathbf{C}.
\end{array}$$

The claim is established by combining together these last two diagrams; using the definitions of the isomorphisms $\Psi_{(l,D_1,D_2)}$, $\widetilde{\Psi}_{(l,D_1,D_2)}$, and δ it is straightforward to check that the resulting diagram is as desired.

Stage 2. Next, we claim that the following diagram commutes:

$$\begin{array}{ccc}
\dfrac{\mathbf{D}(L\otimes M|_{D_1})\otimes\mathbf{D}(\mathcal{O}_{D_2})}{\mathbf{D}(L|_{D_1})\otimes\mathbf{D}(M|_{D_2})} & \xrightarrow{\ \Theta_{(D_1,D_2;\,m)}\ } & \mathbf{C} \\
\downarrow & & \downarrow \\
\dfrac{\mathbf{D}(L\otimes M(-D)|_{D_1})\otimes\mathbf{D}(\mathcal{O}_{D_2})}{\mathbf{D}(L|_{D_1})\otimes\mathbf{D}(M(-D)|_{D_2})}\otimes\mathbf{C} & \xrightarrow{\ \Theta_{(D_1,D_2;\,m\otimes\mathbf{1})}^{\otimes 1}\ } & \mathbf{C}\otimes\mathbf{C}
\end{array}$$

where the rightmost vertical map is the natural isomorphism, and the leftmost vertical map is the rightmost vertical map of the diagram of stage 1.

To see this, consider the following commutative diagram, which obviously commutes:

$$\begin{array}{ccccc}
\dfrac{\mathbf{D}\big(L\otimes M|_{D_1}\big)\otimes\mathbf{D}\big(\mathcal{O}_{D_2}\big)}{\mathbf{D}\big(L|_{D_1}\big)\otimes\mathbf{D}\big(M|_{D_2}\big)} & \longrightarrow & \dfrac{\mathbf{D}\big(L|_{D_1}\big)\otimes\mathbf{D}\big(\mathcal{O}_{D_2}\big)}{\mathbf{D}\big(L|_{D_1}\big)\otimes\mathbf{D}\big(\mathcal{O}_{D_2}\big)} & \longrightarrow & \mathbf{C} \\
\downarrow & & \downarrow & & \downarrow \\
\dfrac{\mathbf{D}\big(L\otimes M(-D)|_{D_1}\big)\otimes\mathbf{D}\big(\mathcal{O}_{D_2}\big)}{\mathbf{D}\big(L|_{D_1}\big)\otimes\mathbf{D}\big(M(-D)|_{D_2}\big)}\otimes\mathbf{C} & \longrightarrow & \dfrac{\mathbf{D}\big(L|_{D_1}\big)\otimes\mathbf{D}\big(\mathcal{O}_{D_2}\big)}{\mathbf{D}\big(L|_{D_1}\big)\otimes\mathbf{D}\big(\mathcal{O}_{D_2}\big)}\otimes\mathbf{C} & \longrightarrow & \mathbf{C}\otimes\mathbf{C}
\end{array}$$

where the leftmost vertical map is as in the previous diagram; the top leftmost horizontal map is formed from the determinants of the isomorphisms

$$L\otimes M|_{D_1} \xrightarrow{\sim} L|_{D_1} \qquad \text{defined by} \qquad l\otimes m|_{D_1}\mapsto l|_{D_1}$$

and

$$M|_{D_2} \xrightarrow{\sim} \mathcal{O}_{D_2} \qquad \text{defined by} \qquad m|_{D_2}\mapsto \mathbf{1};$$

the bottom leftmost horizontal map is formed from the determinants of the isomorphisms

$$L\otimes M(-D)|_{D_1} \xrightarrow{\sim} L|_{D_1} \qquad \text{defined by} \qquad l\otimes m\otimes\mathbf{1}|_{D_1}\mapsto l|_{D_1}$$

and

$$M(-D)|_{D_2} \xrightarrow{\sim} \mathcal{O}_{D_2} \qquad \text{defined by} \qquad m\otimes\mathbf{1}|_{D_2}\mapsto \mathbf{1};$$

the other maps are the obvious isomorphisms.

By the results of (4.7), Chapter 1, we see that the top row of the above diagram is a factorization of $\Theta_{(D_1,D_2;\,m)}$, and the bottom row is a factorization of $\Theta_{(D_1,D_2;\,m\otimes\mathbf{1})}$.

Stage 3. Finally, we point out that the following trivially commutes:

$$\begin{array}{ccc}
\mathbf{C} & \xrightarrow{\bullet\langle l,m\rangle} & \big\langle L,M\big\rangle \\
\downarrow & & \downarrow \\
\mathbf{C}\otimes\mathbf{C} & \xrightarrow{\bullet\langle l,m\otimes\mathbf{1}\rangle\otimes\langle l,\mathbf{1}\rangle} & \big\langle L,M(-D)\big\rangle\otimes\big\langle L,\mathcal{O}_X(D)\big\rangle
\end{array}$$

where the right vertical map is formed from the canonical isomorphism of Proposition 7.4 of Chapter 1.

The overall result follows from combining together the diagrams of the above three stages.

2. The Riemann-Roch Isomorphism for Arithmetic Surfaces

Throughout this section let A be the ring of integers of a number field K. Let X be projective and flat over $S = \operatorname{Spec} A$ with one-dimensional fibers; assume also that (i) every fiber is Cohen-Macaulay and has components all of dimension 1, (ii) the relative dualizing sheaf $\mathcal{K}_{X/S}$ is an invertible sheaf, and (iii) $X_{\overline{K}}$ is connected and non-singular everywhere except for possibly a finite number of ordinary double points; here $\overline{K}$ denotes the algebraic closure of K. As mentioned before in the introduction to Section 4 of Chapter 3, the relative dualizing sheaf is invertible in many important situations including the case when X is semi-stable over S.

Let Σ be the set of embeddings of K into the complex numbers $\mathbf{C}$. For any $\sigma \in \Sigma$, let X_σ be the scheme obtained through base change from K to $\mathbf{C}$ via σ. For each $\sigma \in \Sigma$, choose an intersection function g_σ on X_σ, and let Φ_σ be the order function of X_σ determined by g_σ (see (2.1) of Chapter 4).

Such a scheme X together with an above set of intersection functions $\{g_\sigma\}$ will be called an *arithmetic surface.*

(2.1) Definition. A *normed line bundle* on $\operatorname{Spec} A$ is a line bundle L on $\operatorname{Spec} A$, together with a non-trivial norm $|\,|_\sigma$ on each of the the one-dimensional complex vector spaces $L_\sigma = L \otimes_\sigma \mathbf{C}$, where $\sigma \in \Sigma$.

An isomorphism between normed line bundles is an isomorphism between the underlying line bundles such that, for each $\sigma \in \Sigma$, the associated isomorphism between one-dimensional complex vector spaces is an isometry. The trivial line bundle is made into a normed line bundle in the obvious way. Similarly, the inverse of a normed line bundle and the tensor product of two normed line bundles are normed line bundles under the obvious definitions.

Let $\widehat{L}$ be a normed line bundle with underlying line bundle L, and let l be a meromorphic section of L. It is a basic fact of number theory that the quantity

$$\sum_P \operatorname{ord}_P(l) \cdot \log N_P - \sum_{\sigma \in \Sigma} \log |l|_\sigma$$

is independent of l; we call this quantity the *degree* of $\widehat{L}$. Here the first sum is taken over the closed points of $S = \operatorname{Spec} A$, $\operatorname{ord}_P(l)$ denotes the order vanishing of l at P, and N_P denotes the number of elements of the residue field associated with P.

(2.2) Definition. An *arithmetic line bundle* $\widehat{L}$ on X is a line bundle L together with, for each $\sigma \in \Sigma$, an order function $\Phi_{\widehat{L}_\sigma}$ on L_σ which is compatible with Φ_σ, where L_σ is the pull-back of L to X_σ (see (2.2) of Chapter 4).

The inverse of an arithmetic line bundle and the tensor product of two arithmetic line bundles are made into arithmetic line bundles in the obvious way. The notion of an isomorphism between arithmetic line bundles is the obvious one. The line bundle $\mathcal{O}_X$ is made into an arithmetic line bundle $\widehat{\mathcal{O}}_X$ as follows: the pull-back of $\mathcal{O}_X$ to X_σ is canonically isomorphic to $\mathcal{O}_{X_\sigma}$; we take, as an order function for $\mathcal{O}_{X_\sigma}$, the order function Φ_σ of X_σ.

The relative dualizing sheaf $\mathcal{K}_{X/S}$ is made into an arithmetic line bundle $\widehat{\mathcal{K}}_{X/S}$ as follows: the pull-back of $\mathcal{K}_{X/S}$ to X_σ is canonically isomorphic to $\mathcal{K}_{X_\sigma}$; we take, as an order function for $\mathcal{K}_{X_\sigma}$, the order function mentioned in (2.3) of Chapter 4.

(2.3) Definition. Let $\widehat{L}$ and $\widehat{M}$ be arithmetic line bundles on X with underlying line bundles L and M. For each $\sigma \in \Sigma$, let L_σ and M_σ be the pull-backs of L and M

to X_σ. By Proposition 6.2 of Chapter 2, the intersection pairing is compatible with flat base change, so, for each $\sigma \in \Sigma$, $\langle L, M\rangle \otimes_\sigma \mathbf{C}$ is canonically isomorphic to $\langle L_\sigma, M_\sigma\rangle$. In (2.6) of Section 2, Chapter 4 the one-dimensional complex vector space $\langle L_\sigma, M_\sigma\rangle$ is given a non-trivial norm. This makes $\langle L, M\rangle$ into a normed line bundle, which we will denote by $\langle \widehat{L}, \widehat{M}\rangle$.

The *intersection number* $[\widehat{L}, \widehat{M}]$ of the pair $\widehat{L}$, $\widehat{M}$ is defined to be the degree of the normed line bundle $\langle \widehat{L}, \widehat{M}\rangle$. The canonical isomorphism of Proposition 4.5, Chapter 2 between $\langle L, M\rangle$ and $\langle M, L\rangle$ gives, by Proposition 2.6.1 of Chapter 4, an isomorphism between the normed line bundles $\langle \widehat{L}, \widehat{M}\rangle$ and $\langle \widehat{M}, \widehat{L}\rangle$. This implies that the intersection number is symmetric:

$$[\widehat{L}, \widehat{M}] = [\widehat{M}, \widehat{L}].$$

A similar argument shows that the intersection number is linear in both variable. For example, if $\widehat{L_1}$, $\widehat{L_2}$, and $\widehat{M}$ are arithmetic line bundles of X, then

$$[\widehat{L_1} \otimes \widehat{L_2}, \widehat{M}] = [\widehat{L_1}, \widehat{M}] + [\widehat{L_2}, \widehat{M}].$$

(2.4) Definition. Let $\widehat{L}$ be an arithmetic line bundle on X with underlying line bundle L. For each $\sigma \in \Sigma$, let L_σ be the pull-back of L to X_σ. For each $\sigma \in \Sigma$, the determinant of cohomology $\mathbf{D}(L)$ is compatible with base change via σ, i.e., $\mathbf{D}(L) \otimes_\sigma \mathbf{C}$ is canonically isomorphic to $\mathbf{D}(L_\sigma)$; in the previous section, the one-dimensional vector space $\mathbf{D}(L_\sigma)$ is given a non-trivial norm. This makes $\mathbf{D}(L)$ into a normed line bundle which we denote $\mathbf{D}(\widehat{L})$.

Actually, to be honest, the norm on $\mathbf{D}(L_\sigma)$ is only defined after we have fixed a non-trivial norm for the vector space $\mathbf{D}(\mathcal{O}_{X_\sigma})$. We assume that, for each $\sigma \in \Sigma$, this norm is fixed once and for all. We will not address here the problem of determining a natural choice for such a norm.

We define the *Euler characteristic* $\chi_{X/S}(\widehat{L})$ to be the degree of $\mathbf{D}(\widehat{L})$. Although $\chi_{X/S}(\widehat{L})$ depends on the norms on each $\mathbf{D}(\mathcal{O}_{X_\sigma})$, the difference we are interested in, $\chi_{X/S}(\widehat{L}) - \chi_{X/S}(\widehat{\mathcal{O}}_X)$, does not.

(2.5) Theorem (The Arithmetic Riemann-Roch Isomorphism). *Let $\widehat{L}$ be an arithmetic line bundle on X with underlying line bundle L. The Riemann-Roch isomorphism*

$$\mathbf{D}(L)^{\otimes 2} \otimes \mathbf{D}(\mathcal{O}_X)^{\otimes -2} \xrightarrow{\sim} \langle L, L \otimes \mathcal{K}_{X/S}^{-1}\rangle$$

of Section 4, Chapter 3 induces the following isomorphism between the associated normed line bundles

$$\mathbf{D}(\widehat{L})^{\otimes 2} \otimes \mathbf{D}(\widehat{\mathcal{O}}_X)^{\otimes -2} \xrightarrow{\sim} \langle \widehat{L}, \widehat{L} \otimes \widehat{\mathcal{K}}_{X/S}^{-1}\rangle.$$

Note. For any given line bundle on X there are actually two Riemann-Roch isomorphisms differing from each other by a sign, so there are two arithmetic Riemann-Roch isomorphisms for any given arithmetic line bundle.

Proof. The intersection pairing, the determinant of cohomology, the relative dualizing sheaf, and the Riemann-Roch isomorphism are all well-behaved with respect to pull-back via any $\sigma \in \Sigma$. So the theorem follows from Proposition 1.3.

(2.5.1) Corollary. *Let $\widehat{L}$ be an arithmetic line bundle on X. Then*

$$\chi_{X/S}(\widehat{L}) - \chi_{X/S}(\widehat{\mathcal{O}}_X) = \frac{1}{2}[\widehat{L}, \widehat{L}] - \frac{1}{2}[\widehat{L}, \widehat{\mathcal{K}}_{X/S}].$$

Bibliography

[A] W. Aitken, *An explicit sign formula for the determinant of cohomology* (preprint)

[Ar1] S. J. Arakelov, *An intersection theory for divisors on an arithmetic surface*, Math. USSR Izv. **8** (1974), No. 6, 1167 – 1180.

[Ar2] S. J. Arakelov, *Theory of intersections on an arithmetic surface*, Proceedings, International Congress of Mathematicians, Vancouver (1975), Vol 1, 405 – 408.

[D] P. Deligne, *Le déterminant de la cohomologie*, Contemporary Mathematics **67** (1987), American Mathematical Society.

[E1] R. Elkik, *Intersections relatives de fibrés en droites et intégrales de classes de Chern*, Ann. Sci. École Norm. Sup **22** (1989), 195 – 226.

[E2] R. Elkik, *Métriques sur les fibrés d'intersection*, Duke Math. Journal **61** (1990), 303-328.

[F1] G. Faltings, *Calculus on arithmetic surfaces*, Annals of Math. **119** (1984), 387 – 424.

[F2] G. Faltings, *Lectures on the Arithmetic Riemann-Roch Theorem*, Princeton University Press (1992).

[Fr] J. Franke, *Chow categories*, Compositio Mathematica, **76** (1990), 101-162.

[H] R. Hartshorne, *Algebraic Geometry*, Springer-Verlag (1977).

[K] S. L. Kleiman, *Relative duality for quasi-coherent sheaves*, Compositio Mathematica, **41** (1980).

[L] S. Lang, *Introduction to Arakelov Theory*, Springer-Verlag (1988).

[M] H. Matsumura, *Commutative Ring Theory*, Cambridge University Press (1986).

[MB] L. Moret-Bailly, "Métriques permises", *Séminaire sur les pinceaux arithmétiques: la conjecture de Mordell*, Astérisque, **127** (1985).

[SABK] C. Soulé, D. Abramovich, J.-F. Burnol, J. Kramer, *Lectures on Arakelov Geometry*, Cambridge University Press (1992).

Editors

This journal is designed particularly for long research papers (and groups of cognate papers) in pure and applied mathematics. Papers intended for publication in the *Memoirs* should be addressed to one of the following editors:

Ordinary differential equations, partial differential equations, and applied mathematics to JOHN MALLET-PARET, Division of Applied Mathematics, Brown University, Providence, RI 02912-9000; e-mail: am438000@brownvm.brown.edu.

Harmonic analysis, representation theory, and Lie theory to ROBERT J. STANTON, Department of Mathematics, The Ohio State University, 231 West 18th Avenue, Columbus, OH 43210-1174; electronic mail: stanton@function.mps.ohio-state.edu.

Ergodic theory, dynamical systems, and abstract analysis to DANIEL J. RUDOLPH, Department of Mathematics, University of Maryland, College Park, MD 20742; e-mail: djr@math.umd.edu.

Real and harmonic analysis and elliptic partial differential equations to JILL C. PIPHER, Department of Mathematics, Brown University, Providence, RI 02910-9000; e-mail: jpipher@gauss.math.brown.edu.

Algebra and algebraic geometry to EFIM ZELMANOV, Department of Mathematics, University of Wisconsin, 480 Lincoln Drive, Madison, WI 53706-1388; e-mail: zelmanov@math.wisc.edu

Algebraic topology and cohomology of groups to STEWART PRIDDY, Department of Mathematics, Northwestern University, 2033 Sheridan Road, Evanston, IL 60208-2730; e-mail: s_priddy@math.nwu.edu.

Global analysis and differential geometry to ROBERT L. BRYANT, Department of Mathematics, Duke University, Durham, NC 27706-7706; e-mail: bryant@math.duke.edu.

Probability and statistics to RICHARD DURRETT, Department of Mathematics, Cornell University, White Hall, Ithaca, NY 14853-7901; e-mail: rtd@cornella.cit.cornell.edu.

Combinatorics and Lie theory to PHILIP J. HANLON, Department of Mathematics, University of Michigan, Ann Arbor, MI 48109-1003; e-mail: phil.hanlon@math.lsa.umich.edu.

Logic and universal algebra to GREGORY L. CHERLIN, Department of Mathematics, Rutgers University, Hill Center, Busch Campus, New Brunswick, NJ 08903; e-mail: cherlin@math.rutgers.edu.

Number theory and arithmetic algebraic geometry to ALICE SILVERBERG, Department of Mathematics, Ohio State University, Columbus, OH 43210-1174; e-mail: silver@math.ohio-state.edu.

Complex analysis and complex geometry to DANIEL M. BURNS, Department of Mathematics, University of Michigan, Ann Arbor, MI 48109-1003; e-mail: burns@gauss.stanford.edu.

Algebraic geometry and commutative algebra to LAWRENCE EIN, Department of Mathematics, University of Illinois, 851 S. Morgan (MIC 249), Chicago, IL 60607-7045; email: u22425@uicvm.uic.edu.

All other communications to the editors should be addressed to the Managing Editor, PETER SHALEN, Department of Mathematics, Statistics, and Computer Science, University of Illinois at Chicago, Chicago, IL 60680; e-mail: shalen@math.uic.edu.

Authors with FTP access may retrieve an author package from the Society's Internet node e-MATH.ams.org (130.44.1.100). For those without FTP access, the author package can be obtained free of charge by sending e-mail to pub@math.ams.org (Internet) or from the Publication Division, American Mathematical Society, P.O. Box 6248, Providence, RI 02940-6248. When requesting an author package, please specify $\mathcal{A}\mathcal{M}\mathcal{S}$-TEX or $\mathcal{A}\mathcal{M}\mathcal{S}$-LATEX, Macintosh or IBM (3.5) format, and the publication in which your paper will appear. Please be sure to include your complete mailing address.

Submission of electronic files. At the time of submission, the source file(s) should be sent to the Providence office (this includes any TEX source file, any graphics files, and the DVI or PostScript file).

Before sending the source file, be sure you have proofread your paper carefully. The files you send must be the EXACT files used to generate the proof copy that was accepted for publication. For all publications, authors are required to send a printed copy of their paper, which exactly matches the copy approved for publication, along with any graphics that will appear in the paper.

TEX files may be submitted by email, FTP, or on diskette. The DVI file(s) and PostScript files should be submitted only by FTP or on diskette unless they are encoded properly to submit through e-mail. (DVI files are binary and PostScript files tend to be very large.)

Files sent by electronic mail should be addressed to the Internet address pub-submit@math.ams.org. The subject line of the message should include the publication code to identify it as a Memoir. TEX source files, DVI files, and PostScript files can be transferred over the Internet by FTP to the Internet node e-math.ams.org (130.44.1.100).

Electronic graphics. Figures may be submitted to the AMS in an electronic format. The AMS recommends that graphics created electronically be saved in Encapsulated PostScript (EPS) format. This includes graphics originated via a graphics application as well as scanned photographs or other computer-generated images.

If the graphics package used does not support EPS output, the graphics file should be saved in one of the standard graphics formats—such as TIFF, PICT, GIF, etc.—rather than in an application-dependent format. Graphics files submitted in an application-dependent format are not likely to be used. No matter what method was used to produce the graphic, it is necessary to provide a paper copy to the AMS.

Authors using graphics packages for the creation of electronic art should also avoid the use of any lines thinner than 0.5 points in width. Many graphics packages allow the user to specify a "hairline" for a very thin line. Hairlines often look acceptable when proofed on a typical laser printer. However, when produced on a high-resolution laser imagesetter, hairlines become nearly invisible and will be lost entirely in the final printing process.

Screens should be set to values between 15% and 85%. Screens which fall outside of this range are too light or too dark to print correctly.

Any inquiries concerning a paper that has been accepted for publication should be sent directly to the Editorial Department, American Mathematical Society, P. O. Box 6248, Providence, RI 02940-6248.

Editorial Information

To be published in the *Memoirs*, a paper must be correct, new, nontrivial, and significant. Further, it must be well written and of interest to a substantial number of mathematicians. Piecemeal results, such as an inconclusive step toward an unproved major theorem or a minor variation on a known result, are in general not acceptable for publication. *Transactions* Editors shall solicit and encourage publication of worthy papers. Papers appearing in *Memoirs* are generally longer than those appearing in *Transactions* with which it shares an editorial committee.

As of November 30, 1995, the backlog for this journal was approximately 5 volumes. This estimate is the result of dividing the number of manuscripts for this journal in the Providence office that have not yet gone to the printer on the above date by the average number of monographs per volume over the previous twelve months, reduced by the number of issues published in four months (the time necessary for preparing an issue for the printer). (There are 6 volumes per year, each containing at least 4 numbers.)

A Copyright Transfer Agreement is required before a paper will be published in this journal. By submitting a paper to this journal, authors certify that the manuscript has not been submitted to nor is it under consideration for publication by another journal, conference proceedings, or similar publication.

Information for Authors and Editors

Memoirs are printed by photo-offset from camera copy fully prepared by the author. This means that the finished book will look exactly like the copy submitted.

The paper must contain a *descriptive title* and an *abstract* that summarizes the article in language suitable for workers in the general field (algebra, analysis, etc.). The *descriptive title* should be short, but informative; useless or vague phrases such as "some remarks about" or "concerning" should be avoided. The *abstract* should be at least one complete sentence, and at most 300 words. Included with the footnotes to the paper, there should be the 1991 *Mathematics Subject Classification* representing the primary and secondary subjects of the article. This may be followed by a list of *key words and phrases* describing the subject matter of the article and taken from it. A list of the numbers may be found in the annual index of *Mathematical Reviews*, published with the December issue starting in 1990, as well as from the electronic service e-MATH [**telnet e-MATH.ams.org** (or **telnet 130.44.1.100**). Login and password are **e-math**]. For journal abbreviations used in bibliographies, see the list of serials in the latest *Mathematical Reviews* annual index. When the manuscript is submitted, authors should supply the editor with electronic addresses if available. These will be printed after the postal address at the end of each article.

Electronically prepared papers. The AMS encourages submission of electronically prepared papers in $\mathcal{A}\mathcal{M}\mathcal{S}$-TEX or $\mathcal{A}\mathcal{M}\mathcal{S}$-LATEX. The Society has prepared author packages for each AMS publication. Author packages include instructions for preparing electronic papers, the *AMS Author Handbook*, samples, and a style file that generates the particular design specifications of that publication series for both $\mathcal{A}\mathcal{M}\mathcal{S}$-TEX and $\mathcal{A}\mathcal{M}\mathcal{S}$-LATEX.

Other Titles in This Series

(*Continued from the front of this publication*)

(See the AMS catalog for earlier titles)